Physical Models of Living Systems

Philip Nelson

University of Pennsylvania

*with the assistance of Sarina Bromberg,
Ann Hermundstad, and Jason Prentice*

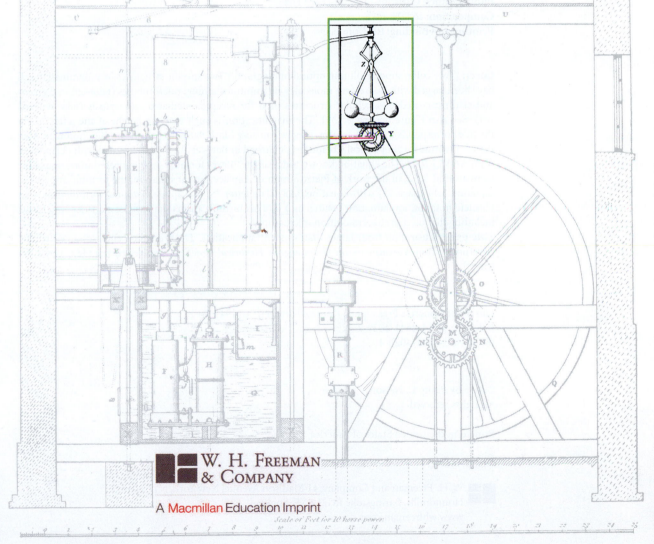

W. H. FREEMAN & COMPANY

A Macmillan Education Imprint

Publisher: Kate Parker
Acquisitions Editor: Alicia Brady
Senior Development Editor: Blythe Robbins
Assistant Editor: Courtney Lyons
Editorial Assistant: Nandini Ahuja
Marketing Manager: Taryn Burns
Senior Media and Supplements Editor: Amy Thorne
Director of Editing, Design, and Media Production: Tracey Kuehn
Managing Editor: Lisa Kinne
Project Editor: Kerry O'Shaughnessy
Production Manager: Susan Wein
Design Manager and Cover Designer: Vicki Tomaselli
Illustration Coordinator: Matt McAdams
Photo Editors: Christine Buese, Richard Fox
Composition: codeMantra
Printing and Binding: RR Donnelley

Cover: [Two-color, superresolution optical micrograph.] Two specific structures in a mammalian cell have been tagged with fluorescent molecules via immunostaining: microtubules (false-colored *green*) and clathrin-coated pits, cellular structures used for receptor-mediated endocytosis (false-colored *red*). See also Figure 6.5 (page 139). The magnification is such that the height of the letter "o" in the title corresponds to about 1.4 μm. [Image courtesy Mark Bates, Dept. of NanoBiophotonics, Max Planck Institute for Biophysical Chemistry, published in Bates et al., 2007. Reprinted with permission from AAAS.] *Inset:* The equation known today as the "Bayes formula" first appeared in recognizable form around 1812, in the work of Pierre Simon de Laplace. In our notation, the formula appears as Equation 3.17 (page 52) with Equation 3.18. (The letter "S" in Laplace's original formulation is an obsolete notation for sum, now written as $\sum$.) This formula forms the basis of statistical inference, including that used in superresolution microscopy.
Title page: Illustration from James Watt's patent application. The green box encloses a centrifugal governor. [From *A treatise on the steam engine: Historical, practical, and descriptive* (1827) by John Farey.]

Library of Congress Preassigned Control Number: 2014949574
ISBN-13: 978-1-4641-4029-7
ISBN-10: 1-4641-4029-4

Printed in the United States of America

First printing

 W. H. Freeman and Company, 41 Madison Avenue, New York, NY 10010
Houndmills, Basingstoke RG21 6XS, England
www.whfreeman.com

For my classmates Janice Enagonio, Feng Shechao, and Andrew Lange.

Whose dwelling is the light of setting suns,
And the round ocean and the living air,
And the blue sky, and in the mind of man:
A motion and a spirit, that impels
All thinking things, all objects of all thought,
And rolls through all things.

– William Wordsworth

Brief Contents

PART III   Control in Cells

Detailed Contents

Chapter 4   Some Useful Discrete Distributions   69

PART III   Control in Cells

Web Resources

The book's Web site (http://www.macmillanhighered.com/physicalmodels1e) contains links to the following resources:

- The *Student's Guide* contains an introduction to some computer math systems, and some guided computer laboratory exercises.
- *Datasets* contains datasets that are used in the problems. In the text, these are cited like this: Dataset 1, with numbers keyed to the list on the Web site.
- *Media* gives links to external media (graphics, audio, and video). In the text, these are cited like this: Media 2, with numbers keyed to the list on the Web site.
- Finally, *Errata* is self-explanatory.

To the Student

Learn from science that you must doubt the experts.
—Richard Feynman

This is a book about physical models of living systems. As you work through it, you'll gain some skills needed to create such models for yourself. You'll also become better able to assess scientific claims without having to trust the experts.

The *living systems* we'll study range in scale from single macromolecules all the way up to complete organisms. At every level of organization, the degree of inherent complexity may at first seem overwhelming, if you are more accustomed to studying physics. For example, the dance of molecules needed for even a single cell to make a decision makes Isaac Newton's equation for the Moon's orbit look like child's play. And yet, the Moon's motion, too, is complex when we look in detail—there are tidal interactions, mode locking, precession, and so on. To study any complex system, we must first make it manageable by adopting a *physical model*, a set of idealizations that focus our attention on the most important features.

Physical models also generally exploit analogies to other systems, which may already be better understood than the one under study. It's amazing how a handful of basic concepts can be used to understand myriad problems at all levels, in both life science and physical science.

Physical modeling seeks to account for experimental data quantitatively. The point is not just to summarize the data succinctly, but also to shed light on underlying mechanisms by testing the different predictions made by various competing models. The reason for insisting on quantitative prediction is that often we can think up a cartoon, either as an actual sketch or in words, that sounds reasonable but fails quantitatively. If, on the contrary, a model's numerical predictions are found to be confirmed in detail, then this is unlikely to be a fluke. Sometimes the predictions have a definite character, stating what should happen every time; such models can be tested in a single experimental trial. More commonly, however, the output of a model is probabilistic in character. This book will develop some of the key ideas of probability, to enable us to make precise statements about the predictions of models and how well they are obeyed by real data.

Perhaps most crucially in practice, a good model not only guides our interpretation of the data we've got, but also suggests what *new* data to go out and *get next*. For example, it may suggest what quantitative, physical intervention to apply when taking those data, in order to probe the model for weaknesses. If weaknesses are found, a physical model may suggest how to improve it by accounting for more aspects of the system, or treating them more realistically. A model that survives enough attempts at falsification eventually earns the label "promising." It may even one day be "accepted."

This book will show you some examples of the modeling process at work. In some cases, physical modeling of quantitative data has allowed scientists to deduce mechanisms whose key molecular actors were at the time unsuspected. These case studies are worth studying, so that you'll be ready to operate in this mode when it's time to make your own discoveries.

Skills

Science is not just a pile of facts for you to memorize. Certainly you need to know many facts, and this book will supply some as background to the case studies. But you also need skills. Skills cannot be gained just by reading through this (or any) book. Instead you'll need to work through at least some of the exercises, both those at the ends of chapters and others sprinkled throughout the text.

Specifically, this book emphasizes

- *Model construction skills:* It's important to find an appropriate level of description and then write formulas that make sense at that level. (Is randomness likely to be an essential feature of this system? Does the proposed model check out at the level of dimensional analysis?) When reading others' work, too, it's important to be able to grasp what assumptions their model embodies, what approximations are being made, and so on.

- *Interconnection skills:* Physical models can bridge topics that are not normally discussed together, by uncovering a hidden similarity. Many big advances in science came about when someone found an analogy of this sort.

- *Critical skills:* Sometimes a beloved physical model turns out to be . . . wrong. Aristotle taught that the main function of the brain was to cool the blood. To evaluate more modern hypotheses, you generally need to understand how raw *data* can give us *information,* and then *understanding.*

- *Computer skills:* Especially when studying biological systems, it's usually necessary to run many trials, each of which will give slightly different results. The experimental data very quickly outstrip our abilities to handle them by using the analytical tools taught in math classes. Not very long ago, a book like this one would have to content itself with telling you things that faraway people had done; you couldn't do the actual analysis yourself, because it was too difficult to make computers do anything. Today you can do industrial-strength analysis on any personal computer.

- *Communication skills:* The biggest discovery is of little use until it makes it all the way into another person's brain. For this to happen reliably, you need to sharpen some communication skills. So when writing up your answers to the problems in this book, imagine that you are preparing a report for peer review by a skeptical reader. Can you take another few minutes to make it easier to figure out what you did and why? Can you label graph axes better, add comments to your code for readability, or justify a step? Can you anticipate objections?

You'll need skills like these for reading primary research literature, for interpreting your own data when you do experiments, and even for evaluating the many statistical and pseudostatistical claims you read in the newspapers.

One more skill deserves separate mention. Some of the book's problems may sound suspiciously vague, for example, "Comment on . . . ." They are intentionally written to make you ask, "What is interesting and worthy of comment here?" There are multiple "right" answers, because there may be more than one interesting thing to say. In your own scientific research, *nobody will tell you the questions*. So it's good to get the habit of asking yourself such things.

Acquiring these skills can be empowering. For instance, some of the most interesting graphs in this book do not actually appear anywhere. You will create them yourself, starting from data on the companion Web site.

What computers can do for you

A model begins in your mind as a proposed mechanism to account for some observations. You may represent those ideas by sketching a diagram on paper. Such diagrams can help you to think clearly about your model, explain it to others, and begin making testable experimental predictions.

Despite the usefulness of such traditional representations, generally you must also carry out some calculational steps before you get predictions that are detailed enough to test the model. Sometimes these steps are easy enough to do with pencil, paper, and a calculator. More often, however, at some point you will need an extremely fast and accurate assistant. Your computer can play this role.

You may need a computer because your model makes a statistical prediction, and a large amount of experimental data is needed to test it. Or perhaps there are a large number of entities participating in your mechanism, leading to long calculations. Sometimes testing the model involves *simulating* the system, including any random elements it contains; sometimes the simulation must be run many times, each time with different values of some unknown parameters, in order to find the values that best describe the observed behavior. Computers can do all these things very rapidly.

To compute responsibly, you also need some insight into what's going on under the hood. Sometimes the key is to write your own simple analysis code from scratch. Many of the exercises in this book ask you to practice this skill.

Finally, you will need to understand your results, and communicate them to others. *Data visualization* is the craft of representing quantitative information in ways that are meaningful, and honest. From the simplest *xy* graph to the fanciest interactive 3D image, computers have transformed data visualization, making it faster and easier than ever before.

This book does not include any chapters explicitly about computer programming or data visualization. The *Student's Guide* contains a brief introduction; your instructor can help you find other resources appropriate for the platform you'll be using.

What computers *can't* do for you

Computers are *not* skilled at formulating imaginative models in the first place. They do not have intuitions, based on analogies to past experience, that help them to identify the important players and their interactions. They don't know what sorts of predictions can be readily measured in the lab. They cannot help you choose which mode of visualization will communicate your results best.

Above all, *a computer doesn't know whether it's appropriate to use a computer* for any phase of a calculation, or whether on the contrary you would be better off with pencil and paper. Nor can it tell you that certain styles of visualization are misleading or cluttered with irrelevant information. Those high-level insights are your job.

Structure and features

- Every chapter contains "Your Turn" questions. Generally these are short and easy (though not always). Beyond these explicit questions, however, most of the formulas are consequences of something said previously, which you should derive yourself. Doing so will greatly improve your understanding of the material—and your fluency when it's time to write an exam.
- Most chapters end with a "Track 2" section. These are generally for advanced students; some of them assume more background knowledge than the main, "Track 1," material. (Others just go into greater detail.) Similarly, there are Track 2 footnotes and homework problems, marked with the glyph $\boxed{T_2}$.
- Appendix A summarizes mathematical notation and key symbols that are used consistently throughout the book. Appendix B discusses some useful tools for solving problems. Appendix C gathers a few constants of Nature for reference.
- Many equations and key ideas are set off and numbered for reference. The notations "Equation x.y" and "Idea x.y" both refer to the same numbered series.
- When a distant figure gets cited, you may or may not need to flip back to see it. To help you decide, many figure references are accompanied by an iconified version of the cited figure in the margin.

Other books

The goal of this book is to help you to teach yourself some of the skills and frameworks you will need in order to become a scientist, in the context of physical models of living systems. A companion book introduces a different slice through the subject (Nelson, 2014), including mechanics and fluid mechanics, entropy and entropic forces, bioelectricity and neural impulses, and mechanochemical energy transduction.

Many other books instead attempt a more complete coverage of the field of biophysics, and would make excellent complements to this one. A few recent examples include
General: Ahlborn, 2004; Franklin et al., 2010; Nordlund, 2011.
Cell biology/biochemistry background: Alberts et al., 2014; Berg et al., 2012; Karp, 2013; Lodish et al., 2012.
Medicine/physiology: Amador Kane, 2009; Dillon, 2012; Herman, 2007; Hobbie & Roth, 2007; McCall, 2010.
Networks: Alon, 2006; Cosentino & Bates, 2012; Vecchio & Murray, 2014; Voit, 2013.
Mathematical background: Otto & Day, 2007; Shankar, 1995.
Probability in biology and physics: Denny & Gaines, 2000; Linden et al., 2014.
Cell and molecular biophysics: Boal, 2012; Phillips et al., 2012; Schiessel, 2013.
Biophysical chemistry: Atkins & de Paula, 2011; Dill & Bromberg, 2010.
Experimental methods: Leake, 2013; Nadeau, 2012.
Computer methods: Computation: DeVries & Hasbun, 2011; Newman, 2013. Other computer skills: Haddock & Dunn, 2011.

Finally, no book can be as up-to-date as the resources available online. Generic sources such as Wikipedia contain many helpful articles, but you may also want to consult `http://bionumbers.hms.harvard.edu/` for specific numerical values, so often needed when constructing physical models of living systems.

To the Instructor

Physicist: "I want to study the brain. Tell me something helpful."
Biologist: "Well, first of all, the brain has two sides . . . ."
Physicist: "Stop! You've told me too much!"
—V. Adrian Parsegian

This book is the text for a course that I have taught for several years to undergraduates at the University of Pennsylvania. The class mainly consists of second- and third-year science and engineering students who have taken at least one year of introductory physics and the associated math courses. Many have heard the buzz about synthetic biology, superresolution microscopy, or something else, and they want a piece of the action.

Many recent articles stress that future breakthroughs in medicine and life science will come from researchers with strong quantitative backgrounds, and with experience at systems-level analysis. Answering this call, many textbooks on "Mathematical Biology," "Systems Biology," "Bioinformatics," and so on have appeared. Few of these, however, seem to stress the importance of physical models. And yet there is something remarkably— unreasonably—effective about physical models. This book attempts to show this using a few case studies.

The book also embodies a few convictions, including[1]

- The study of living organisms is an inspiring context in which to learn many fundamental physical ideas—even for physical-science students who don't (or don't yet) intend to study biophysics further.
- The study of fundamental physical ideas sheds light on the design and functioning of living organisms, and the instruments used to study them. It's important even for life-science students who don't (or don't yet) intend to study biophysics further.

---

[1] See also "To the Student."

In short, this is a book about how *physical science and life science illuminate each other*.

I've also come to believe that

- Whenever possible, we should try to relate our concepts to familiar experience.
- All science students need some intuitions about probability and inference, in order to make sense of methods now in use in many fields. These include likelihood maximization and Bayesian modeling. Other universal topics, often neglected in undergraduate syllabi, include the notion of convolution, long-tail distributions, feedback control, and the Poisson process (and other Markov processes).
- Algorithmic thinking is different from pencil-and-paper analysis. Many students have not yet encountered it by this stage of their careers, yet it's crucial to the daily practice of almost every branch of science. Recent reports have commented on this disconnect and recommended changes in curricula (e.g., Pevzner & Shamir, 2009; National Research Council, 2003). The earlier students come to grips with this mode of thought, the better.
- Students need explicit discussions about Where Theories Come From, in the context of concrete case studies.

This book is certainly not intended as a comprehensive survey of the enormous and protean field of Biophysics. Instead, it's intended to develop the *skills and frameworks* that students need in many fields of science, engineering, and applied math, in the context of understanding how living organisms manage a few of their remarkable abilities. I have tried to tell a limited number of stories with sufficient detail to bring students to the point where they can do research-level analysis for themselves. I have selected stories that seem to fit a single narrative, and that seem to open the most doors to current work. I also tried to stick with stories for which the student can actually do all the calculations, instead of resorting to "Smith has shown . . . ."

Students in the course come from a wide range of majors, with a correspondingly wide range of backgrounds. This can lead to some tricky, yet valuable, cross-cultural moments, like the one in the epigraph to this section. I have found that a little bit of social engineering, to bring together students with different strengths, can start the process of interdisciplinary contact at the moment when it is most likely to become a habit.

Ways to use this book

Most chapters end with "Track 2" sections. Some of these contain material appropriate for students with more advanced backgrounds. Others discuss topics that are at the undergraduate level, but will not be needed later in the book. They can be discussed a la carte, based on your and the students' interests. The main, "Track 1," sections do not rely on any of this material. Also, the *Instructor's Guide* contains many additional bibliographic references, some of which could be helpful for starting projects based on primary literature.

This book could serve as the basis of a course on the science underpinning contemporary biological physics. Or it can be used as a supplement in more specialized courses on physics, biophysics, or several kinds of engineering or applied math. Although Track 1 is meant as an undergraduate course, it contains a lot of material not generally included in undergraduate physics curricula. Thus, it could easily form the basis of a graduate course, if you add all or part of Track 2, and perhaps some reading from your own specialty (or work cited in the *Instructor's Guide*).

This book is not a sequel to my earlier one (Nelson, 2014). Indeed there is very little overlap between these books, which partly explains why certain topics are not covered here. Still other topics will appear in a forthcoming book on light, imaging, and vision. A few of the many other recent books with overlapping goals are listed in "To the Student"; others appear at the ends of chapters.

There are many ways to organize the material: by organism type, by length scale, and so on. I have tried to arrange topics in a way that gradually builds up the framework needed to understand an important and emblematic system in Chapter 11.

Computer-based assignments

> *The difference between a text without problems and a text with problems is like the difference between learning to read a language and learning to speak it.*
> —Freeman Dyson

All of the problems set in this book have been tested on real students. Many ask the student to use a computer. One can learn some of the material without doing this, but I think it's important for students to learn how to write their own short codes, from scratch. It's best to do this not in the vacuum of a course dedicated to programming, but in the context of some problems of independent scientific interest—for example, biophysics. The book's companion Web site features a collection of real experimental datasets to accompany the homework problems. Many reports stress the importance of students working with such data (for example, see National Research Council, 2003).

To do research, students need skills relevant for data visualization, simulation of random variables, and handling of datasets, all of which are covered in this book's problems. Several general-purpose programming environments would work well for this, depending on your own preference, for example, *Mathematica*®, MATLAB®, Octave, Python, R, or Sage. Some of these are free and open source. It's hugely motivating when that beautiful fit to data emerges, and important for students to have this experience early and often.

In my own course, many students arrive with no programming experience. A separate *Student's Guide* gives them some computer laboratory exercises and other suggestions for how to get started. The *Instructor's Guide* gives solutions to these exercises, and to the Problems and Your Turn questions in this book. Keep in mind that programming is very time consuming for beginners; you can probably only assign a few of the longer problems in a semester, and your students may need lots of support.

Classroom demonstrations

One kind of experiential learning is almost unique to physical science classes: We bring a piece of apparatus into the class and show the students some surprising *real* phenomenon—not a simulation, not a metaphor. The *Instructor's Guide* offers some suggestions for where to give demonstrations.

New directions in education

Will life-science students really need this much background in physical science? Although this is not a book about medicine per se, nevertheless many of its goals mesh with recent guidelines for the preparation of premedical students, and specifically for the revised MCAT exam (American Association of Medical Colleges, 2014):[2]

1. "Achieving economies of time spent on science instruction would be facilitated by breaking down barriers among departments and fostering interdisciplinary approaches to science education. Indeed, the need for increased scientific rigor and its relevance to human biology is most likely to be met by more interdisciplinary courses."

2. Premedical students should enter medical school able to

   - "Apply quantitative reasoning and appropriate mathematics to describe or explain phenomena in the natural world."

   - "Demonstrate understanding of the process of scientific inquiry, and explain how scientific knowledge is discovered and validated," as well as "knowledge of basic physical and chemical principles and their applications to the understanding of living systems."

   - "Demonstrate knowledge of how biomolecules contribute to the structure and function of cells."

   - "Apply understanding of principles of how molecular and cell assemblies, organs, and organisms develop structure and carry out function."

   - "Explain how organisms sense and control their internal environment and how they respond to external change."

3. At the next level, students *in* medical school need another set of core competencies, including an understanding of technologies used in medicine.

4. Finally, practicing physicians need to explain to patients the role of complexity and variability, and must be able to communicate approaches to quantitative evidence.

This book may be regarded as showing one model for how physical science and engineering departments can address these goals in their course offerings.

Standard disclaimers

This is a textbook, not a monograph. Many fine points have been intentionally banished to Track 2, to the *Instructor's guide*, or even farther out into deep space. The experiments described here were chosen simply because they illustrated points I needed to make. The citation of original works is haphazard. No claim is made that anything in this book is original. No attempt at historical completeness is implied.

---

[2] See also American Association of Medical Colleges / Howard Hughes Medical Institute, 2009. Similar competencies are listed in the context of biology education in another recent report (American Association for the Advancement of Science, 2011), for example, "apply concepts from other sciences to interpret biological phenomena," "apply physical laws to biological dynamics," and "apply imaging technologies."

Prolog:
A Breakthrough on HIV

Los Alamos, 1994

Alan Perelson was frustrated. For some years, he, and many other researchers, had been staring at an enigmatic graph (Figure 0.1). Like any graph, it consisted of dry, unemotional squiggles. But like any graph, it also told a story.

   The enigmatic feature of the graph was precisely what made HIV so dangerous: After a brief spike, the concentration of virus particles in the blood fell to a low, steady level. Thus, after a short, flu-like episode, the typical patient had no serious symptoms, but remained

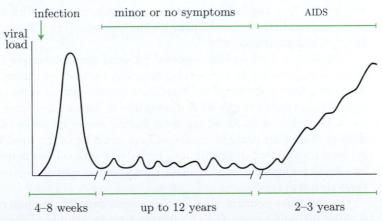

Figure 0.1 [Sketch graph.] **The time course of HIV infection,** representing the progression of the disease as it was understood in the early 1990s. After a brief, sharp peak, the concentration of virus particles in the blood ("viral load") settled down to a low, nearly steady level for up to ten years. During this period, the patient showed no symptoms. Ultimately, however, the viral load increased and the symptoms of full AIDS appeared. [After Weiss, 1993.]

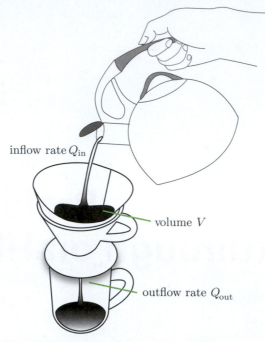

inflow rate Q_{in}

volume V

outflow rate Q_{out}

Figure 0.2 [Metaphor.] **Steady state in a leaky container.** Inflow at a rate Q_{in} replenishes the container, compensating outflow at a rate Q_{out}. If we observe that the volume V of liquid in the container is steady, we can conclude that Q_{out} matches Q_{in}, but we can't determine the actual value of either quantity without more information. In the analogy to viral dynamics, Q_{in} corresponds to the body's production of virus particles and Q_{out} to the immune system's rate of virus clearance (see Chapter 1).

contagious, for up to ten years. Inevitably, however, the virus level eventually rose again, and the patient died.

In the early 1990s, many researchers believed that these facts implied that HIV was a slow virus, which remained in the body, nearly dormant, for years before rising sharply in number. But how could such a long latency period be possible? What was happening during those ten years? How could the patient's immune system fight the virus effectively at first, and then ultimately succumb?

Perelson and others had suspected for some time that maybe HIV was not slow or dormant at all during the apparent latent period. He made an analogy to a physical system: If we see a leaky container that nevertheless retains water at some constant level, we can conclude that there must be water flowing into it (Figure 0.2). But we can't determine *how fast* water is flowing in. All we can say is that the rate of inflow equals the rate of outflow. Both of those rates could be small—or both could be large. Applying this idea to HIV, Perelson realized that, during the long period of low blood concentration, the virus might actually be multiplying rapidly, but after the brief initial episode, it could be eliminated by the body just as rapidly.

A real leaky container has another simple property reminiscent of the HIV data: Because the outflow rate $Q_{out}(V)$ increases as the volume of the water (and hence its pressure at the exit point) goes up, the system can *self-adjust* to a steady state, no matter what inflow rate Q_{in} we select. Similarly, different HIV-infected patients have quite different steady levels of virus concentration, but all maintain that steady level for long periods.

Perelson was head of the Theoretical Biology and Biophysics Group at Los Alamos National Laboratory. By 1994, he had already developed a number of elaborate mathematical models in an attempt to see if they could describe clinical reality. But his models were full of unknown parameters. The available data (Figure 0.1) didn't help very much. How could he make progress without some better knowledge of the underlying cellular events giving rise to the aggregate behavior?

New York City, 1994

David Ho was puzzled. As the head of the Aaron Diamond AIDS Research Center, he had the resources to conduct clinical trials. He also had access to the latest anti-HIV drugs and had begun tests with ritonavir, a "protease inhibitor" designed to stop the replication of the HIV virus.

Something strange was beginning to emerge from these trials: The effect of treatment with ritonavir seemed to be a very *sudden* drop in the patient's total number of virus particles. This was a paradoxical result, because it was known that ritonavir by itself didn't destroy existing virus particles, but simply stopped the creation of new ones. If HIV were really a slow virus, as many believed, wouldn't it also *stay around* for a long time, even once its replication was stopped? What was going on?

Also, it had been known for some time that patients treated with antiviral drugs got much better, but only temporarily. After a few months, ritonavir and other such drugs always lost their effectiveness. Some radically new viewpoint was needed.

Hilton Head Island, 1994

Perelson didn't know about the new drugs; he just knew he needed quantitative data. At a conference on HIV, he heard a talk by one of Ho's colleagues, R. Koup, on a different topic. Intrigued, he later phoned to discuss Koup's work. The conversation turned to the surprising results just starting to emerge with ritonavir. Koup said that the group was looking for a collaborator to help make sense of the strange data they had been getting. Was Perelson interested? He was.

Ho and his colleagues suspected that simply measuring viral populations before and after a month of treatment (the usual practice at the time) was not showing enough detail. The crucial measurement would be one that examined an asymptomatic patient, not one with full AIDS, and that monitored the blood virus concentration *every day* after administering the drug.

More clinical trials followed. Measurements from patient after patient told the same story (Figure 0.3): *Shutting down the replication of virus particles brought a hundredfold drop in their population in 2–3 weeks.*

Perelson and Ho were stunned. The rapid drop implied that the body was constantly clearing the virus at a tremendous rate; in the language of Figure 0.2, Q_{out} was huge. That could only mean that, without the drug, the production rate Q_{in} was also huge. Similar results were soon obtained with several other types of antiviral drugs. The virus wasn't dormant at all; it was replicating like mad. Analysis of the data yielded a numerical value for Q_{out}, as we'll see in Chapter 1. Using this measurement, the researchers estimated that the typical asymptomatic patient's body was actually making at least *a billion* new virus particles each day.[3]

As often happens, elsewhere another research group, led by George Shaw, independently pursued a similar program. This group, too, contained an "outsider" to AIDS

---

[3]Later, more refined estimates showed that the average production rate was actually even larger than this initial lower bound.

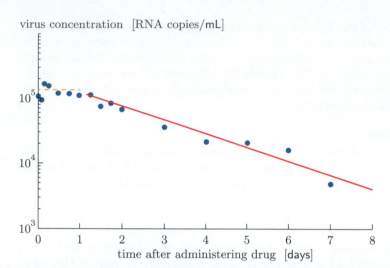

Figure 0.3 [Experimental data with preliminary fit.] **Virus concentration in a patient's blood ("viral load") after treatment with a protease inhibitor,** showing the rapid decline after treatment. In this semilog plot, the *solid line* shows the time course corresponding to elimination of half the total viral population every 1.4 days. The *dashed line* highlights a deviation from this behavior at early times (the "initial plateau"); see Chapter 1. [Data from Perelson, 2002; see Dataset 1.]

research, a mathematician named Martin Nowak. Both groups published their findings simultaneously in *Nature*. The implications of this work were profound. Because the virus is replicating so rapidly, it can easily mutate to find a form resistant to any given drug.[4] Indeed, as we'll see later, the virus mutates often enough to generate every possible single-base mutation every few hours. Hence, every infected patient *already* has some resistant mutant viruses before the drug is even administered; in a couple of weeks, this strain takes over and the patient is sick again. The same observation also goes to the heart of HIV's ability to evade total destruction by the body: It is constantly, furiously, playing cat-and-mouse with the patient's immune system.

But what if we simultaneously administer *two* antiviral drugs? It's not so easy for a virus to sample every possible *pair* of mutations, and harder still to get three or more. And in fact, subsequent work showed that "cocktails" of three different drugs can halt the progression of HIV infection, apparently indefinitely. The patients taking these drugs have not been cured; they still carry low levels of the virus. But they are alive, thanks to the treatment.

The message

This book is about basic science. It's not about AIDS, nor indeed is it directly about medicine at all. But the story just recounted has some important lessons.

The two research groups mentioned above made significant progress against a terrible disease. They did this by following some general steps:

1. Assemble (or join) an interdisciplinary team to look at the problem with different sets of tools;
2. Apply simple physical metaphors (the leaky container of water) and the corresponding disciplines (dynamical systems theory, an area of physics) to make a hypothesis; and

---

[4]Actually the *fact* of mutation had already been established a few years earlier. Prior to the experiments described here, however, it was difficult to understand how mutation could lead to fast evolution.

3. Perform experiments specifically designed to give new, quantitative data to support or refute the hypothesis.

This strategy will continue to yield important results in the future.

The rest of the book will get a bit dry in places. There will be many abstract ideas. But abstract ideas do matter when you understand them well enough to find their concrete applications. In fact, sometimes their abstractness just reflects the fact that they are so widely applicable: Good ideas can jump like wildfires from one discipline to another. Let's get started.

PART I

First Steps

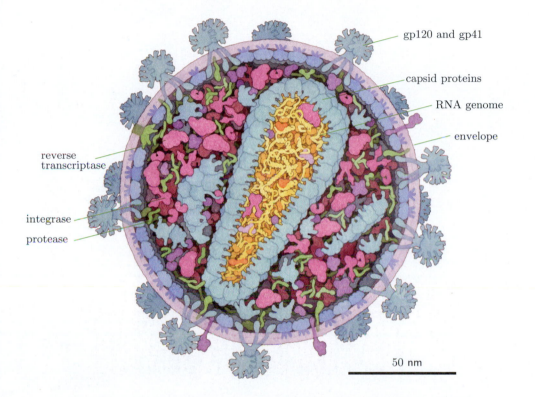

gp120 and gp41

capsid proteins

RNA genome

envelope

reverse
transcriptase

integrase

protease

50 nm

[Artist's reconstructions based on structural data.] **A human immunodeficiency virus particle** (virion), surrounded by its lipid membrane envelope. The envelope is studded with gp120, the protein that recognizes human T cells. The envelope encloses several enzymes (proteins that act as molecular machines), including HIV protease, reverse transcriptase, and integrase. Two RNA strands carrying the genome of HIV are packaged in a cone-shaped protein shell called the capsid. See also Media 1. [Courtesy David S Goodsell.]

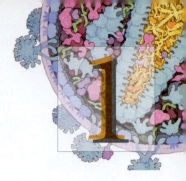

Virus Dynamics

We all know that Art is not truth. Art is a lie that makes us realize the truth.
—Pablo Picasso

1.1 First Signpost

The Prolog suggested a three-step procedure to make headway on a scientific problem (see page 4). Unfortunately, the experiment that can be performed usually does not directly yield the information we desire, and hence does not directly confirm or disprove our original hypothesis. For example, this chapter will argue that testing the viral mutation hypothesis in the Prolog actually requires information not directly visible in the data that were available in 1995.

Thus, a fourth step is almost always needed:

4. Embody the physical metaphor (or **physical model**) in mathematical form, and attempt to fit it to the experimental data.

In this statement, **fit** means "adjust one or more numbers appearing in the model." For each set of these **fit parameter** values that we choose, the model makes a prediction for some experimentally measurable quantity, which we compare with actual observations. If a successful fit can be found, then we may call the model "promising" and begin to draw tentative conclusions from the parameter values that yield the best fit. This chapter will take a closer look at the system discussed in the Prolog, illustrating how to construct a physical model, express it in mathematical form, fit it to data, evaluate the adequacy of the fit, and draw conclusions. The chapter will also get you started with some of the basic computer skills needed to carry out these steps.

Each chapter of this book begins with a biological question to keep in mind as you read, and an idea that will prove relevant to that question.
This chapter's Focus Question is
Biological question: Why did the first antiviral drugs succeed briefly against HIV, but then fail?
Physical idea: A physical model, combined with a clinical trial designed to test it, established a surprising feature of HIV infection.

1.2 Modeling the Course of HIV Infection

We begin with just a few relevant facts about HIV, many of which were known in 1995. It will not be necessary to understand these in detail, but it is important to appreciate just how much was already known at that time.

1.2.1 Biological background

In 1981, the US Centers for Disease Control noticed a rising incidence of rare diseases characterized by suppression of the body's immune system. As the number of patients dying of normally nonlethal infections rose, it became clear that some new disease, with an unknown mechanism, had appeared. Eventually, it was given the descriptive name acquired immune deficiency syndrome (AIDS).

Two years later, research teams in France and the United States showed that a virus was present in lymph fluid taken from AIDS patients. The virus was named human immunodeficiency virus (HIV). To understand why HIV is so difficult to eradicate, we must very briefly outline its mechanism as it was later understood.

HIV consists of a small package (the virus particle, or **virion**) containing some nucleic acid (two copies of the genome), a protective protein shell (or **capsid**), a few other protein molecules needed for the initial steps of infection, and a surrounding **envelope** (see the figure on page 7). In a **retrovirus** like HIV, the genome takes the form of RNA molecules, which must be converted to DNA during the infection.[1]

The genome of HIV is extremely short, roughly 10 000 bases. It contains nine genes, which direct the synthesis of just 19 different proteins. Three of the genes code for proteins that perform the following functions:[2]

- *gag* generates the four proteins that make the virion's capsid.
- *pol* generates three protein machines (enzymes): A **reverse transcriptase** converts the genome to DNA, an **integrase** helps to insert the viral DNA copy into the infected cell's own genome, and a **protease** cleaves (cuts) the product of *gag* (and of *pol* itself) into separate proteins (Figure 1.1).
- *env* generates a protein that embeds itself in the envelope (called gp41), and another (called gp120; see page 7) that attaches to gp41, protrudes from the envelope, and helps the virus to target and enter its host.

HIV targets some of the very immune cells that normally protect us from disease. Its gp120 protein binds to a receptor found on a human immune cell (the **CD4+ helper T cell**,

---

[1] The normal direction of information transmission is from DNA to RNA; a *retro*virus is so named because it reverses this flow.
[2] That is, they are "structural" genes. The other six genes code for transcription factors; see Chapter 9.

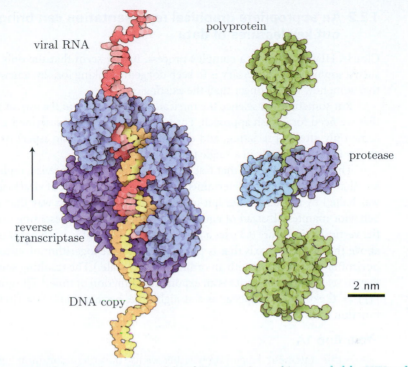

Figure 1.1 [Artist's reconstructions based on structural data.] **Two protein machines needed for HIV replication.** *Left:* Reverse transcriptase is shown transcribing the viral RNA into a DNA copy. This molecular machine moves along the RNA (*arrow*), destroying it as it goes and synthesizing the corresponding DNA sequence. *Right:* HIV protease cleaves the polyprotein generated by a viral gene (in this case, *gag*) into individual proteins. Many antiviral drugs block the action of one of these two enzymes. [Courtesy David S Goodsell.]

or simply "T cell"). Binding triggers fusion of the virion's envelope with the T cell's outer membrane, and hence allows entry of the viral contents into the cell. The virion includes some ready-to-go copies of reverse transcriptase, which converts the viral genome to DNA, and integrase, which incorporates the DNA transcript into the host cell's own genome. Later, this rogue DNA directs the production of more viral RNA. Some of the new RNA is translated into new viral proteins. The rest gets packaged with those proteins, to form several thousand new virions from each infected cell. The new virions escape from the host cell, killing it and spreading the infection. The immune system can keep the infection in check for many years, but eventually the population of T cells falls. When it drops to about 20% of its normal value, then immune response is seriously compromised and the symptoms of AIDS appear.

The preceding summary is simplified, but it lets us describe some of the drugs that were available by the late 1980s. The first useful anti-HIV drug, zidovudine (or AZT), blocks the action of the reverse transcriptase molecule. Other **reverse transcriptase inhibitors** have since been found. As mentioned in the Prolog, however, their effects are generally short lived. A second approach targets the protease molecule; **protease inhibitors** such as ritonavir result in defective (not properly cleaved) proteins within the virion. The effects of these drugs also proved to be temporary.

1.2.2 An appropriate graphical representation can bring out key features of data

Clearly, HIV infection is a complex process. It may seem that the only way to checkmate such a sophisticated adversary is to keep doggedly looking for, say, a new protease inhibitor that somehow works longer than the existing ones.

But sometimes in science the intricate details can *get in the way* of the viewpoint shift that we need for a fresh approach. For example, the Prolog described a breakthrough that came only after appreciating, and documenting, a very basic aspect of HIV infection: its ability to evolve rapidly in a single patient's body.

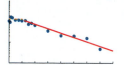

Figure 0.3 (page 4)

The Prolog suggested that fast mutation is possible if the viral replication rate is high, and that this rate could be determined by examining the falloff of viral load when production was halted by a drug. The graph in Figure 0.3 is drawn in a way that makes a particular behavior manifest: Instead of equally spaced tick marks representing uniform intervals on the vertical axis, Figure 0.3 uses a logarithmic scale. That is, each point is drawn at a height above the horizontal axis that is proportional to the logarithm of virus population. (The horizontal axis is drawn with an ordinary linear scale.) The resulting **semilog plot** makes it easy to see when a data series is an exponential function of time:[3] The graph of the function $f(t) = C \exp(kt)$ will appear as a straight line, because $\ln f(t) = (\ln C) + kt$ is a linear function of t.

Your Turn 1A

Does this statement depend on whether we use natural or common logarithms?

Log axes are usually labeled at each power of 10, as shown in Figure 0.3. The unequally spaced "minor" tick marks between the "major" ones in the figure are a visual cue, alerting the reader to the log feature. Most computer math packages can create such axes for you. Note that the tick marks on the vertical axis in Figure 0.3 represent 1000, 2000, 3000, . . . , 9000, 10 000, 20 000, 30 000, . . . , 900 000, 1 000 000. In particular, *the next tick after* 10^4 *represents* $2 \cdot 10^4$, *not* 11 000, and so on.

1.2.3 Physical modeling begins by identifying the key actors and their main interactions

Although the data shown in Figure 0.3 are suggestive, they are not quite what we need to establish the hypothesis of rapid virus mutation. The main source of mutations is reverse transcription.[4] Reverse transcription usually occurs only once per T cell infection. So if we wish to establish that there are many opportunities for mutation, then we need to show that many *new T cell infections* are occurring per unit time. This is not the same thing as showing that new virions are being created rapidly, so we must think more carefully about what we can learn from the data. Simply making a graph is not enough.

Moreover, the agreement between the data (dots in Figure 0.3) and the simple expectation of exponential falloff (line) is actually not very good. Close inspection shows that the rapid fall of virus population does not begin at the moment the drug is administered

---

[3]The prefix "semi-" reminds us that only one axis is logarithmic. A "log-log plot" uses logarithmic scales on *both* axes; we'll use this device later to bring out a different feature in other datasets.

[4]The later step of making new viral genomes from the copy integrated into the T cell's DNA is much more accurate; we can neglect mutations arising at that step.

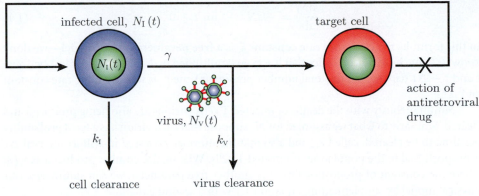

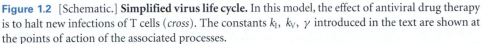

Figure 1.2 [Schematic.] **Simplified virus life cycle.** In this model, the effect of antiviral drug therapy is to halt new infections of T cells (*cross*). The constants k_I, k_V, γ introduced in the text are shown at the points of action of the associated processes.

("time 0"). Instead, the data shown in the figure (and similar graphs from other patients) show an initial pause in virus population at early times, prior to the exponential drop. It's not surprising, really—so far we haven't even attempted to write any quantitative version of our initial intuition.

To do better than this, we begin by identifying the relevant processes and quantities affecting virus population. Infected T cells produce free virions, which in turn infect new cells. Meanwhile, infected T cells eventually die, and the body's defenses also kill them; we will call the combined effect **clearance** of infected cells. The immune system also destroys free virions, another kind of clearance. To include these processes in our model, we first assign names to all the associated quantities. Thus, let t be time after administering the antiviral drug. Let $N_I(t)$ be the number of infected T cells at time t, and $N_V(t)$ the number of free virions in the blood (the **viral load**).[5]

Before drug treatment (that is, at time $t < 0$), production and removal processes roughly balance, leading to a nearly steady (**quasi-steady**) state, the long period of low virus population seen in Figure 0.1. In this state, the rate of new infections must balance T cell clearance—so finding their rate of clearance will tell us the rate of new infections in the quasi-steady state, which is what we are seeking.

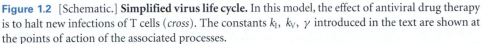

Figure 0.1 (page 1)

Let's simplify by assuming that the antiviral drug completely stops new infections of T cells. From that moment, uninfected T cells become irrelevant—they "decouple" from infected T cells and virions. We also simplify by assuming that each infected T cell has some fixed chance of being cleared in any short time interval. That chance depends on the duration Δt of the interval, and it becomes zero if $\Delta t = 0$, so it's reasonable to suppose that it's the product of Δt times a constant, which we'll call k_I.[6] These assumptions imply that, after administering the drug, N_I changes with time according to a simple equation:

---

[5]The concentration of virions is therefore $N_V(t)$ divided by the blood volume. Because the blood volume is constant in any particular patient, we can work with either concentration or total number.

[6] $\boxed{T_2}$ We are ignoring the possibility of saturation, that is, decrease in clearance probability per infected cell per time when the concentration of infected cells is so high that the immune system is overloaded. This assumption will not be valid in the late stages of infection. We also assume that an infected T cell is unlikely to divide before being cleared.

$$\frac{dN_I}{dt} = -k_I N_I \qquad \text{for } t \geq 0. \tag{1.1}$$

In this formula, the **clearance rate constant** k_I is a **free parameter** of the model—we don't know its value in advance. Notice that it appears multiplied by N_I: The number lost between t and $t + \Delta t$ depends on the total number present at time t, as well as on the rate constant and Δt.

Simultaneously with the deaths of infected T cells, virions are also being produced and cleared. Similarly to what we assumed for N_I, suppose that each virion has a fixed probability per time to be cleared, called k_V, and also that the number *produced* in a short interval Δt is proportional to the population of infected T cells. Writing the number produced as γN_I, where the constant of proportionality γ is another free parameter, we can summarize our physical model by supplementing Equation 1.1 with a second equation:

$$\frac{dN_V}{dt} = -k_V N_V + \gamma N_I. \tag{1.2}$$

Figure 1.2 summarizes the foregoing discussion in a cartoon. For reference, the following quantities will appear in our analysis:

t	time since administering drug
$N_I(t)$	population of infected T cells; its initial value is N_{I0}
$N_V(t)$	population of virions; its initial value is N_{V0}
k_I	clearance rate constant for infected T cells
k_V	clearance rate constant for virions
γ	rate constant for virion production per infected T cell
β	an abbreviation for γN_{I0}

1.2.4 Mathematical analysis yields a family of predicted behaviors

With the terminology in place, we can describe our plan more concretely than before:

- We want to test the hypothesis that the virus evolves within a single patient.
- To this end, we'd like to find the rate at which T cells get infected in the quasi-steady state, because virus mutation is most likely to happen at the error-prone reverse transcription step, which happens once per infection.
- But the production rate of newly infected T cells was not directly measurable in 1995. The measurable quantity was the viral load N_V as a function of time after administering the drug.
- Our model, Equations 1.1–1.2, connects what we've got to what we want, because (*i*) in the quasi-steady state, the infection rate is equal to the rate k_I at which T cells are lost, and (*ii*) k_I is a parameter that we can extract by fitting the model in Equations 1.1–1.2 to data about the *non*-steady state after administering an antiviral drug.

Notice that Equation 1.1 doesn't involve N_V at all; it's one equation in one unknown function, and a very famous one too. Its solution is exponential decay: $N_I(t) = N_{I0}e^{-k_I t}$. The constant N_{I0} is the initial number of infected T cells at time zero. We can just substitute that solution into Equation 1.2 and then forget about Equation 1.1. That is,

$$\frac{dN_V}{dt} = -k_V N_V + \gamma N_{I0}e^{-k_I t}. \tag{1.3}$$

Here k_I, k_V, γ, and N_{I0} are four unknown quantities. But we can simplify our work by noticing that two of them enter the equation only via their product. Hence, we can replace γ and N_{I0} by a single unknown, which we'll abbreviate as $\beta = \gamma N_{I0}$. Because the experiments did not actually measure the population of infected T cells, we need not predict it, and hence we won't need the separate values of γ and N_{I0}.

We could directly solve Equation 1.3 by the methods of calculus, but it's usually best to try for an intuitive understanding first. Think about the metaphor in Figure 0.2, where the volume of water in the middle chamber at any moment plays the role of N_V. For a real leaky container, the rate of outflow depends on the pressure at the bottom, and hence on the level of the water; similarly, Equation 1.2 specifies that the clearance (outflow) rate at time t depends on $N_V(t)$. Next, consider the inflow: Instead of being a constant equal to the outflow, as it is in the steady state, Equation 1.3 gives the inflow rate as $\beta e^{-k_I t}$ (see Figure 0.2).

$\beta e^{-k_I t}$

N_V

$-k_V N_V$

Figure 0.2 (page 2)

Our physical metaphor now lets us guess some general behavior. If $k_I \gg k_V$, then we get a burst of inflow that quickly shuts off, before much has had a chance to run out. So after this brief **transient** behavior, N_V falls exponentially with time, in a way controlled by the decay rate constant k_V. In the opposite extreme case, $k_V \gg k_I$, the container drains nearly as fast as it's being filled;[7] the water level simply tracks the inflow. Thus, again the water level falls exponentially, but this time in a way controlled by the *inflow* decay rate constant k_I.

Our intuition from the preceding paragraph suggests that the long-time behavior of the solution to Equation 1.3 is proportional either to $e^{-k_I t}$ or $e^{-k_V t}$, depending on which rate constant is smaller. We can now try to guess a **trial solution** with this property. In fact, the function

$$N_V(t) \overset{?}{=} Xe^{-k_I t} + (N_{V0} - X)e^{-k_V t}, \tag{1.4}$$

where X is any constant value, has the desired behavior. Moreover, Equation 1.4 equals N_{V0} at time zero. We now ask if we can choose a value of X that makes Equation 1.4 a solution to Equation 1.3. Substitution shows that indeed it works, if we make the choice $X = \beta/(k_V - k_I)$.

Your Turn 1B

a. Confirm the last statement.
b. The trial solution, Equation 1.4, seems to be the sum of two terms, each of which decreases in time. So how could it have the initial pause that is often seen in data (Figure 0.3)?

Figure 0.3 (page 4)

$\boxed{T_2}$ *Section 1.2.4′ (page 21) discusses the hypothesis of viral evolution within a single patient in greater detail.*

1.2.5 Most models must be fitted to data

What has been accomplished so far? We proposed a physical model with three unknown parameters, k_I, k_V, and β. One of these, k_I, is relevant to our hypothesis that virus is rapidly infecting T cells, so we'd like to know its numerical value. Although the population of infected T cells was not directly measurable in 1995, we found that the model makes a

---

[7]The water never drains completely, because in our model the rate of outflow goes to zero as the height goes to zero. This may not be a good description of a real bucket, but it's reasonable for a virus, which becomes harder for the immune system to find as its concentration decreases.

virus concentration [RNA/mL]

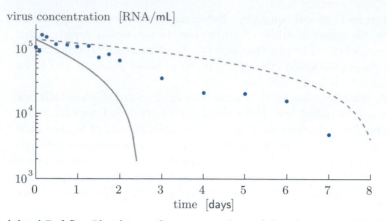

Figure 1.3 [Experimental data.] **Bad fits.** *Blue dots* are the same experimental data that appeared in Figure 0.3 (page 4). The *solid curve* shows the trial solution to our model (Equation 1.4), with a bad set of parameter values. Although the solution starts out at the observed value, it quickly deviates from the data. However, a different choice of parameters does lead to a function that works (see Problem 1.4). The *dashed curve* shows a fit to a different functional form, one not based on any physical model. Although it starts and ends at the right values, in fact, *no* choice of parameters for this model can fit the data.

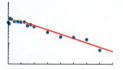

Figure 0.3 (page 4)

prediction for the virus number $N_V(t)$, which *was* observable then. Fitting the model to experimentally measured virus concentration data can thus help us determine the desired rate constant k_I.

The math gave a prediction for $N_V(t)$ in terms of the initial viral load N_{V0} and the values of k_I, k_V, and β. The prediction does have the observed qualitative behavior that after an initial transient, $N_V(t)$ falls exponentially, as seen in Figure 0.3. The remaining challenges are that

- We haven't yet determined the unknowns k_I, k_V, and β.
- The model itself needs to be evaluated critically; all of the assumptions and approximations that went into it are, in principle, suspect.

To gain some confidence in the model, and find the unknown parameter values, we must attempt a detailed comparison with data. We would especially hope to find that different patients, despite having widely different initial viral load N_{V0}, nevertheless all have similar values of k_I. Then the claim that this rate is "very large" (and hence may allow for virus evolution in a single patient) may have some general validity.

Certainly it's not difficult to find things that *don't* work! Figure 1.3 shows the experimental data, along with two functions that don't fit them. One of these functions belongs to the class of trial solutions that we constructed in the preceding section. It looks terrible, but in fact you'll show in Problem 1.4 that a function of this form can be made to look like the data, with appropriate choices of the fitting parameters. The other function shown in the figure is an attempt to fit the data with a simple linear function, $N_V(t) \overset{?}{=} A - Bt$. We can make this function pass through our data's starting and ending points, but there is no way to make it fit all of the data. It's not surprising—we have no physical model leading us to expect a linear falloff with time. If we *were* considering such a model, however, our inability to fit it would have to be considered as strong evidence that it is wrong.

After you work through Problem 1.4, you'll have a graph of the best-fitting version of the model proposed in the preceding section. Examining it will let you evaluate the main hypothesis proposed there, by drawing a conclusion from the fit parameter values.

1.2.6 Overconstraint versus overfitting

Our physical model includes crude representations of some processes that we knew must be present in our system, although the model neglects others. It's not perfect, but the agreement between model and data that you'll find in Problem 1.4 is detailed enough to make the model seem "promising." Fitting the model requires that we adjust three unknown parameters, however. There is always the possibility that a model is fundamentally wrong (omits some important features of reality) but nevertheless can be made to *look* good by tweaking the values of its parameters. It's a serious concern, because in that case, the parameter values that gave the fortuitously good fit can be meaningless. Concerns like these must be addressed any time we attempt to model any system.

In our case, however, Problem 1.4 shows that you can fit *more than three* data points by an appropriate choice of three parameters. We say that the data **overconstrain** the model, because there are more conditions to be met than there are parameters to adjust. When a model can match data despite being overconstrained, that fact is unlikely to be a coincidence. The opposite situation is often called **overfitting**; in extreme cases, a model may have so many free fit parameters that it can fit almost any data, regardless of whether it is correct.

Successfully fitting an overconstrained model increases our confidence that it reflects reality, even if it proposes the existence of *hidden actors*, for which we may have little or no direct evidence. In the HIV example, these actors were the T cells, whose population was not directly observable in the original experiments.

T_2 | *Section 1.2.6′ (page 21) discusses in more detail the sense in which our model is overdetermined, and outlines a more realistic model for virus dynamics.*

1.3 Just a Few Words About Modeling

Sometimes we examine experimental data, and their form immediately suggests some simple mathematical function. We can write down that function with some parameters, plot it alongside the data, and adjust the parameters to optimize the fit. In Problem 1.5 you'll use this approach, which is sometimes called "blind fitting." Superficially, it resembles what we did with HIV data in this chapter, but there is a key difference.

Blind fitting is often a convenient way to *summarize* existing data. Because many systems respond continuously as time progresses or a parameter is changed, choosing a simple smooth function to summarize data also lets us **interpolate** (predict what would have been measured at points lying between actual measurements). But, as you'll see in Problem 1.5, blind fitting often fails spectacularly at **extrapolation** (predicting what would have been measured at points lying *outside* the range of actual measurements).[8] That's because the mathematical function that we choose may not have any connection to any underlying mechanism giving rise to the behavior.

This chapter has followed a very different procedure. We first imagined a plausible mechanism consistent with other things we knew about the world (a physical model), and then embodied it in mathematical formulas. The physical model may be wrong; for example, it may neglect some important players or interactions. But if it gives nontrivial successful predictions about data, then we are encouraged to test it outside the range of conditions studied in the initial experiments. If it passes those tests as well, then it's "promising,"

---

[8] Even interpolation can fail: We may have so few data points that a simple function seems to fit them, even though no such relation exists in reality. A third pitfall with blind fitting is that, in some cases, a system's behavior does *not* change smoothly as parameters are changed. Such "bifurcation" phenomena are discussed in Chapter 10.

and we are justified in trying to apply its results to other situations, different from the first experiments. Thus, successful fitting of a model to HIV data suggested a successful treatment strategy (the multidrug treatment described in the Prolog). Other chapters in this book will look at different stories.

In earlier times, a theoretical "model" often just meant some words, or a cartoon. Why must we clutter such images with math? One reason is that equations force a model to be precise, complete, and self-consistent, and they allow its full implications to be worked out, including possible experimental tests. Some "word models" sound reasonable but, when expressed precisely, turn out to be self-inconsistent, or to depend on physically impossible values for some parameters. In this way, modeling can formulate, explore, and often reject potential mechanisms, letting us focus only on experiments that test the promising ones.

Finally, skillful modeling can tell us in advance whether an experiment is likely to be able to discriminate two or more of the mechanisms under consideration, or point to what changes in the experimental design will enhance that ability. For that reason, modeling can also help us make more efficient use of the time and money needed to perform experiments.

THE BIG PICTURE

This chapter has explored a minimal, reductionist approach to a complex biological system. Such an approach has strengths and weaknesses. One strength is generality: The same sort of equations that are useful in understanding the progression of HIV have proven to be useful in understanding other infections, such as hepatitis B and C.

More broadly, we may characterize a good scientific experience as one in which a puzzling result is explained in quantitative detail by a simplified model, perhaps involving some sort of equations. But a *great* scientific experience can arise when you find that some totally different-seeming system or process obeys the *same* equations. Then you gain insights about one system from your experience with the other. In this chapter, the key equations had associations with systems like leaky containers, which helped lead us to the desired solution.

Our physical model of HIV dynamics still has a big limitation, however. Although the fit to data supports the picture of a high virus production rate, this does not completely validate the overall picture proposed in the Prolog (page 4). That proposal went on to assert that *high production somehow leads to the evolution of drug resistance.* In fact, although this connection is intuitive, the framework of this chapter cannot establish it quantitatively. When writing down Equations 1.1–1.2, we tacitly made the assumption that T cell and virus populations were quantities that changed continuously in time. This assumption allowed us to apply some familiar techniques from calculus. But really, those populations are *integers,* and so must change discontinuously in time.

In many situations, the difference is immaterial. Populations are generally huge numbers, and their graininess is too small to worry about. But in our problem we are interested in the rare, chance mutation of *just one* virion from susceptible to resistant (Figure 1.4). We will need to develop some new methods, and intuition, to handle such problems.

The word "chance" in the preceding paragraph highlights another gap in our understanding so far: Our equations, and calculus in general, describe *deterministic* systems, ones for which the future follows inevitably once the present is sufficiently well known. It's an approach that works well for some phenomena, like predicting eclipses of the Sun. But clockwork determinism is not very reminiscent of Life. And even many purely physical phenomena, we will see, are inherently probabilistic in character. Chapters 3–7 will develop the ideas we need to introduce randomness into our physical models of living systems.

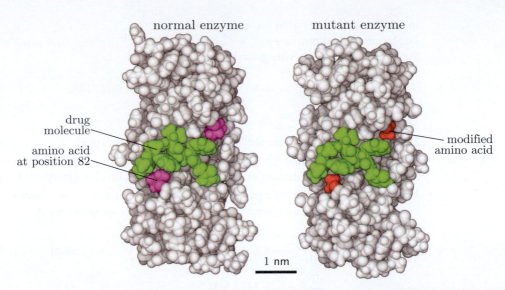

Figure 1.4 [Artist's reconstructions based on structural data.] **Antiviral drugs are molecules that bind tightly to HIV's enzymes** and block their action. HIV protease can become resistant to drugs by mutating certain of its amino acids; such a mutation changes its shape slightly, degrading the fit of the drug molecule to its usual binding site. The drug ritonavir is shown in *green*. *Left:* The amino acids at position 82 in each of the enzyme's two protein chains are normally valines (*magenta*); they form close contacts with the drug, stabilizing the binding. The bound drug molecule then obstructs the enzyme's active site, preventing it from carrying out its function (see Figure 1.1). *Right:* In the mutant enzyme, this amino acid has been changed to the smaller alanine (*red*), weakening the contact slightly. Ritonavir then binds poorly, and so does not interfere with the enzyme's activity even when it is present. [Courtesy David S Goodsell.]

KEY FORMULAS

Throughout the book, closing sections like this one will collect useful formulas that appeared in each chapter. In this chapter, however, the section also includes formulas from your previous study of math that will be needed later on.

- *Mathematical results:* Make sure you recall these formulas and how they follow from Taylor's theorem. Some are valid only when x is "small" in some sense.

$$\exp(x) = 1 + x + \cdots + \frac{1}{n!}x^n + \cdots$$

$$\cos(x) = 1 - \frac{1}{2!}x^2 + \frac{1}{4!}x^4 - \cdots$$

$$\sin(x) = x - \frac{1}{3!}x^3 + \frac{1}{5!}x^5 \cdots$$

$$1/(1-x) = 1 + x + \cdots + x^n + \cdots$$

$$\ln(1-x) = -x - \frac{1}{2}x^2 - \cdots - \frac{1}{n}x^n - \cdots$$

$$\sqrt{1+x} = 1 + \frac{1}{2}x - \frac{1}{8}x^2 + \cdots$$

In addition, we'll later need these formulas:

The binomial theorem: $(x+y)^M = C_{M,0}x^M y^0 + C_{M,1}x^{M-1}y^1 + \cdots + C_{M,M}x^0 y^M$, where the binomial coefficients are given by

$$C_{M,\ell} = M!/\big(\ell!(M-\ell)!\big) \text{ for } \ell = 0, \ldots, M.$$

The Gaussian integral: $\int_{-\infty}^{\infty} dx \, \exp(-x^2) = \sqrt{\pi}$.

The compound interest formula:[9] $\lim_{M\to\infty} \left(1 + \frac{a}{M}\right)^M = \exp(a)$.

- *Continuous growth/decay:* The differential equation $dN_I/dt = kN_I$ has solution $N_I(t) = N_{I0}\exp(kt)$, which displays exponential decay (if k is negative) or growth (if k is positive).
- *Viral dynamics model:* After a patient begins taking antiviral drugs, we proposed a model in which the viral load and population of infected T cells are solutions to

$$\frac{dN_I}{dt} = -k_I N_I \qquad \text{for } t \geq 0, \tag{1.1}$$

$$\frac{dN_V}{dt} = -k_V N_V + \gamma N_I. \tag{1.2}$$

FURTHER READING

Semipopular:
On overfitting: Silver, 2012, chapt. 5.

Intermediate:
HIV: Freeman & Herron, 2007.
Modeling and HIV dynamics: Ellner & Guckenheimer, 2006, §6.6; Nowak, 2006; Otto & Day, 2007, chapt. 1; Shonkwiler & Herod, 2009, chapt. 10.

Technical:
Ho et al., 1995; Nowak & May, 2000; Perelson & Nelson, 1999; Wei et al., 1995. Equations 1.1 and 1.2 appeared in Wei et al., 1995, along with an analysis equivalent to Section 1.2.4.

---

[9]The left side of this formula is the factor multiplying an initial balance on a savings account after one year, if interest is compounded M times a year at an annual interest rate a.

T_2 **Track 2**

1.2.4′ Exit from the latency period

Prior to the events of 1995, Nowak, May, and Anderson had already developed a theory for the general behavior shown in Figure 0.1 (see Nowak, 2006, chapt. 10). According to this theory, during the initial spike in viral load, one particular strain of HIV becomes dominant, because it reproduces faster than the others. The immune system manages to control this one strain, but over time it mutates, generating diversity. Eventually, the immune system gets pushed to a point beyond which it is unable to cope simultaneously with all the strains that have evolved by mutation, and the virus concentration rises rapidly. Meanwhile, each round of mutation stimulates a new class of T cells to respond, more and more of which are already infected, weakening their response.

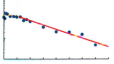

Figure 0.1 (page 1)

T_2 **Track 2**

1.2.6′a Informal criterion for a falsifiable prediction

The main text stated that our result was significant because we could fit many (more than three) data points by adjusting only three unknown parameters: k_I, k_V, and β. It's a bit more precise to say that the data in Figure 0.3 have several independent "visual features": the slope and intercept of the final exponential decay line, the initial slope and value N_{V0}, and the sharpness of the transition from initial plateau to exponential decay. Of these five features, N_{V0} was already used in writing the solution, leaving four that must be fit by parameter choice. But we have only three parameters to adjust, which in principle makes our trial solution a **falsifiable prediction**: There is no mathematical guarantee that *any* choice of parameters can be found that will fit such data. If we do find a set of values that fit, we may at least say that the data have missed an opportunity to falsify the model, increasing our confidence that the model may be correct. In this particular case, none of the visual features are very precisely known, due to scatter in the data and the small number of data points available. Thus, we can only say that (*i*) the data are not qualitatively inconsistent with the model, but (*ii*) the data *are* inconsistent with the value $(k_I)^{-1} \approx 10\,\text{years}$ suggested by the hypothesis of a slow virus.

Figure 0.3 (page 4)

1.2.6′b More realistic viral dynamics models

Many improvements to the model of HIV dynamics in the main text have been explored. For example, we supposed that no new infections occur after administering the drug, but neither of the drug classes mentioned in the text work in exactly this way. Some instead block reverse transcription after virus entry; such a drug may be only partially effective, so that new infections of T cells continue, at a reduced rate, after administration of the drug. Other drugs seek to stop the production of "competent" virions; these, too, may be only partly effective. A more complex set of equations incorporating these ideas appeared in Perelson (2002). Letting $N_U(t)$ be the population of uninfected T cells and $N_X(t)$ that of inactive virions, the model becomes

$$\frac{dN_U}{dt} = \lambda - k_V N_U - \epsilon N_V N_U, \tag{1.5}$$

$$\frac{dN_I}{dt} = \epsilon N_V N_U - k_I N_I, \tag{1.6}$$

$$\frac{dN_V}{dt} = \epsilon' \gamma N_I - k_V N_V, \tag{1.7}$$

$$\frac{dN_X}{dt} = (1 - \epsilon') \gamma N_I - k_V N_X. \tag{1.8}$$

Here, the normal birth and death of T cells are described by the constants λ and k_V, respectively, the residual infectivity by ϵ, and the fraction of competent virions produced by ϵ'.

Even this more elaborate model makes assumptions that are not easy to justify. But it can account nicely for a lot of data, including the fact that at longer times than those shown in Figure 0.3, virus concentration stops dropping exponentially.

Figure 0.3 (page 4)

1.2.6′c Eradication of HIV

Unfortunately, not all infected cells are short lived. The cell death rate that we found from the math reflects only the subset of infected cells that actively add virions to the blood; but some infected cells do not. Some of the other, nonproductive infected cells have a latent "provirus" in their genome, which can be activated later. That's one reason why complete eradication of HIV remains difficult.

PROBLEMS

1.1 Molecular graphics

You can make your own molecular graphics. Access the Protein Data Bank (Media 3[10]). The simplest way to use it is to enter the name or code of a macromolecule in the search box. Then click "3D view" on the right and manipulate the resulting image. Alternatively, you can download coordinate data for the molecule to your own computer and then visualize it by using one of the many free software packages listed in Media 3. To find some interesting examples, you can explore the past Molecules of the Month or see page 321 for the names of entries used in creating images in this book.

a. Make images based on the following entries, which are relevant to the present chapter:

- 1jlb (HIV-1 reverse transcriptase in complex with nevirapine)
- 1hsg (HIV-2 protease complexed with a protease inhibitor)
- 1rl8 (resistant strain of HIV-1 protease with ritonavir)

b. Now try these entries, which are molecules to be discussed in Chapter 10:

- 1lbh (*lac* repressor bound to the gratuitous inducer IPTG)
- 3cro (Cro transcription factor, bound to a segment of DNA)
- 1pv7 (lactose permease bound to the lactose-like molecule TDG)

c. These entries are also interesting:

- 1mme (hammerhead ribozyme)
- 2f8s (small interfering RNA)

1.2 Semilog and log-log plots

a. Use a computer to plot the functions $f_1(x) = \exp(x)$ and $f_2(x) = x^{3.5}$ on the range $2 \le x \le 7$. These functions may appear qualitatively similar.

b. Now make semilogarithmic graphs of the same two functions. What outstanding feature of the exponential function jumps out in this representation?

c. Finally, make log-log graphs of the two functions and comment.

1.3 Half-life

Sometimes instead of quoting a rate constant like k_I in Equation 1.1 (page 14), scientists will quote a **half-life**, the time after which an exponentially falling population has decreased to half its original value. Derive the relation between these two quantities.

1.4 Model action of antiviral drug

Finish the analysis of the time course of HIV infection after administering an antiviral drug. For this problem, you may assume that virus clearance is faster than T cell death (though not necessarily much faster). That is, assume $k_V > k_I$.

a. Follow Section 1.2.4 (page 14) to write down the trial solution for $N_V(t)$, the observable quantity, in terms of the initial viral load N_{V0} and three unknown constants k_I, k_V, and β.

b. Obtain Dataset 1,[11] and use a computer to make a semilog plot. Don't join the points by line segments; make each point a symbol, for example, a small circle or plus sign. Label the axes of your plot. Give it a title, too. Superimpose a graph of the trial solution, with

---

[10] References of this form refer to Media links on the book's Web site.
[11] References of this form refer to Dataset links on the book's Web site.

some arbitrary values of k_I, k_V, and β, on your graph of the actual data. Then try to fit the trial solution to the data, by choosing better values of the parameters.

c. You may quickly discover that it's difficult to find the right parameter values just by guessing. Rather than resort to some black-box software to perform the search, however, try to choose parameter values that make certain features of your graph coincide with the data, as follows. First, note that the experimental data approach a straight line on a semilog plot at long times (apart from some experimental scatter). The trial solution Equation 1.4 also approaches such a line, namely, the graph of the function $N_{V,\text{asymp}}(t) = Xe^{-k_I t}$, so you can match that function to the data. Lay a ruler along the data, adjust it until it seems to match the long-time trend of the data, and find two points on that straight line. From this information, find values of k_I and X that make $N_{V,\text{asymp}}(t)$ match the data.

d. Substitute your values of k_I and X into Equation 1.4 (page 15) to get a trial solution with the right initial value N_{V0} and the right long-time behavior. This is still not quite enough to give you the value of k_V needed to specify a unique solution. However, the model suggests another constraint. Immediately after administering the drug, the number of infected T cells has not yet had a chance to begin falling. Thus, in this model both viral production and clearance are the same as they were prior to time zero, so the solution is initially still quasi-steady:

$$\left.\frac{dN_V}{dt}\right|_{t=0} = 0.$$

Use this constraint to determine all parameters of the trial solution from your answer to (c), and plot the result along with the data. (You may want to tweak the approximate values you used for N_{V0}, and other parameters, in order to make the fit look better.)

e. The hypothesis that we have been exploring is that the reciprocal of the T cell infection rate is much shorter than the typical latency period for the infection, or in other words that

$$(1/k_I) \text{ is much smaller than } 10 \text{ years.}$$

Do the data support this claim?

f. Use your result from Problem 1.3 to convert your answers to (d) into half-life values for virions and infected T cells. These represent half-lives in a hypothetical system with clearance, but no new virion production nor new infections.

1.5 Blind fitting

Obtain Dataset 2. This file contains an array consisting of two columns of data. The first is the date in years after an arbitrary starting point. The second is the estimated world population on that date.

a. Use a computer to plot the data points.

b. Here is a simple mathematical function that roughly reproduces the data:

$$f(t) = \frac{100\,000}{2050 - (t/1 \text{ year})}. \tag{1.9}$$

Have your computer draw this function, and superimpose it on your graph of the actual data. Now play around with the function, trying others of the same form

$f(t) = A/(B - t)$, for some constants A and B. You can get a pretty good-looking fit in this way. (There are automated ways to do this, but it's instructive to try it "by hand" at least once.)

c. Do you think this is a good model for the data? That is, does it tell us anything interesting beyond roughly reproducing the data points? Explain.

1.6 $\boxed{T_2}$ Special case of a differential equation system

Consider the following system of two coupled linear differential equations, simplified a bit from Equations 1.1 and 1.2 on page 14:

$$dA/dt = -k_A A \quad \text{and} \quad dB/dt = -k_B B + A.$$

This set of equations has two linearly independent solutions, which can be added in any linear combination to get the general solution.[12] So the general solution has two free parameters, the respective amounts of each independent solution.

Usually, a system of linear differential equations with constant coefficients has solutions in the form of exponential functions. However, there is an exceptional case, which can arise even in the simplest example of two equations. Remember the physical analogy for this problem: The first equation determines a function $A(t)$, which is the flow rate into a container B, which in turn has a hole at the bottom.

Section 1.2.4 argued that if $k_A \ll k_B$, then B can't accumulate much, and its outflow is eventually determined by k_A. In the opposite case, $k_B \ll k_A$, B fills up until A runs out, and then trickles out its contents in a way controlled by k_B. Either way, the long-time behavior is an exponential, and in fact we found that at *all* times the behavior is a combination of two exponentials.

But what if $k_A = k_B$? The above reasoning is inapplicable in that case, so we can't be sure that every solution falls as an exponential at long times. In fact, there is *one* exponential solution, which corresponds to the situation where $A(0) = 0$, so we have only B running out. But there must also be a second, independent solution.

a. Find the other solution when $k_A = k_B$, and hence find the complete general solution to the system. Get an analytic result (a formula), not a numerical one. [*Hint:* Solve container A's behavior explicitly: $A(t) = A_0 \exp(-k_A t)$, and substitute into the other equation to get

$$dB/dt = -k_A B + A_0 \exp(-k_A t).$$

If $A_0 \neq 0$, then no solution of the form $B(t) = \exp(-k_A t)$ works. Instead, play around with multiplying that solution by various powers of t until you find something that solves the equation.]

b. Why do you suppose this case is *not* likely to be relevant to a real-life problem like our HIV story?

1.7 $\boxed{T_2}$ Infected cell count

First, work Problem 1.4. Then continue as follows to get an estimate of the population of infected T cells in the quasi-steady state. Chapter 3 will argue that this number is needed in order to evaluate the hypothesis of viral evolution in individual patients.

---

[12]In general, a system of N first-order linear differential equations in N unknowns has N independent solutions.

Figure 0.3 (page 4)

a. The human body contains about 5 L of blood. Each HIV virion carries two copies of its RNA genome. Thus, the total virion population is about $2.5 \cdot 10^3$ mL times the quantity plotted in Figure 0.3. Express the values of N_{v0} and X that you found in Problem 1.4 in terms of total virion population.

b. Obtain a numerical estimate for β from your fit. (You found the value of k_I in Problem 1.4.)

c. The symbol β is an abbreviation for the product γN_{I0}, where $\gamma \approx 100 k_I$ is the rate of virion release by an infected T cell and N_{I0} is the quantity we want to find. Turn your results from (a) and (b) into an estimate of N_{I0}.

Physics and Biology

It is not the strongest of the species that survives, nor the most intelligent, but rather the one most responsive to change.
—Charles Darwin

2.1 Signpost

The rest of this book will explore two broad classes of propositions:

1a. Living organisms use physical mechanisms to gain information about their surroundings, and to respond to that information. Appreciating the basic science underlying those mechanisms is critical to understanding how they work.

1b. Scientists also use physical mechanisms to gain information about the systems they are studying. Here, too, appreciating some basic science allows us to extend the range and validity of our measurements (and, in turn, of the models that those measurements support).

2a. Living organisms must make inferences (educated guesses) about the best response to make, because their information is partly diluted by noise.[1]

2b. In many cases scientists, too, must reason probabilistically to extract the meaning from our measurements.

In fact, the single most characteristic feature of living organisms, cutting across their immense diversity, is their adaptive response to the opportunities and constraints of their ever-fluctuating physical environment. Organisms must gather *information* about the world, make *inferences* about its present and future states based on that information, and *modify behavior* in ways that optimize some outcome.

---

[1] The everyday definition of **noise** is "uninformative audio stimuli," or more specifically "music enjoyed by people in any generation other than my own." But in this book, the word is a synonym for "randomness," defined in Chapter 3.

Each of the propositions above has been written in two parallel forms, in order to highlight a nice symmetry:

> *The same sort of probabilistic inference needed to understand your lab data must also be used by the lion and the gazelle as they integrate their own data and make their decisions.*

2.2 The Intersection

At first sight, Physics may seem to be almost at the opposite intellectual pole from Biology. On one side, we have Newton's three simple laws;[2] on the other, the seemingly arbitrary Tree of Life. On one side, there is the relentless search for simplicity; on the other, the appearance of irreducible complexity. One side's narrative stresses universality; the other's seems dominated by historical accidents. One side stresses determinism, the prediction of the future from measurement of the present situation; the other's reality is highly unpredictable.

But there is more to Physics than Newton's laws. Gradually during the 19th century, scientists came to accept the lumpy (molecular) character of all matter. At about the same time, they realized that if the air in a room consists of tiny particles independently flying about, that motion must be *random*—not deterministic. We can't see this motion directly, but by the early 20th century it became clear that it was the cause of the incessant, random motion of any micrometer-size particle in water (called **Brownian motion**; see Media 2). A branch of Physics accordingly arose to describe such purely physical, yet random, systems. It turned out that conclusions can be drawn from intrinsically random behavior, essentially because every sort of "randomness" actually has characteristics that we can measure quantitatively, and even try to predict. Similar methods apply to the randomness found in living systems.

Another major discovery of early 20th century Physics was that light, too, has a lumpy character. Just as we don't perceive that a stream of water consists of discrete molecules, so too in normal circumstances we don't notice the granular character of a beam of light. Nevertheless, that aspect will prove essential in our later study of localization microscopy.

Turning now to Biology, the great advance of the late 19th century was the principle of *common descent:* Because all living organisms partially share their family tree, we can learn about any of them by studying any other one. Just as physicists can study simple atoms and hope to find clues about the most complex molecule, so too could biologists study bacteria with reasonable expectations of learning clues about butterflies and giraffes. Moreover, inheritance along that vast family tree has a particulate character: It involves discrete lumps of information (genes), which are either copied exactly or else suffer random, discrete changes (mutation or recombination). This insight was extensively developed in the 20th century; it became clear that, although the long-run *outcomes* of inheritance are subtle and gorgeously varied, many of the underlying *mechanisms* are universal throughout the living world.

Individuals within a species meet, compete, and mate at least partially at random, in ways that may remind us of the physical processes of chemical reactions. Within each individual, too, each cell's life processes are literally chemical reactions, again involving discrete molecules. Some of the key actors in this inner dance appear in only a very few copies. Some respond to external signals that involve only a few discrete entities. For example,

---

[2] And Maxwell's less simple, but still concise, equations.

olfactory (smell) receptors can respond to just a few odorant molecules; visual receptors can respond to the absorption of even *one* unit of light. At this deep level, the distinction between biological and physical science melts away.

In short, 20th century scientists gained an ever increasing appreciation of the fact that

> *Discreteness and randomness lie at the roots of many physical and biological phenomena.*

They developed mathematical techniques appropriate to both realms, continuing to the present.

2.3 Dimensional Analysis

Appendix B describes an indispensable tool for organizing our thoughts about physical models. Here are two quick exercises in this area that are relevant to topics in this book.

Your Turn 2A

Go back to Equation 1.3 (page 14) and check that it conforms to the rules for units given in Appendix B. Along the way, find the appropriate units for the quantities k_1, k_V, β, and γ. From that, show that statements like "k_V is much larger than $1/(10\,\text{years})$" indeed make sense dimensionally.

Your Turn 2B

Find the angular diameter of a coin held at a distance of 3 m from your eye. (Take the coin's diameter to be 1 cm.) Express your answer in radians and in arc minutes. Compare your answer to the angular diameter of the Moon when viewed from Earth, which is about 32 arcmin.

In this book, the names of units are set in a special typeface, to help you distinguish them from named quantities. Thus cm denotes "centimeters," whereas *cm* could denote the product of a concentration times a mass, and "cm" could be an abbreviation for some ordinary word. Symbols like $\mathbb{L}$ denote dimensions (in this case, length); see Appendix B.

Named quantities are generally single italicized letters. We can assign them arbitrarily, but we must use them consistently, so that others know what we mean. Appendix A collects definitions of many of the named quantities, and other symbols, used in this book.

THE BIG PICTURE

In physics classes, "error analysis" is sometimes presented as a distasteful chore needed to overcome some tiresome professor's (or peer reviewer's) objections to our work. In the biology curriculum, it's sometimes relegated to a separate course on the design of clinical trials. This book will instead try to integrate probabilistic reasoning directly into the study of how living organisms manage their amazing trick of responding to their environment.

Looking through this lens, we will study some case histories of responses at many levels and on many time scales. As mentioned in the Prolog, even the very most primitive life forms (viruses) respond at the population level by evolving responses to novel challenges, and so do all higher organisms. Moving up a step, individual bacteria have genetic and metabolic circuits that endow them with faster responses to change, enabling them to turn on certain capabilities only when they are needed, become more adaptable in hard times, and even search for food. Much more elaborate still, we vertebrates have exceedingly fast neural circuits that let us hit a speeding tennis ball, or snag an insect with our long, sticky tongue (as appropriate). Every level involves physical ideas. Some of those ideas may be new to you; some seem to fly in the face of common sense. (You may need to change and adapt a bit yourself to get a working understanding of them.)

KEY FORMULAS

See also Appendix B.

• *Angles:* To find the angle between two rays that intersect at their end points, draw a circular arc, centered on the common end point, that starts on one ray and ends on the other one. The angle in radians (rad) is the ratio of the length of that arc to its radius. Thus, angles, and the unit rad, are dimensionless. Another dimensionless unit of angle is the degree, defined as $\pi/180$ radians.

FURTHER READING

Here are four books that view living organisms as information-processing machines:

Semipopular:
Bray, 2009.

Intermediate:
Alon, 2006; Laughlin & Sterling, 2015.

Technical:
Bialek, 2012.

PROBLEMS

2.1 Greek to me

We'll be using a lot of letters from the Greek alphabet. Here are the letters most often used by scientists. The following list gives both lowercase and uppercase (but omits the uppercase when it looks just like a Roman letter):

$$\alpha, \ \beta, \ \gamma/\Gamma, \ \delta/\Delta, \ \epsilon, \ \zeta, \ \eta, \ \theta/\Theta, \ \kappa, \ \lambda/\Lambda, \ \mu, \ \nu,$$

$$\xi/\Xi, \ \pi/\Pi, \ \rho, \ \sigma/\Sigma, \ \tau, \ \upsilon/\Upsilon, \ \phi/\Phi, \ \chi, \ \psi/\Psi, \text{and } \omega/\Omega$$

When writing computer code, we often spell them out as alpha, beta, gamma, delta, epsilon, zeta, eta, theta, kappa, lambda, mu, nu, xi (pronounced "k'see"), pi, rho, sigma, tau, upsilon, phi, chi (pronounced "ky"), psi, and omega, respectively.

Practice by examining the following quote:

> Cell and tissue, shell and bone, leaf and flower, are so many portions of matter, and it is in obedience to the laws of physics that their particles have been moved, moulded, and conformed. They are no exception to the rule that Θεὸς ἀεὶ γεωμετρεῖ. – D'Arcy Thompson

From the sounds made by each letter, can you guess what Thompson was trying to say? [*Hint:* ς is an alternate form of σ.]

2.2 Unusual units

In the United States, automobile fuel consumption is usually quantified by stating the car's "miles per gallon" rating. In some ways, the reciprocal of this quantity, called "fuel efficiency," is more meaningful. State the dimensions of fuel efficiency, and propose a natural SI unit with those dimensions. Give a physical/geometrical interpretation of the fuel efficiency of a car that gets 30 miles per gallon of gasoline.

2.3 Quetelet index

It's straightforward to diagnose obesity if a subject's percentage of body fat is known, but this quantity is not easy to measure. Frequently the "body mass index" (BMI, or "Quetelet index") is used instead, as a rough proxy. BMI is defined as

$$\mathrm{BMI} = \frac{\text{body mass in kilograms}}{(\text{height in meters})^2},$$

and BMI > 25 is sometimes taken as a criterion for overweight.

a. Re-express this criterion in terms of the quantity m/h^2. Instead of the pure number 25, your answer will involve a number with dimensions.

b. What's wrong with the simpler criterion that a subject is overweight if body mass m exceeds some fixed threshold?

c. Speculate why the definition above for BMI might be a better, though not perfect, proxy for overweight.

2.4 Mechanical sensitivity

Most people can just barely feel a single grain of salt dropped on their skin from a height $h = 10\,\mathrm{cm}$. Model a grain of salt as a cube of length about 0.2 mm made of a material of mass density about $10^3 \, \mathrm{kg \, m^{-3}}$. How much gravitational potential energy does that grain release when it falls from 10 cm? [*Hint:* If you forget the formula, take the parameters given in the problem and use dimensional analysis to find a quantity with the units $\mathrm{J = kg \, m^2 s^{-2}}$. Recall that the acceleration due to gravity is $g \approx 10 \, \mathrm{m \, s^{-2}}$.]

2.5 Do the wave

a. Find an approximate formula for the speed of a wave on the surface of the deep ocean. Your answer may involve the mass density of water ($\rho \approx 10^3 \, \mathrm{kg \, m^{-3}}$), the wavelength λ of the wave, and/or the acceleration of gravity ($g \approx 10 \, \mathrm{m \, s^{-2}}$). [*Hints:* Don't work hard; don't write or solve any equation of motion. The depth of the ocean doesn't matter (it's essentially infinite), nor do the surface tension or viscosity of the water (they're negligible).]

b. Evaluate your answer numerically for a wavelength of one meter to see if your result is reasonable.

2.6 Concentration units

Appendix B introduces a unit for concentration called "molar," abbreviated M. To practice dimensional analysis, consider a sugar solution with concentration 1 mM. Find the average number of sugar molecules in one cubic micrometer of such a solution.

2.7 Atomic energy scale

Read Appendix B.

a. Using the same logic as in Section B.6, try to construct an *energy* scale as a combination of the force constant k_e defined there, the electron mass m_e, and Planck's constant $\hbar$. Get a numerical answer in joules. What are the values of the exponents a, b, and c analogous to Equation B.1 (page 314)?

b. We know that chemical reactions involve a certain amount of energy per molecule, which is generally a few eV, where the **electron volt** unit is $1.6 \times 10^{-19} \, \mathrm{J}$. (For example, the energy needed to remove the electron from a hydrogen atom is about 14 eV.) How well does your estimate in (a) work?

PART II

Randomness in Biology

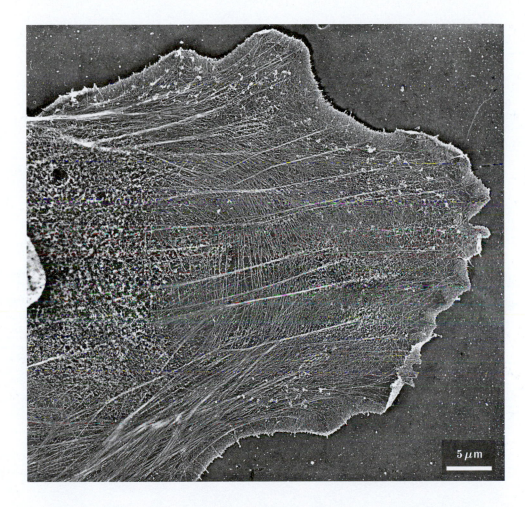

[Electron micrograph.] **The leading edge of a crawling cell** (a fibroblast from the frog *Xenopus laevis*). An intricate network of filaments (the cytoskeleton) has been highlighted. Although it is not perfectly regular, neither is this network perfectly random—in any region, the distribution of filament orientations is well defined and related to the cell's function. [Courtesy Tatyana Svitkina, University of Pennsylvania.]

Discrete Randomness

Chance is a more fundamental conception than causality.
—Max Born

3.1 Signpost

Suppose that 30 people are gathered for a meeting. We organize them alphabetically by first name, then list each person's height in that order. It seems intuitive that the resulting list of numbers is "random," in the sense that there is no way to predict any of the numbers. But certainly the list is not "totally random"—we can predict in advance that there will be no heights exceeding, say, 3 m. No series of observations is ever totally unpredictable.

It also makes intuitive sense to say that if we sort the list in ascending order of height, it becomes "less random" than before: Each entry is known to be no smaller than its predecessor. Moreover, if the first 25 heights in the alphabetical list are all under 1 m, then it seems reasonable to draw some tentative conclusions about those people (probably they are children), and even about person number 26 (probably also a child).

This chapter will distill intuitions like these in a mathematical framework general enough for our purposes. This systematic study of randomness will pay off as we begin to construct physical models of living systems, which must cope with (*i*) randomness coming from their external environment and (*ii*) intrinsic randomness from the molecular mechanisms that implement their decisions and actions.

Many of our discussions will take the humble coin toss as a point of departure. This may not seem like a very biological topic. But the coin toss will lead us directly to some less trivial random distributions that do pervade biology, for example, the Binomial, Poisson, and Geometric distributions. It also gives us a familiar setting in which to frame more general ideas, such as likelihood; starting from such a concrete realization will keep our feet on the ground as we generalize to more abstract problems.

This chapter's Focus Question is
Biological question: If each attempt at catching prey is an independent random trial, how long must a predator wait for its supper?
Physical idea: Distributions like this one arise in many physical contexts, for example, in the waiting times between enzyme turnovers.

3.2 Avatars of Randomness

3.2.1 Five iconic examples illustrate the concept of randomness

Let's consider five concrete physical systems that yield results commonly described as "random." Comparing and contrasting the examples will help us to build a visual vocabulary to describe the kinds of randomness arising in Nature:

1. We flip a coin and record which side lands upward (<u>heads</u> or <u>tails</u>).
2. We evaluate the integer random number function defined in a computer math package.
3. We flip a coin m times and use the results to construct a "random, m-bit binary fraction," a number analogous to the familiar decimal fractions:

$$ x = \frac{1}{2}s_1 + \frac{1}{4}s_2 + \cdots + \frac{1}{2^m}s_m, \qquad (3.1) $$

   where $s_i = 1$ for <u>heads</u> or 0 for <u>tails</u>. x is always a number between 0 and 1.
4. We observe a very dim light source using a sensitive light detector. The detector responds with individual electrical impulses ("blips"), and we record the elapsed <u>waiting</u> time t_w between each blip and its predecessor.[1]
5. We observe the successive positions of a micrometer-size particle undergoing free motion in water (Brownian motion) by taking video frames every few seconds.[2]

Let's look more closely at these examples, in turn. We'd like to extract a general idea of randomness, and also learn how to characterize different *kinds* of randomness.

1a. Actually, coin flipping is not intrinsically unpredictable: We can imagine a precision mechanical coin flipper, isolated from air currents, that reliably results in <u>heads</u> landing up every time. Nevertheless, when a human flips a coin, we do get a series $s_1, s_2, \ldots$ that has no *discernible, relevant* structure: Apart from the constraint that each s_i has only two allowed values, it is essentially unpredictable. Even if we construct an unfair coinlike object that lands <u>heads</u> some fraction ξ of the time, where $\xi \neq 1/2$, nevertheless that one number completely characterizes the resulting series. We will refer often to this kind of randomness, which is called a **Bernoulli trial**. As long as ξ does not equal 1 or 0, we cannot completely predict any result from its predecessors.

2a. A computer-generated series of "random" numbers also cannot literally be random—computers are designed to give perfectly deterministic answers to mathematical calculations. But the computer's algorithm yields a sequence so complex that, for nearly any practical purpose, it too has no discernible, relevant structure.

3a. Turning to the binary fraction example, consider the case of double flips of a fair coin ($m = 2$ and $\xi = 1/2$). There are four possible outcomes, namely, *TT, TH, HT,* and

---

[1] You can listen to a sample of these blips (Media 5) and contrast it with a sample of regularly spaced clicks with the same average rate (Media 6).
[2] See Media 2.

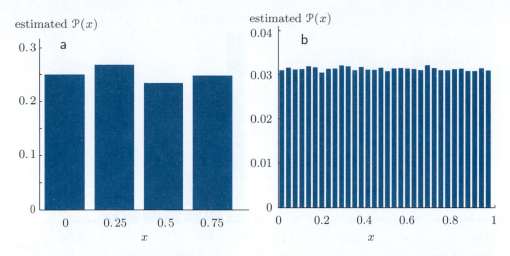

Figure 3.1 [Computer simulations.] **Uniformly distributed random variables.** Empirical distributions of (a) 500 two-bit random binary fractions (that is, $m = 2$ in Equation 3.1) and (b) 250 000 five-bit binary fractions ($m = 5$). The symbol $\mathcal{P}(x)$ refers to the probabilities of various outcomes; it will be defined precisely later in this chapter.

HHH, which yield the numbers $x = 0$, 1/4, 1/2, and 3/4, respectively. If we make a lot of double flips and draw a histogram of the results (Figure 3.1a), we expect to find four bars of roughly equal height: The successive x values are drawn from a **discrete Uniform distribution**. If we choose a larger value of m, say, $m = 5$, we find many more possible outcomes; the allowed x values are squeezed close to one another, staying always in the range from $x = 0$ to 1 (Figure 3.1b). Again, the bars in the resulting histogram are all roughly equal in height (if we have measured a large enough number of instances).[3] All we can say about the next number drawn from this procedure is that $0 \le x < 1$, but even that is *some* knowledge.

4a. For the light detector example, no matter how hard we work to improve our apparatus, we always get irregularly spaced blips at low light intensity. However, we do observe a certain amount of structure in the intervals between successive light detector blips: These waiting times t_w are always greater than zero, and, moreover, short waits are more common than long ones (Figure 3.2a). That is, the thing that's predictable is, again, a *distribution*—in this case, of the waiting times. The "cloud" representation shown in the figure makes it hard to say anything more precise than that the outcomes are not all equally probable. But a histogram-like representation (Figure 3.2b) reveals a definite form. Unlike example **3** above, the figures show that this time the distribution is *non-Uniform:* It's an example of an "Exponential distribution."[4]

So in this case, we again have limited knowledge, obtained from our experience with many previous experiments, that helps us to guess the waiting time before the next blip. We can learn a bit more by examining, say, the first 100 entries in the series of waiting

---

[3]We can even imagine a limit of very large m; now the tops of the histogram bars nearly form a continuous line, and we are essentially generating a sequence of real numbers drawn from the **continuous Uniform distribution** on the interval. Chapter 5 will discuss continuous distributions.
[4]Chapter 7 will discuss this distribution in detail.

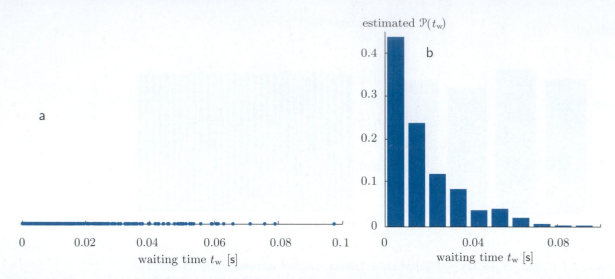

Figure 3.2 [Experimental data.] **Two ways to visualize a distribution.** (a) Cloud diagram showing the waiting times between 290 successive light detector blips as a cloud of 289 dots. The dot density is higher at the left of the diagram. (b) The same data, presented as a histogram. Taller bars correspond to greater density of dots in (a). The data have been subdivided into 10 discrete bins. [Data courtesy John F Beausang (see Dataset 3).]

times and finding their average, which is related to the intensity of the light source. But there is a limit to the information we can obtain in this way. Once we know that the general form of the distribution is Exponential, then it is completely characterized by the average waiting time; there is nothing more that any number of initial measurements can tell us about the next one, apart from refining our estimate of that one quantity.

5a. The positions of a particle undergoing Brownian motion reflect the unimaginably complex impacts of many water molecules during the intervals between snapshots ("video frames"), separated by equal time intervals Δt. Nevertheless, again we know at least a little bit about them: Such a particle will never jump, say, 1 cm from one video frame to the next, although there is no limit to how far it could move if given enough time. That is, position at video frame i does give us some partial information about position at frame $i + 1$: Successive observed positions are "correlated," in contrast to examples **1–3**.[5] However, once we know the position at frame i, then also knowing it at any frame *prior* to i gives us no additional help predicting it at $i + 1$; we say the Brownian particle "forgets" everything about its past history other than its current position, then takes a random step whose distribution of outcomes depends only on that one datum. A random system that generates a series of steps with this "forgetful" property is called a **Markov process**.

Brownian motion is a particularly simple kind of Markov process, because the distribution of positions at frame $i + 1$ has a simple form: It's a universal function, common to all steps, simply *shifted* by the position at i. Thus, if we subtract the vector position x_{i-1} from x_i, the resulting displacements Δx_i are *uncorrelated*, with a distribution peaked at zero displacement (see Figure 3.3).

---

[5] Section 3.4.1 will give a more precise definition of correlation.

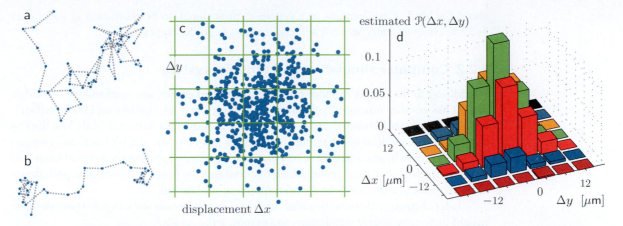

Figure 3.3 [Experimental data.] **Progressively more abstract representations of Brownian motion.** (a) *Dots* show successive observations of the position of a micrometer-size particle, taken at 30-second intervals. The *lines* merely join the dots; the particle did not really undergo straight-line motion between observations. (b) Similar data from another trial. (c) Cloud representation of the displacement vectors $\Delta\boldsymbol{x}$ joining successive positions, for 508 such position observations. Thus, a dot at the exact center would correspond to zero displacement. The grid lines are separated by about $6\,\mu$m. Thus, on a few occasions, the particle was observed to have moved more than $16\,\mu$m, though it usually moved much less than that. (d) The same data, presented as a histogram. The data have been subdivided into 49 discrete bins. [Data from Perrin, 1909; see Dataset 4.]

We will return to these key examples often as we study more biologically relevant systems. For now, we simply note that they motivate a pragmatic definition of randomness:

> *A system that yields a series of outcomes is effectively random if a list of those outcomes has no discernible, relevant structure beyond what we explicitly state.*   (3.2)

The examples above described quantities that are, for many purposes, effectively random after acknowledging these characteristics:

1b–3b. Coin flips (or computer-generated random integers) are characterized by a finite list of allowed values, and the Uniform distribution on that list.

4b. Blip waiting times are characterized by a range of allowed values ($t_w > 0$), and a particular distribution on that range (in this case, not Uniform).

5b. Brownian motion is characterized by a non-Uniform distribution of positions in each video frame, which depends in a specific way on the position in the previous frame.

In short, each random system that we study has its own structure, which characterizes "what kind of randomness" it has.

The definition in Idea 3.2 may sound suspiciously imprecise. But the alternative—a precise-sounding mathematical definition—is often not helpful. How do we know for sure that any particular biological or physical system really fits the precise definition? Maybe there is some unsuspected seasonal fluctuation, or some slow drift in the electricity powering the apparatus. In fact, very often in science, a tentative identification of a supposedly random system turns out to omit some structure hiding in the actual observations (for example, correlations). Later we may discern that extra structure, and discover something new. It's best to avoid the illusion that we know everything about a system, and treat all our statements about the kind of randomness in a system as provisional, to be sharpened as more data become available. Later sections will explain how to do this in practice.

3.2.2 Computer simulation of a random system

In addition to the integer random function mentioned earlier, any mathematical software system has another function that simulates a sample from the continuous Uniform distribution in the range between 0 and 1. We can use it to simulate a Bernoulli trial (coin flip) by drawing such a random number and comparing it to a constant ξ; if it's less than ξ, we can call the outcome <u>heads</u>, and otherwise <u>tails</u>. That is, we can partition the interval from 0 to 1 into two subintervals and report which one contains the number drawn. The probability of each outcome is the width of the corresponding subinterval.

Later chapters will extend this idea considerably, but already you can gain some valuable insight by writing simple simulations and graphing the results.[6]

3.2.3 Biological and biochemical examples

Here are three more examples with features similar to the ones in Section 3.2.1:

1c. Many organisms are **diploid**; that is, each of their cells contains two complete copies of the genome. One copy comes from the male parent, the other from the female parent. Each parent forms germ cells (egg or sperm/pollen) via **meiosis**, in which *one* copy of each gene ends up in each germ cell. That copy is essentially chosen at random from the two that were initially present. That is, for many purposes, inheritance can be thought of as a Bernoulli trial with $\xi = 1/2$, where <u>heads</u> could mean that the germ cell receives the copy originally given by the grandmother, and <u>tails</u> means the other copy.[7]

4c. Many chemical reactions can be approximated as "well mixed." In this case, the probability that the reaction will take a step in a short time interval dt depends only on the total number of reactant molecules present at the start of that interval, and not on any earlier history of the system.[8] For example, a single **enzyme** molecule wanders in a bath of other molecules, most of them irrelevant to its function. But when a particular molecular species, the enzyme's **substrate**, comes near, the enzyme can bind it, transform it chemically, and release the resulting **product** without any net change to itself. Over a time interval in which the enzyme does not significantly change the ambient concentrations of substrate, the individual appearances of product molecules have a distribution of waiting times similar to that in Figure 3.2b.

5c. Over longer times, or if the initial number of substrate molecules is not huge, it may be necessary to account for changes in population. Nevertheless, in well-mixed solutions each reaction step depends only on the *current* numbers of substrate and product molecules, not on the prior history. Even huge systems of many simultaneous reactions, among many species of molecules, can often be treated in this way. Thus, many biochemical reaction networks have the same Markov property as in example **5a** on page 38.

---

[6]See Problems 3.3 and 3.5.

[7] $\boxed{T_2}$ This simplified picture gets complicated by other genetic processes, including gene transposition, duplication, excision, and point mutation. In addition, the Bernoulli trials corresponding to different genes are not necessarily independent of one another, due to physical linkage of the genes on chromosomes.

[8]Chapter 1 assumed that this was also the case for clearance of HIV virions by the immune system.

3.2.4 False patterns: Clusters in epidemiology

Look again at Figure 3.1. These figures show estimated probabilities from finite samples of data known to have been drawn from Uniform distributions. If we were told that these figures represented experimental data, however, we might wonder whether there is additional structure present. Is the extra height in the second bar in panel (a) significant?

    Human intuition is not always a reliable guide to questions like this one when the available data are limited. For example, we may have the home addresses of a number of people diagnosed with a particular disease, and we may observe an apparent geographical cluster in those addresses. But even Uniformly distributed points will show apparent clusters, if our sample is not very large. Chapter 4 will develop some tools to assess questions like these.[9]

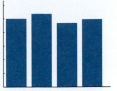

Figure 3.1a (page 37)

Figure 3.1b (page 37)

3.3 Probability Distribution of a Discrete Random System

3.3.1 A probability distribution describes to what extent a random system is, and is not, predictable

Our provisional definition of randomness, Idea 3.2, hinged on the idea of "structure." To make this idea more precise, note that the examples given so far all yield measurements that are, at least in principle, **replicable**. That is, they come from physical systems that are simple enough to reproduce in many copies, each one identical to the others in all relevant aspects and each unconnected to all the others. Let's recall some of the examples in Section 3.2.1:

1d. We can make many identical coins and flip them all using similar hand gestures.

4d. We can construct many identical light sources and shine them on identical light detectors.

5d. We can construct many chambers, each of which releases a micrometer-size particle at a particular point at time zero and observes its subsequent Brownian motion.

We can then use such repeated measurements to learn what discernible, relevant structure our random system may have. Here, "structure" means *everything that we can learn from a large set of measurements that can help us to predict the next one.* Each example given had a rather small amount of structure in this sense:

1e. All that we can learn from a large number of coin tosses that's helpful for guessing the result of the next one is the single number ξ characterizing a Bernoulli trial.

2e–3e. Similarly in these examples, once we list the allowed outcomes and determine that each is equally probable (Uniform distribution), nothing more can be gleaned from past outcomes.

4e. For light detection, each blip waiting time is independent of the others, so again, the distribution of those times is all that we can measure that is useful for predicting the next one. Unlike examples **2e** and **3e**, however, in this case the distribution is non-Uniform (Figure 3.2).

5e. In Brownian motion, the successive positions of a particle are not independent. However, we may still take the *entire trajectory* of the particle throughout our trial as the "outcome" (or observation), and assert that each is independent of the other trials. Then, what we can say about the various trajectories is that the ones with lots

---

[9]See Problem 4.10.

of large jumps are less probable than the ones with few: There is a distribution on the *space of trajectories*. It's not easy to visualize such a high-dimensional distribution, but Figure 3.3 simplified it by looking only at the net displacement Δx after a particular elapsed time.

With these intuitive ideas in place, we can now give a formal definition of a random system's "structure," or probability distribution, starting with the special case of discrete outcomes.

Suppose that a random system is replicable; that is, it can be measured repeatedly, independently, and under the same conditions. Suppose also that the outcomes are always items drawn from a discrete list, indexed by ℓ.[10] Suppose that we make many measurements of the outcome (N_{tot} of them) and find that $\ell = 1$ on N_1 occasions, and so on. Thus, N_ℓ is an integer, the number of times that outcome ℓ was observed; it's called the **frequency** of outcome ℓ.[11] If we start all over with another N_{tot} measurements, we'll get different frequencies N_ℓ', but for large enough N_{tot} they should be about equal to the corresponding N_ℓ. We say that the discrete **probability distribution** of the outcome ℓ is the fraction of trials that yielded ℓ, or[12]

$$\mathcal{P}(\ell) = \lim_{N_{\text{tot}} \to \infty} N_\ell / N_{\text{tot}}. \tag{3.3}$$

Note that $\mathcal{P}(\ell)$ is always nonnegative. Furthermore,

Any discrete probability distribution function is dimensionless,

because for any ℓ, $\mathcal{P}(\ell)$ involves the ratio of two *integers*. The N_ℓ's must add up to N_{tot} (every observation is assigned to *some* outcome). Hence, any discrete distribution must have the property that

$$\sum_\ell \mathcal{P}(\ell) = 1. \quad \text{normalization condition, discrete case} \tag{3.4}$$

Equation 3.3 can also be used with a finite number of observations, to obtain an *estimate* of $\mathcal{P}(\ell)$. When drawing graphs, we often indicate such estimates by representing the values by bars. Then the heights of the bars must all add up to 1, as they do in Figures 3.1a,b, 3.2b, and 3.3d. This representation looks like a histogram, and indeed it differs from an ordinary histogram only in that each bar has been scaled by $1/N_{\text{tot}}$.

We'll call the list of all possible distinct outcomes the **sample space**. If ℓ_1 and ℓ_2 are two distinct outcomes, then we may also ask for the probability that "either ℓ_1 or ℓ_2 was observed." More generally, an **event** E is any subset of the sample space, and it "occurs" whenever we draw from our random system an outcome ℓ that belongs to the subset E. The

---

[10]That list may be infinite, but "discrete" means that we can at least label the entries by an integer. Chapter 5 will discuss continuous distributions.

[11]There is an unavoidable collision of language associated with this term from probability. ΔN is an integer, with no units. But in physics, "frequency" usually refers to a different sort of quantity, with units $\mathbb{T}^{-1}$ (for example, the frequency of a musical note). We must rely on context to determine which meaning is intended.

[12] $\boxed{T_2}$ Some authors call $\mathcal{P}(\ell)$ a **probability mass function** and, unfortunately, assign a completely different meaning to the words "probability distribution function." (That meaning is what we will instead call the "cumulative distribution.")

probability of an event E is the fraction of all draws that yield an outcome belonging to E. The quantity $\mathcal{P}(\ell)$ is just the special case corresponding to an event containing only one point in sample space.

We can also regard events as *statements:* The event E corresponds to the statement "The outcome was observed to be in the set E," which will be true or false every time we draw from (observe) the random system. We can then interpret the logical operations **or**, **and**, and **not** as the usual set operations of union, intersection, and complement.

3.3.2 A random variable has a sample space with numerical meaning

Section 3.3.1 introduced a lot of abstract jargon; let's pause here to give some more concrete examples.

A Bernoulli trial has a sample space with just two points, so $\mathcal{P}_{\text{bern}}(\ell)$ consists of just two numbers, namely, $\mathcal{P}_{\text{bern}}(\text{heads})$ and $\mathcal{P}_{\text{bern}}(\text{tails})$. We have already set the convention that

$$\mathcal{P}_{\text{bern}}(\text{heads}; \xi) = \xi \quad \text{and} \quad \mathcal{P}_{\text{bern}}(\text{tails}; \xi) = (1 - \xi), \tag{3.5}$$

where ξ is a number between 0 and 1. The semicolon followed by ξ reminds us that "the" Bernoulli trial is really a *family* of distributions depending on the *parameter* ξ. Everything before the semicolon specifies an outcome; everything after is a parameter. For example, the fair-coin flip is the special case $\mathcal{P}_{\text{bern}}(s; 1/2)$, where the outcome label s is a variable that ranges over the two values {heads, tails}.

We may have an interpretation for the sample space in which the outcome label is literally a number (like the number of petals on a daisy flower or the number of copies of an RNA molecule in a cell at some moment). Or it may simply be an index to a list of outcomes without any special numerical significance; for example, our random system may consist of successive cases of a disease, and the outcome label s indexes a list of towns where the cases were found. Even in such a situation, there may be one or more interesting numerical *functions of s*. For example, $f(s)$ could be the distance of each town from a particular point, such as the location of a power-generating station. Any such numerical function on the sample space is called a **random variable**. If the outcome label itself can be naturally interpreted as a number, we'll usually call it ℓ; in this case, $f(\ell)$ is any ordinary function, or even ℓ itself.

We already know a rather dull example of a random variable: the Uniformly distributed discrete random variable on some range (Figure 3.1a,b). For example, if ℓ is restricted to integer values between 3 and 6, then this Uniform distribution may be called $\mathcal{P}_{\text{unif}}(\ell; 3, 6)$. It equals $1/4$ if $\ell = 3, 4, 5$, or 6, and 0 otherwise. The semicolon again separates the potential value of a random variable (here, ℓ) from some parameters specifying which distribution function in the family is meant.[13] We will eventually define several such parametrized families of idealized distribution functions.

Another example, which we'll encounter in later chapters, is the "Geometric distribution." To motivate it, imagine a frog that strikes at flies, not always successfully. Each attempt is an independent Bernoulli trial with some probability of success ξ. How long must the frog wait for its next meal? Clearly there is no unique answer to this question, but we can nevertheless ask about the *distribution* of answers. Letting j denote the number of attempts

---

[13]Often we will omit parameter values to shorten our notation, for example, by writing $\mathcal{P}_{\text{unif}}(\ell)$ instead of the cumbersome $\mathcal{P}_{\text{unif}}(\ell; 3, 6)$, if the meaning is clear from context.

needed to get the next success, we'll call this distribution $\mathcal{P}_{\mathrm{geom}}(j;\xi)$. Section 3.4.1.2 will work out an explicit formula for this distribution. For now just note that, in this case, the random variable j can take any positive integer value—the sample space is discrete (although infinite). Also note that, as with the previous examples, "the" Geometric distribution is really a *family* depending on the value of a parameter ξ, which can be any number between 0 and 1.

3.3.3 The addition rule

The probability that the next measured value of ℓ is *either ℓ_1 or ℓ_2* is simply $\mathcal{P}(\ell_1) + \mathcal{P}(\ell_2)$ (unless $\ell_1 = \ell_2$). More generally, if two events E_1 and E_2 have no overlap, we say that they are **mutually exclusive**; then Equation 3.3 implies that

$$\mathcal{P}(\mathsf{E}_1 \text{ or } \mathsf{E}_2) = \mathcal{P}(\mathsf{E}_1) + \mathcal{P}(\mathsf{E}_2). \qquad \begin{array}{l}\textbf{addition rule} \text{ for} \\ \text{mutually exclusive events}\end{array} \qquad (3.6)$$

If the events do overlap, then just adding the probabilities will overstate the probability of (E_1 **or** E_2), because some outcomes will be counted twice.[14] In this case, we must modify our rule to say that

$$\mathcal{P}(\mathsf{E}_1 \text{ or } \mathsf{E}_2) = \mathcal{P}(\mathsf{E}_1) + \mathcal{P}(\mathsf{E}_2) - \mathcal{P}(\mathsf{E}_1 \text{ and } \mathsf{E}_2). \quad \text{general } \textbf{addition rule} \qquad (3.7)$$

Your Turn 3A

Prove Equation 3.7 starting from Equation 3.3.

3.3.4 The negation rule

Let **not-E** be the statement that "the outcome is not included in event E." Then E and **not-E** are mutually exclusive, and, moreover, either one or the other is true for every outcome. In this case, Equation 3.3 implies that

$$\mathcal{P}(\textbf{not-E}) = 1 - \mathcal{P}(\mathsf{E}). \quad \textbf{negation rule} \qquad (3.8)$$

This obvious-seeming rule can be surprisingly helpful when we want to understand a complex event.[15]

If E is one of the outcomes in a Bernoulli trial, then **not-E** is the other one, and Equation 3.8 is the same as the normalization condition. More generally, suppose that we have many events $\mathsf{E}_1, \ldots, \mathsf{E}_n$ with the property that any two are mutually exclusive. Also suppose that together they cover the entire sample space. Then Equation 3.8 generalizes to the statement that the sum of all the $\mathcal{P}(\mathsf{E}_i)$ equals one—a more general form of the normalization condition. For example, each of the bars in Figure 3.2b corresponds to an

---

[14]In logic, "E_1 **or** E_2" means either E_1, E_2, *or both*, is true.
[15]See Problem 3.13.

event defined by a range of possible waiting times; we say that we have **binned the data**, converting a lot of observations of a continuous quantity into a discrete set of bars, whose heights must sum to 1.[16]

3.4 Conditional Probability

3.4.1 Independent events and the product rule

Consider two scenarios:

a. A friend rolls a six-sided die but doesn't show you the result, and then asks you if you'd like to place a bet that wins if the die landed with 5 facing up. Before you reply, a bystander comes by and adds the information that the die is showing some *odd* number. Does this change your assessment of the risk of the bet?

b. A friend rolls a die and flips a coin, doesn't show you the results, and then asks you if you'd like to place a bet that wins if the die landed with 5 facing up. Before you reply, the friend suddenly adds the information that the coin landed with <u>heads</u> up. Does this change your assessment of the risk of the bet?

The reason you changed your opinion in scenario **a** is that the additional information you gained eliminated some of the sample space (all the outcomes corresponding to even numbers). If we roll a die many times but disregard all rolls that came up even, then Equation 3.3 says that the probability of rolling 5, *given that we rolled an odd number*, is 1/3. Letting E_5 be the event "roll a 5" and E_{odd} the event "roll an odd number," we write this quantity as $\mathcal{P}(E_5 \mid E_{odd})$ and call it "the conditional probability of E_5 given E_{odd}." More generally, the conditional probability $\mathcal{P}(E \mid E')$ accounts for partial information by restricting the denominator in Equation 3.3 to only those measurements for which E' is true and restricting the numerator to only those measurements for which *both* E and E' are true:

$$\mathcal{P}(E \mid E') = \lim_{N_{tot} \to \infty} \frac{N(E \text{ and } E')}{N(E')}.$$

We can give a useful rule for computing conditional probabilities by dividing both numerator and denominator by the same thing, the total number of all measurements made:

$$\mathcal{P}(E \mid E') = \lim_{N_{tot} \to \infty} \frac{N(E \text{ and } E')/N_{tot}}{N(E')/N_{tot}}, \text{ or} \tag{3.9}$$

$$\mathcal{P}(E \mid E') = \frac{\mathcal{P}(E \text{ and } E')}{\mathcal{P}(E')}. \quad \textbf{conditional probability} \tag{3.10}$$

Equivalently,

$$\mathcal{P}(E \text{ and } E') = \mathcal{P}(E \mid E') \times \mathcal{P}(E'). \quad \text{general } \textbf{product rule} \tag{3.11}$$

---

[16] $\boxed{T_2}$ Binning isn't always necessary nor desirable; see Section 6.2.4′ (page 142).

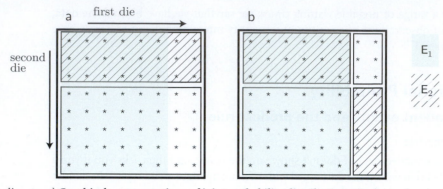

Figure 3.4 [Box diagrams.] **Graphical representations of joint probability distributions.** Each panel represents a distribution graphically as a partitioning of a square. Each consists of 64 equally probable outcomes. (a) Within this sample space, event E_1 corresponds to rolls of two eight-sided dice for which the number on the first die was less than 7; those events lie in the *colored part* of the square. Event E_2 corresponds to rolls for which the second number was less than 4; those events are represented by the *hatched region*. In this situation, E_1 and E_2 are statistically independent. We can see this geometrically because, for example, (E_1 **and** E_2) occupies the upper left rectangle, whose width is that of E_1 and height is that of E_2. (b) A different choice of two events in the same sample space. (E_1 **and** E_2) again occupies the upper left rectangle, but this time its area is *not* the product of $\mathcal{P}(E_1)$ and $\mathcal{P}(E_2)$. Thus, these two events are not independent.

A special case of this rule is particularly important. Sometimes knowing that E' is true tells us nothing relevant to predicting E. (That's why you didn't change your bet in scenario **b** above.) That is, suppose that the additional information you were given was *irrelevant* to what you were trying to predict: $\mathcal{P}(E_5 \mid E_{\text{heads}}) = \mathcal{P}(E_5)$. The product rule then implies that, in this case, $\mathcal{P}(E_5 \text{ and } E_{\text{heads}}) = \mathcal{P}(E_5) \times \mathcal{P}(E_{\text{heads}})$. More generally, we say that two events are **statistically independent**[17] if

$$\mathcal{P}(E \text{ and } E') = \mathcal{P}(E) \times \mathcal{P}(E'). \quad \text{statistically independent events} \qquad (3.12)$$

Two events that are *not* statistically independent are said to be **correlated**.

Equation 3.12 is very useful because often we have a physical model of a random system that states a priori that two events are independent.

It's good to have a pictorial representation of any abstract concept. To represent a random system, we can draw a unit square (sides of length 1), then divide it into boxes corresponding to all of the possible outcomes. For example, suppose that we roll a pair of fair eight-sided dice, so that our sample space consists of 64 elementary events, each of which is equally probable. Figure 3.4a shows the elementary events as asterisks. Symbols set in a shaded background correspond to an event we'll call E_1; the unshaded region is **not**-E_1. The hatched region corresponds to another event called E_2; its complement is **not**-E_2. Because every outcome has the same probability, the probabilities of various events are simply the number of outcomes they contain, times 1/64; equivalently, the probabilities correspond to the *areas* of the various regions in the unit square.

In Figure 3.4a, both blocks on the left side are colored; both on the right are not. Both blocks on the top are hatched; both on the bottom are not. This arrangement implies

---

[17]The abbreviation "independent," or the synonym "uncorrelated," is frequently used instead of "statistically independent."

that $\mathcal{P}(\mathsf{E}_1 \text{ and } \mathsf{E}_2) = \mathcal{P}(\mathsf{E}_1)\mathcal{P}(\mathsf{E}_2)$, and similarly for the other three blocks. Thus, the joint distribution has the product form that implies independence according to the product rule. In contrast, Figure 3.4b graphically represents a different arrangement, in which the events are not independent.

Your Turn 3B

Make this argument more explicit. That is, calculate $\mathcal{P}(\mathsf{E}_1)$ and $\mathcal{P}(\mathsf{E}_1 \mid \mathsf{E}_2)$ for each of Figures 3.4a,b, and comment.

$\boxed{T_2}$ *Section 3.4.1′ (page 60) develops more general forms of the product and negation rules.*

3.4.1.1 Crib death and the prosecutor's fallacy

Officials in the United Kingdom prosecuted hundreds of women, mainly in the 1990s, for the murder of their own infants, who died in their cribs. The families of the targeted women had suffered multiple crib deaths, and the arguments made to juries often took the form that "one is a tragedy, two is suspicious, and three is murder." In one case, an expert justified this claim by noting that, at that time, about one infant in 8500 died in its crib for no known cause in the United Kingdom. The expert then calculated the probability of two such deaths occurring naturally in a family as $(1/8500)^2$, which is a tiny number.

It is true that the observed occurrence of multiple crib deaths in one family, in a population of fewer than 8500^2 families, strongly suggests that successive instances are not statistically independent. The logical flaw in the argument is sometimes called the "prosecutor's fallacy"; it lies in the assumption that *the only possible source* of this nonindependence is willful murder. For example, there could instead be a genetic predisposition to crib death, a noncriminal cause that would nevertheless be correlated within families. After an intervention from the Royal Statistical Society, the UK attorney general initiated legal review of every one of the 258 convictions.

Crib death could also be related to ignorance or custom, which tends to remain constant within each family (hence correlated between successive children). Interestingly, after a vigorous informational campaign to convince parents to put their babies to sleep on their back or side, the incidence of crib death in the United Kingdom dropped by 70%. Had the earlier crib deaths been cases of willful murder, it would have been a remarkable coincidence that they suddenly declined at exactly the same time as the information campaign!

3.4.1.2 The Geometric distribution describes the waiting times for success in a series of independent trials

Section 3.3.2 introduced a problem (frog striking at flies) involving repeated, independent attempts at a yes/no goal, each with probability ξ of success. Let j denote the number of attempts made from one success to the next. For example, $j = 2$ means one failure followed by success (two attempts in all). Let's find the probability distribution of the random variable j.

Once a fly has been caught, there is probability ξ of succeeding again, on the very next attempt: $\mathcal{P}_{\text{geom}}(1; \xi) = \xi$. The outcome of exactly one failure followed by success is then a product: $\mathcal{P}_{\text{geom}}(2; \xi) = \xi(1 - \xi)$, and so on. That is,

$$\mathcal{P}_{\text{geom}}(j; \xi) = \xi(1 - \xi)^{j-1}, \text{ for } j = 1, 2, \ldots . \quad \textbf{Geometric distribution} \qquad (3.13)$$

This family of discrete probability distribution functions is called "Geometric" because each value is a constant times the previous one—a geometric sequence.

Your Turn 3C

a. Graph this probability distribution function for fixed values of $\xi = 0.15, 0.5,$ and 0.9. Because the function $\mathcal{P}_{\text{geom}}(j)$ is defined only at integer values of j, be sure to indicate this by drawing dots or some other symbols at each point—not just a set of line segments.
b. Explain the features of your graphs in terms of the underlying situation being described. What is the most probable value of j in each graph? Think about why you got that result.

Your Turn 3D

Because the values $j = 1, 2, \ldots$ represent a complete set of mutually exclusive possibilities, we must have that $\sum_{j=1}^{\infty} \mathcal{P}_{\text{geom}}(j; \xi) = 1$ for any value of ξ. Confirm this by using the Taylor series for the function $1/(1 - x)$, evaluated near $x = 0$ (see page 19).

You'll work out the basic properties of this family of distributions in Problem 7.2.

3.4.2 Joint distributions

Sometimes a random system yields measurements that each consist of *two* pieces of information; that is, the system's sample space can be naturally labeled by pairs of discrete variables. Consider the combined act of rolling an ordinary six-sided die and also flipping a coin. The sample space consists of all pairs (ℓ, s), where ℓ runs over the list of all allowed outcomes for the die and s runs over those for the coin. Thus, the sample space consists of a total of 12 outcomes. The probability distribution $\mathcal{P}(\ell, s)$, still defined by Equation 3.3, is called the **joint distribution** of ℓ and s. It can be thought of as a table, whose entry in row ℓ and column s is $\mathcal{P}(\ell, s)$. Two-dimensional Brownian motion is a more biological example: Figure 3.3d shows the joint distribution of the random variables Δx and Δy, the components of the displacement vector $\Delta \boldsymbol{x}$ after a fixed elapsed time.

Figure 3.3d (page 39)

Your Turn 3E

Suppose that we roll two six-sided dice. What's the probability that the numbers on the dice add up to 2? To 6? To 12? Think about how you used both the addition and product rules for this calculation.

We may not be interested in the value of s, however. In that case, we can usefully reduce the joint distribution by considering the event $\mathsf{E}_{\ell=\ell_0}$, which is the statement that the random system generated any outcome for which ℓ has the particular value ℓ_0, with no restriction on s. The probability $\mathcal{P}(\mathsf{E}_{\ell=\ell_0})$ is often written simply as $\mathcal{P}_\ell(\ell_0)$, and is called the **marginal distribution** over s; we also say that we obtained it from the joint distribution by "marginalizing" s.[18] We implicitly did this when we reduced the entire path of Brownian motion over 30 s to just the final displacement in Figure 3.3.

---

[18]The subscript "ℓ" serves to distinguish this distribution from other functions of one variable, for example, the distribution $\mathcal{P}_s$ obtained by marginalizing ℓ. When the meaning is clear, we may sometimes drop this subscript. The notation $\mathcal{P}(\ell_0, s_0)$ does not need any subscript, because (ℓ_0, s_0) completely specify a point in the coin/die sample space.

We may find that each of the events $E_{\ell=\ell_0}$ is statistically independent of each of the $E_{s=s_0}$. In that case, we say that the random variables ℓ and s are themselves independent.

Your Turn 3F

a. Show that $\mathcal{P}_\ell(\ell_0) = \sum_s P(\ell_0, s)$.
b. If ℓ and s are independent, then show that $P(\ell, s) = \mathcal{P}_\ell(\ell) \times \mathcal{P}_s(s)$.
c. Imagine a random system in which each "observation" involves drawing a card from a shuffled deck, and then, without replacing it, drawing a second card. If ℓ is the first card's name and s the second one's, are these independent random variables?

The next idea is simple, but subtle enough to be worth stating carefully. Suppose that ℓ corresponds to the roll of a four-sided die and s to a coin flip. We often want to sum over all the possibilities for ℓ, s—for example, to check normalization or compute some average. Let's symbolically call the terms of the sum $[\ell, s]$. We can group the sum in two ways:

$$\Big([1, \underline{\text{tails}}] + [2, \underline{\text{tails}}] + [3, \underline{\text{tails}}] + [4, \underline{\text{tails}}]\Big) + \Big([1, \underline{\text{heads}}] + [2, \underline{\text{heads}}] + [3, \underline{\text{heads}}] + [4, \underline{\text{heads}}]\Big)$$

or

$$\Big([1, \underline{\text{tails}}] + [1, \underline{\text{heads}}]\Big) + \Big([2, \underline{\text{tails}}] + [2, \underline{\text{heads}}]\Big)$$
$$+ \Big([3, \underline{\text{tails}}] + [3, \underline{\text{heads}}]\Big) + \Big([4, \underline{\text{tails}}] + [4, \underline{\text{heads}}]\Big).$$

Either way, it's the same eight terms, just grouped differently. But one of these versions may make it easier to see a point than the other, so often it's helpful to try both.

The first formula above can be expressed in words as "Hold s fixed to $\underline{\text{tails}}$ while summing ℓ, then hold s fixed to $\underline{\text{heads}}$ while again summing ℓ." The second formula can be expressed as "Hold ℓ fixed to 1 while summing s, and so on." The fact that these recipes give the same answer can be written symbolically as

$$\sum_{\ell,s}(\cdots) = \sum_s\Big(\sum_\ell(\cdots)\Big) = \sum_\ell\Big(\sum_s(\cdots)\Big). \tag{3.14}$$

Use this insight to work the following problem:

Your Turn 3G

a. Show that the joint distribution for two independent sets of outcomes will automatically be correctly normalized if the two marginal distributions (for example, our $\mathcal{P}_{\text{die}}$ and $\mathcal{P}_{\text{coin}}$) each have that property.
b. This time, suppose that we are given a properly normalized joint distribution, *not* necessarily for independent outcomes, and we compute the marginal distribution by using the formula you found in Your Turn 3F. Show that the resulting $\mathcal{P}_\ell(\ell)$ is automatically properly normalized.

3.4.3 The proper interpretation of medical tests requires an understanding of conditional probability

Statement of the problem

Let's apply these ideas to a problem whose solution surprises many people. This problem *can* be solved accurately by using common sense, but many people perceive alternate, wrong solutions to be equally reasonable. The concept of conditional probability offers a more sure-footed approach to problems of this sort.

Suppose that you have been tested for some dangerous disease. You participated in a mass random screening; you do not feel sick. The test comes back "positive," that is, indicating that you in fact have the disease. Worse, your doctor tells you the test is "97% accurate." That sounds bad.

Situations like this one are very common in science. We measure something; it's not precisely what we wanted to know, but neither is it irrelevant. Now we must attempt an *inference:* What can we say about the question of interest, based on the available new information? Returning to the specific question, you want to know, "Am I sick?" The ideas of conditional probability let us phrase this question precisely: We wish to know $\mathcal{P}(\text{sick} \mid \text{positive})$, the *probability* of being sick, given one positive test result.

To answer the question, we need some more precise information. The accuracy of a yes/no medical test actually has two distinct components:

- The **sensitivity** is the fraction of truly sick people who test positive. A sensitive test catches almost every sick person; that is, it yields very few **false-negative** results. For illustration, let's assume that the test has 97% sensitivity (a false-negative rate of 3%).
- The **selectivity** is the fraction of truly healthy people who test negative. High selectivity means that the test gives very few **false-positive** results. Let's assume that the test also has 97% selectivity (a false-positive rate of 3%).

In practice, false-positive and -negative results can arise from human error (a label falls off a test tube), intrinsic fluctuations, sample contamination, and so on. Sometimes sensitivity and selectivity depend on a threshold chosen when setting a lab protocol, so that one of them can be increased, but only at the expense of lowering the other one.

Analysis

Let E_{sick} be the event that a randomly chosen member of the population is sick, and E_{pos} the event that a randomly chosen member of the population tests positive. Now, certainly, these two events are *not* independent—the test does tell us *something*—in fact, quite a lot, according to the data given. But the two events are not quite synonymous, because neither the sensitivity nor the selectivity is perfect. Let's abbreviate $\mathcal{P}(S) = \mathcal{P}(E_{\text{sick}})$, and so on. In this language, the sensitivity is $\mathcal{P}(P \mid S) = 0.97$.

Before writing any more abstract formulas, let's attempt a pictorial representation of the problem. We represent the complete population by a 1×1 square containing evenly spaced points, with a point for every individual in the very large population under study. Then the probability of being in any subset is simply the *area* it fills on the square. We segregate the population into four categories based on sick/healthy status (S/H) and test result (P/N), and give names to their areas. For example, let $\mathcal{P}_{HN}$ denote $\mathcal{P}(\text{healthy } \textbf{and} \text{ negative result})$, and so on. Because the test result is not independent of the health of the patient, the figure is similar to Figure 3.4b.

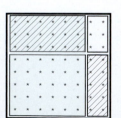

Figure 3.4b (page 46)

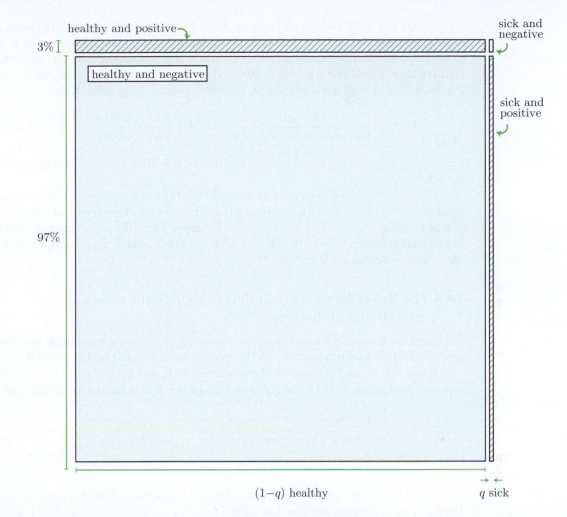

Figure 3.5 [Box diagram.] **A joint distribution representing a medical test.** The labels refer to healthy (*colored*) versus sick patients, and to positive (*hatched*) versus negative test results. The events E_{sick} and E_{pos} are highly (though not perfectly) correlated, so the labels resemble those in Figure 3.4b, not Figure 3.4a (page 46).

Figure 3.5 illustrates the information we have been given. It is partitioned horizontally in such a way that

$$\text{sensitivity} = \mathcal{P}(P \mid S) = \frac{\mathcal{P}(S \textbf{ and } P)}{\mathcal{P}(S)} = \frac{\mathcal{P}_{SP}}{\mathcal{P}_{SN} + \mathcal{P}_{SP}} = 97\%.$$

Your Turn 3H

Confirm that the figure also depicts a selectivity of 97%.

In our imagined scenario, you know you tested positive, but you wish to know whether you're sick. The probability, given what you know, is then

$$\mathcal{P}(S \mid P) = \frac{\mathcal{P}(S \textbf{ and } P)}{\mathcal{P}(P)} = \frac{\mathcal{P}_{SP}}{\mathcal{P}_{SP} + \mathcal{P}_{HP}}. \tag{3.15}$$

So, are you sick? Perhaps surprisingly, there is *no way to answer with the given information*. Figure 3.5 makes it clear that one additional, crucial bit of information is still missing: the fraction of the overall population that is sick. Suppose that you go back to your doctor and find that it's $\mathcal{P}(S) = 0.9\%$. This quantity is called q in the figure. Then $\mathcal{P}_{HP} = 0.03 \times (1 - q)$ and so on, and we can finish evaluating Equation 3.15:

$$\mathcal{P}(S \mid P) = \frac{0.97 \times 0.009}{0.97 \times 0.009 + 0.03 \times 0.991} = \left[1 + \frac{0.03 \times 0.991}{0.97 \times 0.009}\right]^{-1} \approx \frac{1}{4}. \qquad (3.16)$$

Remarkably, although you tested positive, and the test was "97% accurate," you are *probably not sick.*[19]

What just happened? Suppose that we could test the *entire* population. The huge majority of healthy people would generate some false-positive results—more than the number of true positives from the tiny minority of sick people. That's the commonsense analysis. In graphical language, region *HP* of Figure 3.5 is not much smaller than the region *SP* that concerns us—instead, it's *larger* than *SP*.

3.4.4 The Bayes formula streamlines calculations involving conditional probability

Situations like the one in the previous subsection arise often, so it is worthwhile to create a general tool. Consider two events E_1 and E_2, which may or may not be independent.

Notice that (E_1 **and** E_2) is exactly the same event as (E_2 **and** E_1). Equation 3.11 (page 45) therefore implies that $\mathcal{P}(E_1 \mid E_2) \times \mathcal{P}(E_2) = \mathcal{P}(E_2 \mid E_1) \times \mathcal{P}(E_1)$. Rearranging slightly gives

$$\mathcal{P}(E_1 \mid E_2) = \mathcal{P}(E_2 \mid E_1)\frac{\mathcal{P}(E_1)}{\mathcal{P}(E_2)}. \quad \textbf{Bayes formula} \qquad (3.17)$$

In everyday life, people often confuse the conditional probabilities $\mathcal{P}(E_1 \mid E_2)$ and $\mathcal{P}(E_2 \mid E_1)$. The Bayes formula quantifies how they differ.

Equation 3.17 formalizes a procedure that we all use informally. When evaluating a claim E_1, we generally have some notion of how probable it is. We call this our **prior** assessment because it's how strongly we believed E_1 prior to obtaining some new information. After obtaining the new information that E_2 is true, we *update* our prior $\mathcal{P}(E_1)$ to a new **posterior** assessment $\mathcal{P}(E_1 \mid E_2)$. That is, the posterior is the probability of E_1, given the new information that E_2 is true. The Bayes formula tells us that we can compute the posterior in terms of $\mathcal{P}(E_2 \mid E_1)$, which in this context is called the **likelihood**. The formula is useful because sometimes we know the likelihood a priori.

For example, the Bayes formula lets us automate the reasoning of Section 3.4.3.[20] In this context, Equation 3.17 says that $\mathcal{P}(S \mid P) = \mathcal{P}(P \mid S)\mathcal{P}(S)/\mathcal{P}(P)$. The numerator equals the sensitivity times q. The denominator can be expressed in terms of the two ways of getting

---

[19]A one-in-four chance of being sick may nevertheless warrant medical intervention, or at least further testing. "Decision theory" seeks to weight different proposed courses of action according to the probabilities of various outcomes, as well as the severity of the consequences of each action.

[20]The general framework also lets us handle other situations, such as those in which selectivity is not equal to sensitivity (see Problem 3.10).

a positive test result:

$$\mathcal{P}(P) = \mathcal{P}(P \mid S)\mathcal{P}(S) + \mathcal{P}(P \mid \mathbf{not}\text{–}S)\mathcal{P}(\mathbf{not}\text{–}S). \tag{3.18}$$

Your Turn 3I

Prove Equation 3.18 by using Equation 3.11. Then combine it with the Bayes formula to recover Equation 3.16.

Your Turn 3J

a. Suppose that our test has *perfect* sensitivity and selectivity. Write the Bayes formula for this case, and confirm that it connects with what you expect.
b. Suppose that our test is *worthless*; that is, the events E_{sick} and E_{pos} are statistically independent. Confirm that in this case, too, the math connects with what you expect.

T_2 *Section 3.4.4′ (page 60) develops an extended form of the Bayes formula.*

3.5 Expectations and Moments

Suppose that two people play a game in which each move is a Bernoulli trial. Nick pays Nora a penny each time the outcome $s =$ tails; otherwise, Nora pays Nick two pennies. A "round" consists of N_{tot} moves. Clearly, Nick can expect to win about $N_{tot}\xi$ times and lose $N_{tot}(1-\xi)$ times. Thus, he can expect his bank balance to have changed by about $N_{tot}(2\xi - (1 - \xi))$ pennies, although in every round the exact result will be different.

But players in a game of chance have other concerns besides the "typical" net outcome—for example, each will also want to know, "What is the risk of doing substantially *worse* than the typical outcome?" Other living creatures also play games like these, often with higher stakes.

3.5.1 The expectation expresses the average of a random variable over many trials

To make the questions more precise, let's begin by introducing a random variable $f(s)$ that equals $+2$ for $s =$ heads and -1 for $s =$ tails. Then, one useful descriptor of the game is the average of the values that f takes when we make N_{tot} measurements, in the limit of large N_{tot}. This quantity is called the **expectation** of f, and is denoted by the symbol $\langle f \rangle$. But we *don't* mean that we "expect" to observe this exact value in any real measurement; for example, in a discrete distribution, $\langle f \rangle$ generally falls between two allowed values of f, and so will *never* actually be observed.

Example Use Equation 3.3 to show that the expectation of f can be re-expressed by the formula

$$\langle f \rangle = \sum_s f(s)\mathcal{P}(s). \tag{3.19}$$

In this formula, the sum runs only over the list of possible outcomes (not over all N_{tot} repeated measurements); but each term is *weighted* by that outcome's probability.

Solution In the example of a coin-flipping game, suppose that N_1 of the flips yielded <u>heads</u> and N_2 yielded <u>tails</u>. To find the average of $f(s)$ over all of these $N_{\mathrm{tot}} = N_1 + N_2$ trials, we sum all the f values and divide by N_{tot}. Equivalently, however, we can rearrange the sum by first adding up all N_1 trials with $f(\underline{\mathrm{heads}}) = +2$, then adding all N_2 trials with $f(\underline{\mathrm{tails}}) = -1$:

$$\langle f \rangle = (N_1 f(\underline{\mathrm{heads}}) + N_2 f(\underline{\mathrm{tails}}))/N_{\mathrm{tot}}.$$

In the limit of large N_{tot}, this expression is equal to $f(\underline{\mathrm{heads}})\mathcal{P}(\underline{\mathrm{heads}}) + f(\underline{\mathrm{tails}})\mathcal{P}(\underline{\mathrm{tails}})$, which is the same as Equation 3.19. A similar approach proves the formula for any discrete probability distribution.

The left side of Equation 3.19 introduces an abbreviated notation for the expectation.[21] But brevity comes at a price; if we are considering several different distributions—for example, a set of several coins, each with a different value of ξ—then we may need to write something like $\langle f \rangle_\xi$ to distinguish the answers for the different distributions.

Some random systems generate outcomes that are not numbers. For example, if you ask each of your friends to write down a word "at random," then there's no meaning to questions like "What is the average word chosen?" But we have seen that in many cases, the outcome index does have a numerical meaning. As mentioned in Section 3.3.2, we'll usually use the symbol ℓ, not s, for such situations; then it makes sense to discuss the average value of many draws of ℓ itself, sometimes called the **first moment** of $\mathcal{P}(\ell)$. (The word "first" sets the stage for higher moments, which are expectations of higher powers of ℓ.)

Equation 3.19 gives the first moment of a random variable as $\langle \ell \rangle = \sum_\ell \ell \mathcal{P}(\ell)$. Notice that $\langle \ell \rangle$ is a specific number characterizing the distribution, unlike ℓ itself (which is a random value drawn from that distribution), or $\mathcal{P}(\ell)$ (which is a *function* of ℓ). The expectation may not be equal to the **most probable value**, which is the value of ℓ where $\mathcal{P}(\ell)$ attains its maximum.[22] For example, in Figure 3.2b, the most probable value of the waiting time is *zero*, but clearly the average waiting time is greater than that.

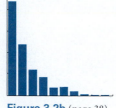

Figure 3.2b (page 38)

Your Turn 3K

Show that $\langle 3 \rangle = 3$. That is, consider a "random variable" whose value on every draw is always exactly equal to 3. More generally, the expectation of any *constant* is simply that constant, regardless of what distribution we use. So, in particular, think about why $\langle\langle f \rangle\rangle$ is the same as $\langle f \rangle$.

3.5.2 The variance of a random variable is one measure of its fluctuation

If you measure ℓ just once, you are not guaranteed to observe exactly the most probable value. We use words like "spread," "jitter," "noise," "dispersion," and "fluctuation" to describe

---

[21] The notations $\langle f \rangle$, $\mathbb{E}(f)$, μ_f, "expectation of f," "expected value of f," and "expectation value of f" are all synonyms in various cultures for "the mean of an infinitely replicated set of measurements of a random variable." This concept is different from "the mean of a particular, finite set of measurements," which we will call the "sample mean."

[22] The most probable value of a discrete distribution is also called its **mode**. If $\mathcal{P}(\ell)$ attains its maximum value at two or more distinct outcomes, then its most probable value is not defined. A Uniform distribution is an extreme example of this situation.

this phenomenon. It is closely related to the "risk" that Nick and Nora wanted to assess in their coin-toss game. For a Uniform distribution, the "spread" clearly has something to do with how *wide* the range of reasonably likely ℓ values is. Can we make this notion precise, for any kind of distribution?

One way to make these intuitions quantitative is to define[23]

$$\mathrm{var}\, f = \langle (f - \langle f \rangle)^2 \rangle. \quad \textbf{variance} \text{ of a random variable} \qquad (3.20)$$

The right side of Equation 3.20 essentially answers the question, "How much does f deviate from its expectation, on average?" But notice that in this definition, it was crucial to square $(f - \langle f \rangle)$. Had we computed the expectation of $(f - \langle f \rangle)$, we'd have found that the answer was always zero, which doesn't tell us much about the spread of f! By squaring the deviation, we ensure that variations above and below the expectation make reinforcing, not canceling, contributions to the variance.

Like the expectation, the variance depends both on which random variable $f(\ell)$ we are studying and also on the distribution $\mathcal{P}(\ell)$ being considered. Thus, if we study a family of distributions with a parameter, such as ξ for the coin flip, then $\mathrm{var}\, f$ will be a *function of* ξ. It is not, however, a function of ℓ, because that variable is summed in Equation 3.19.

Another variation on the same idea is the **standard deviation** of f in the given distribution,[24] defined as $\sqrt{\mathrm{var}\, f}$. The point of taking the square root is to arrive at a quantity with the same dimensions as f.

Example Here's another motivation for introducing the square root into the definition of standard deviation. Imagine a population of Martian students, each exactly twice as tall as a corresponding student in your class. Surely the "spread" of the second distribution should be twice the "spread" of the first. Which descriptor has that property?

Solution The variance for Martian students is $\mathrm{var}(2\ell) = \langle ((2\ell) - \langle 2\ell \rangle)^2 \rangle = 2^2 \langle (\ell - \langle \ell \rangle)^2 \rangle$. Thus, the variance of the Martians' height distribution is *four* times as great as ours. We say that the factor of 2 "inside" the variance became 2^2 when we moved it "outside." The standard deviation, not the variance, scales with a factor of 2.

Example a. Show that $\mathrm{var}\, f = \langle f^2 \rangle - (\langle f \rangle)^2$. (If f is ℓ itself, we say, "The variance is the second moment minus the square of the first moment.")
b. Show that, if $\mathrm{var}\, f = 0$, Equation 3.20 implies that every measurement of f actually does give exactly $\langle f \rangle$.

Solution a. Expand Equation 3.20 to find $\mathrm{var}\, f = \langle f^2 \rangle - 2\langle f \langle f \rangle \rangle + \langle (\langle f \rangle)^2 \rangle$. Now remember that $\langle f \rangle$ is itself a constant, not a random variable. So it can be pulled out of expectations

---

[23] Section 5.2 will introduce a class of distributions for which the variance is *not* useful as a descriptor of the spread. Nevertheless, the variance is simple, widely used, and appropriate in many cases.
[24] The standard deviation is also called the "root-mean-square" or **RMS deviation** of f. Think about why that's a good name for it.

(see also Your Turn 3K), and we get

$$\operatorname{var} f = \langle f^2 \rangle - 2(\langle f \rangle)^2 + (\langle f \rangle)^2,$$

which reduces to what was to be shown.

b. Let $f_* = \langle f \rangle$. We are given that $0 = \langle (f - f_*)^2 \rangle = \sum_\ell \mathcal{P}(\ell)(f(\ell) - f_*)^2$. Every term on the right side is ≥ 0, yet their sum equals zero. So every term is separately zero. For each outcome ℓ, then, we must either have $\mathcal{P}(\ell) = 0$, or else $f(\ell) = f_*$. The outcomes with $\mathcal{P} = 0$ never happen, so every measurement of f yields the value f_*.

Suppose that a discrete random system has outcomes that are labeled by an integer ℓ. We can construct a new random variable m as follows: Every time we are asked to produce a sample of m, we draw a sample of ℓ and add the constant 2. (That is, $m = \ell + 2$.) Then the distribution $\mathcal{P}_{\mathrm{m}}(m)$ has a graph that looks exactly like that of $\mathcal{P}_\ell(\ell)$, but *shifted* to the right by 2, so not surprisingly $\langle m \rangle = \langle \ell \rangle + 2$. Both distributions are equally wide, so (again, not surprisingly) both have the same variance.

Your Turn 3L

a. Prove those two claims, starting from the relevant definitions.
b. Suppose that another random system yields *two* numerical values on every draw, ℓ and s, and the expectations and variances of both are given to us. Find the expectation of $2\ell + 5s$. Express what you found as a general rule for the expectation of a linear combination of random variables.
c. Continuing (b), can you determine the variance of $2\ell + 5s$ from the given information?

Example Find the expectation and variance of the Bernoulli trial distribution, $\mathcal{P}_{\mathrm{bern}}(s; \xi)$, as functions of the parameter ξ.

Solution The answer depends on what numerical values $f(s)$ we assign to heads and tails; suppose these are 1 and 0, respectively. Summing over the sample space just means adding two terms. Hence, $\langle f \rangle = 0 \times (1 - \xi) + 1 \times \xi = \xi$, and

$$\operatorname{var} f = \langle f^2 \rangle - (\langle f \rangle)^2 = (0^2 \times (1 - \xi) + 1^2 \times \xi) - \xi^2,$$

or

$$\langle f \rangle = \xi, \qquad \operatorname{var} f = \xi(1 - \xi). \quad \text{for Bernoulli trial} \tag{3.21}$$

Think about why these results are reasonable: The extreme values of ξ (0 and 1) correspond to certainty, or no spread in the results. The trial is most unpredictable when $\xi = \frac{1}{2}$, and that's exactly where the function $\xi(1 - \xi)$ attains its maximum. Try the derivation again, with different values for $f(\text{heads})$ and $f(\text{tails})$.

Your Turn 3M

Suppose that f and g are two independent random variables in a discrete random system.

a. Show that $\langle fg \rangle = \langle f \rangle \langle g \rangle$. Think about how you had to use the assumption of independence and Equation 3.14 (page 49); give a counterexample of two *non*independent random variables that *don't* obey this rule.

b. Find the expectation and variance of $f + g$ in terms of the expectations and variances of f and g separately.

c. Repeat (b) for the quantity $f - g$.

d. Suppose that the expectations of f and g are both greater than zero. Define the **relative standard deviation** (RSD) of a random variable x as $(\text{var } x)/|\langle x \rangle|$, a dimensionless quantity. Compare the RSD of $f + g$ with the corresponding quantity for $f - g$.

We can summarize part of what you just found by saying,

*The difference of two noisy variables is a **very** noisy variable.*   (3.22)

$\boxed{T_2}$ *Section 3.5.2′ (page 60) discusses some other moments that are useful as reduced descriptions of a distribution, and some tests for statistical independence of two random variables.*

3.5.3 The standard error of the mean improves with increasing sample size

Suppose that we've got a replicable random system: It allows repeated, independent measurements of a quantity f. We'd like to know the expectation of f, but we don't have time to make an infinite set of measurements; nor do we know a priori the distribution $\mathcal{P}(\ell)$ needed to evaluate Equation 3.19. So we make a finite set of M measurements and average over that, obtaining the **sample mean** $\bar{f}$. This quantity is itself a random variable, because when we make another batch of M measurements and evaluate it, we won't get exactly the same answer.[25] Only in the limit of an infinitely big sample do we expect the sample mean to become a specific number. Because we never measure infinitely big samples in practice, we'd like to know: How good an estimate of the true expectation is $\bar{f}$?

Certainly $\langle \bar{f} \rangle$ is $1/M$ times the sum of M terms, each of which has the same expectation (namely, $\langle f \rangle$). Thus, $\langle \bar{f} \rangle = \langle f \rangle$. But we also need an estimate of how much $\bar{f}$ *varies* from one batch of samples to the next, that is, its variance:

$$\text{var}(\bar{f}) = \text{var}\left(\frac{1}{M}(f_1 + \cdots + f_M) \right).$$

Here, f_i is the value that we measured in the ith measurement of a batch. The random variables f_i are all assumed to be independent of one another, because each copy of a replicable system is unaffected by every other one. The constant $1/M$ inside the variance can be replaced by a factor of $1/M^2$ outside.[26] Also, in Your Turn 3M(b), you found that

---

[25] $\boxed{T_2}$ More precisely, $\bar{f}$ is a random variable on the joint distribution of batches of M independent measurements.

[26] See page 55.

the variance of the sum of independent variables equals the sum of their variances, which in this case are all equal. So,

$$\mathrm{var}(\bar{f}) = \left(\frac{1}{M^2}M\right)\left(\mathrm{var}\,f\right) = \frac{1}{M}\,\mathrm{var}\,f. \tag{3.23}$$

The factor $1/M$ in this answer means that

> *The sample mean becomes a better estimate of the true expectation as we average over more measurements.* $\tag{3.24}$

The square root of Equation 3.23 is called the **standard error of the mean**, or **SEM**.

The SEM illustrates a broader idea: A **statistic** is something we compute from a finite sample of data by following a standard recipe. An **estimator** is a statistic that is useful for inferring some property of the underlying distribution of the data. Idea 3.24 says that the sample mean is a useful estimator for the expectation.

THE BIG PICTURE

Living organisms are inference machines, constantly seeking patterns in their world and ways to exploit those regularities. Many of these patterns are veiled by partial randomness. This chapter has begun our study of how to extract whatever discernible, relevant structure can be found from a limited number of observations.

Chapters 4–8 will extend these ideas, but already we have obtained a powerful tool, the Bayes formula (Equation 3.17, page 52). In a strictly mathematical sense, this formula is a trivial consequence of the definition of conditional probability. But we have seen that conditional probability itself is a subtle concept, and one that arises naturally in certain questions that we need to understand (see Section 3.4.3); the Bayes formula clarifies how to apply it.

More broadly, randomness is often a big component of a physical model, and so that model's prediction will in general be a probability distribution. We need to learn how to confront such models with experimental data. Chapter 4 will develop this idea in the context of a historic experiment on bacterial genetics.

KEY FORMULAS

- *Probability distribution of a discrete, replicable random system:* $\mathcal{P}(\ell) = \lim_{N_{\mathrm{tot}}\to\infty} N_\ell/N_{\mathrm{tot}}$. For a finite number of draws N_{tot}, the integers N_ℓ are sometimes called the frequencies of the various possible outcomes; the numbers N_ℓ/N_{tot} all lie between 0 and 1 and can be used as estimates of $\mathcal{P}(\ell)$.
- *Normalization of discrete distribution:* $\sum_\ell \mathcal{P}(\ell) = 1$.
- *Bernoulli trial:* $\mathcal{P}_{\mathrm{bern}}(\underline{\mathrm{heads}};\xi) = \xi$ and $\mathcal{P}_{\mathrm{bern}}(\underline{\mathrm{tails}};\xi) = 1 - \xi$. The parameter ξ, and $\mathcal{P}$ itself, are dimensionless. If $\underline{\mathrm{heads}}$ and $\underline{\mathrm{tails}}$ are assigned numerical values $s = 1$ and 0, respectively, then the expectation of the random variable is $\langle s \rangle = \xi$ and the variance is $\mathrm{var}\,s = \xi(1 - \xi)$.
- *Addition rule:* $\mathcal{P}(\mathsf{E}_1 \text{ or } \mathsf{E}_2) = \mathcal{P}(\mathsf{E}_1) + \mathcal{P}(\mathsf{E}_2) - \mathcal{P}(\mathsf{E}_1 \text{ and } \mathsf{E}_2)$.
- *Negation rule:* $\mathcal{P}(\mathbf{not}\text{-}\mathsf{E}) = 1 - \mathcal{P}(\mathsf{E})$.
- *Product rule:* $\mathcal{P}(\mathsf{E}_1 \text{ and } \mathsf{E}_2) = \mathcal{P}(\mathsf{E}_1 \mid \mathsf{E}_2) \times \mathcal{P}(\mathsf{E}_2)$. (This formula is actually the definition of the conditional probability.)

- *Independence:* Two events are statistically independent if $\mathcal{P}(\mathsf{E} \text{ and } \mathsf{E}') = \mathcal{P}(\mathsf{E}) \times \mathcal{P}(\mathsf{E}')$, or equivalently $\mathcal{P}(\mathsf{E}_1 \mid \mathsf{E}_2) = \mathcal{P}(\mathsf{E}_1 \mid \mathbf{not}\text{-}\mathsf{E}_2) = \mathcal{P}(\mathsf{E}_1)$.

- *Geometric distribution:* $\mathcal{P}_{\mathrm{geom}}(j; \xi) = \xi(1 - \xi)^{(j-1)}$ for discrete, independent attempts with probability of "success" equal to ξ on any trial. The probability $\mathcal{P}$, the random variable $j = 1, 2, \ldots$, and the parameter ξ are all dimensionless. The expectation of j is $1/\xi$, and the variance is $(1 - \xi)/(\xi^2)$.

- *Marginal distribution:* For a joint distribution $\mathcal{P}(\ell, s)$, the marginal distributions are $\mathcal{P}_\ell(\ell_0) = \sum_s \mathcal{P}(\ell_0, s)$ and $\mathcal{P}_s(s_0) = \sum_\ell \mathcal{P}(\ell, s_0)$. If ℓ and s are independent, then $\mathcal{P}(\ell \mid s) = \mathcal{P}_\ell(\ell)$ and conversely; equivalently, $\mathcal{P}(\ell, s) = \mathcal{P}_\ell(\ell)\mathcal{P}_s(s)$ in this case.

- *Bayes:* $\mathcal{P}(\mathsf{E}_1 \mid \mathsf{E}_2) = \mathcal{P}(\mathsf{E}_2 \mid \mathsf{E}_1)\mathcal{P}(\mathsf{E}_1)/\mathcal{P}(\mathsf{E}_2)$. In the context of inferring a model, we call $\mathcal{P}(\mathsf{E}_1)$ the prior distribution, $\mathcal{P}(\mathsf{E}_1 \mid \mathsf{E}_2)$ the posterior distribution in the light of new information E_2, and $\mathcal{P}(\mathsf{E}_2 \mid \mathsf{E}_1)$ the likelihood function.

  Sometimes the formula can usefully be rewritten by expressing the denominator as $\mathcal{P}(\mathsf{E}_2) = \mathcal{P}(\mathsf{E}_2 \mid \mathsf{E}_1)\mathcal{P}(\mathsf{E}_1) + \mathcal{P}(\mathsf{E}_2 \mid \mathbf{not}\text{–}\mathsf{E}_1)\mathcal{P}(\mathbf{not}\text{–}\mathsf{E}_1)$.

- *Moments:* The expectation of a discrete random variable f is its first moment: $\langle f \rangle = \sum_\ell f(\ell)\mathcal{P}(\ell)$. The variance is the mean-square deviation from the expected value: $\mathrm{var}\,\ell = \langle (\ell - \langle \ell \rangle)^2 \rangle$. Equivalently, $\mathrm{var}\,\ell = \langle \ell^2 \rangle - (\langle \ell \rangle)^2$. The standard deviation is the square root of the variance. $\boxed{T_2}$ Skewness and kurtosis are defined in Section 3.5.2' (page 60).

- $\boxed{T_2}$ *Correlation and covariance:* $\mathrm{cov}(\ell, s) = \langle (\ell - \langle \ell \rangle)(s - \langle s \rangle) \rangle$. $\mathrm{corr}(\ell, s) = \mathrm{cov}(\ell, s)/\sqrt{(\mathrm{var}\,\ell)(\mathrm{var}\,s)}$.

FURTHER READING

Semipopular:
Conditional probability and the Bayes formula: Gigerenzer, 2002; Mlodinow, 2008; Strogatz, 2012; Woolfson, 2012.

Intermediate:
Bolker, 2008; Denny & Gaines, 2000; Dill & Bromberg, 2010, chapt. 1; Otto & Day, 2007, §P3.

Technical:
Gelman et al., 2014.

$\boxed{T_2}$ **Track 2**

3.4.1′a Extended negation rule
Here is another useful fact about conditional probabilities:

> **Your Turn 3N**
>
> a. Show that $\mathcal{P}(\textbf{not-}E_1 \mid E_2) = 1 - \mathcal{P}(E_1 \mid E_2)$.
> b. More generally, find a normalization rule for $\mathcal{P}(\ell \mid E)$, where ℓ is a discrete random variable and E is any event.

3.4.1′b Extended product rule
Similarly,

$$\mathcal{P}(E_1 \textbf{ and } E_2 \mid E_3) = \mathcal{P}(E_1 \mid E_2 \textbf{ and } E_3) \times \mathcal{P}(E_2 \mid E_3). \tag{3.25}$$

> **Your Turn 3O**
>
> Prove Equation 3.25.

3.4.1′c Extended independence property
We can extend the discussion in Section 3.4.1 by saying that E_1 and E_2 are "independent under condition E_3" if knowing E_2 gives us no additional information about E_1 beyond what we already had from E_3; that is,

$$\mathcal{P}(E_1 \mid E_2 \textbf{ and } E_3) = \mathcal{P}(E_1 \mid E_3). \qquad \text{independence under condition } E_3$$

Substituting into Equation 3.25 shows that, if two events are independent under a third condition, then

$$\mathcal{P}(E_1 \textbf{ and } E_2 \mid E_3) = \mathcal{P}(E_1 \mid E_3) \times \mathcal{P}(E_2 \mid E_3). \tag{3.26}$$

$\boxed{T_2}$ **Track 2**

3.4.4′ Generalized Bayes formula
There is a useful extension of the Bayes formula that states

$$\mathcal{P}(E_1 \mid E_2 \textbf{ and } E_3) = \mathcal{P}(E_2 \mid E_1 \textbf{ and } E_3) \times \mathcal{P}(E_1 \mid E_3)/\mathcal{P}(E_2 \mid E_3). \tag{3.27}$$

> **Your Turn 3P**
>
> Use your result in Your Turn 3O to prove Equation 3.27.

$\boxed{T_2}$ **Track 2**

3.5.2′a Skewness and kurtosis
The first and second moments of a distribution, related to the location and width of its peak, are useful summary statistics, particularly when we repackage them as

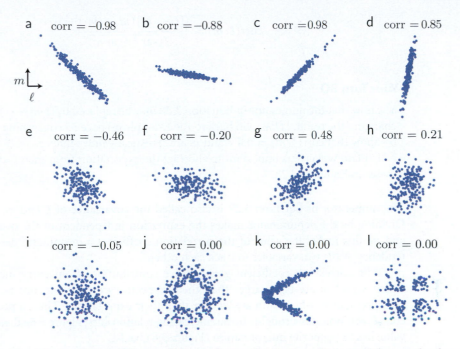

Figure 3.6 [Simulated datasets.] **Correlation coefficients of some distributions.** Each panel shows a cloud representation of a joint probability distribution, as a set of points in the ℓ-m plane; the corresponding value for $\text{corr}(\ell, m)$ is given above each set. Note that the correlation coefficient reflects the noisiness and direction of a linear relationship (a–h), and it's zero for independent variables (i), but it misses other kinds of correlation (j–l). In each case, the correlation coefficient was estimated from a sample of 5000 points, but only the first 200 are shown.

expectation and variance. Two other moments are often used to give more detailed information:

- Some distributions are asymmetric about their peak. The asymmetry can be quantified by computing the **skewness**, defined by $\langle (\ell - \langle \ell \rangle)^3 \rangle / (\text{var } \ell)^{3/2}$. This quantity equals zero for any symmetric distribution.
- Even if two distributions each have a single, symmetric peak, and both have the same variance, nevertheless their peaks may not have the same shape. The **kurtosis** further specifies the peak shape; it is defined as $\langle (\ell - \langle \ell \rangle)^4 \rangle / (\text{var } \ell)^2$.

3.5.2′b Correlation and covariance

The product rule for independent events (Equation 3.12) can also be regarded as a *test* for whether two events are statistically independent. This criterion, however, is not always easy to evaluate. How can we tell from a joint probability distribution $\mathcal{P}(\ell, m)$ whether it can be written as a product? One way would be to evaluate the conditional probability $\mathcal{P}(\ell \mid m)$ and see, for every value of ℓ, whether it depends on m. But there is a short-cut that can at least show that two variables are *not* independent (that is, that they are correlated).

Suppose that ℓ and m both have numerical values; that is, both are random variables. Then we can define the **correlation coefficient** as an expectation:

$$\text{corr}(\ell, m) = \frac{\langle(\ell - \langle\ell\rangle)(m - \langle m\rangle)\rangle}{\sqrt{(\text{var }\ell)(\text{var }m)}}. \tag{3.28}$$

Your Turn 3Q

a. Show that the numerator in Equation 3.28 may be replaced by $\langle\ell m\rangle - \langle\ell\rangle\langle m\rangle$ without changing the result. [*Hint:* Go back to the Example on page 55 concerning variance.]

b. Show that $\text{corr}(\ell, m) = 0$ if ℓ and m are statistically independent.

c. Explain why it was important to subtract the expectation from each factor in parentheses in Equation 3.28.

The numerator of Equation 3.28 is also called the **covariance** of ℓ and m, or $\text{cov}(\ell, m)$. Dividing by the denominator makes the expression independent of the overall scale of ℓ and m; this makes the value of the correlation coefficient a meaningful descriptor of the tendency of the two variables to track each other.

The correlation coefficient gets positive contributions from every measurement in which ℓ and m are both larger than their respective expectations, but also from every measurement in which both are *smaller* than their expectations. Thus, a positive value of $\text{corr}(\ell, m)$ indicates a roughly linear, increasing relationship (Figure 3.6c,d,g,h). A negative value has the opposite interpretation (Figure 3.6a,b,e,f).

When we flip a coin repeatedly, there's a natural linear ordering according to time: Our data form a **time series**. We don't expect the probability of flipping heads on trial i to depend on the results of the previous trials, and certainly not on those of future trials. But many other time series do have such dependences.[27] To spot them, assign numerical values $f(s)$ to each flip outcome and consider each flip $f_1, \ldots, f_M$ in the series to be a different random variable, in which the f_i's may or may not be independent. If the random system is stationary (all probabilities are unchanged if we shift every index by the same amount), then we can define its **autocorrelation function** as $C(j) = \text{cov}(f_i, f_{i+j})$ for any starting point i. If this function is nonzero for any j (other than $j = 0$), then the time series is correlated.

3.5.2′c Limitations of the correlation coefficient

Equation 3.28 introduced a quantity that equals zero if two random variables are statistically independent. It follows that if the correlation coefficient of two random variables is nonzero, then they are correlated. However, the converse statement is not always true: It is possible for two nonindependent random variables to have correlation coefficient equal to zero. Panels (j–l) of Figure 3.6 show some examples. For example, panel (j) represents a distribution with the property that ℓ and m are never both close to zero; thus, knowing the value of one tells something about the value of the other, even though there is no linear relation.

---

[27] For example, the successive positions of a particle undergoing Brownian motion (example **5** on page 38).

PROBLEMS

3.1 Complex time series

Think about the weather—for example, the daily peak temperature. It's proverbially unpredictable. Nevertheless, there are several kinds of structure to this time series. Name a few and discuss.

3.2 Medical test

Look at Figures 3.4a,b. If E_2 is the outcome of a medical test and E_1 is the statement that the patient is actually sick, then which of these figures describes a better test?

Figures 3.4a (page 46)

3.3 Six flips

Write a few lines of computer code to make distributions like Figures 3.1a,b, but with $m = 6$ and 6000 total draws. (That is, generate 6000 six-bit random binary fractions.) If you don't like what you see, explain and then fix it.

3.4 Random walk end point distribution

This problem introduces the "random walk," a physical model for Brownian motion. Get Dataset 4, which contains experimental data. The two columns represent the x and y coordinates of the displacements of a particle undergoing Brownian motion, observed at periodic time intervals (see Figure 3.3a,b).

Figure 3.4b (page 46)

a. Tabulate the values of $x^2 + y^2$ in the experimental data, and display them as a histogram.

Suppose that a chess piece is placed on a line, initially at a point labeled 0. Once per second, the chess piece is moved a distance $d = 1\ \mu\text{m}$ along either the $+$ or $-$ direction. The choice is random, each direction is equally probable, and each step is statistically independent of all the others. We imagine making many trajectories, all starting at 0.

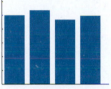

Figures 3.1a (page 37)

b. Simulate 1000 two-dimensional random walks, all starting at the origin. To do this, at each step randomly choose $\Delta x = \pm 1\ \mu\text{m}$ and also $\Delta y = \pm 1\ \mu\text{m}$. Display the histogram of $x^2 + y^2$ after 500 steps, and compare your answer qualitatively with the experimental result.

c. Do your results suggest a possible mathematical form for this distribution? How could you replot your data to check this idea?

Figure 3.1b (page 37)

3.5 Gambler's fallacy

There seems to be a hardwired misperception in the human brain that says, "If I've flipped heads five times in a row, that increases the probability that I'll get tails the next time." Intellectually, we know that's false, but it's still hard to avoid disguised versions of this error in our daily lives.

If we call heads $+1$ and tails -1, we can let X be the sum of, say, 10 flips. On any given 10-flip round we probably won't get exactly zero. But if we keep doing 10-flip rounds, then the long-term average of X is zero.

Suppose that one trial starts out with five heads in a row. We wish to check the proposition

"My next five flips will be more than half tails, in order to pull X closer to zero, because X 'wants' to be zero on average."

Figure 3.3a,b (page 39)

a. Use a computer to simulate 200 ten-flip sequences, pull out those few that start with five heads in a row, and find the average value of X among only those few sequences.

[*Hint:* Define variables called Ntrials and Nflips; use those names throughout your code. At the top, say Ntrials=200, Nflips=10.]

b. Repeat (a) with Ntrials = 2000 and 8000 sequences. Does the answer seem to be converging to zero, as predicted in the quoted text above? Whatever your answer, give some explanation for why the answer you found should have been expected.

c. To understand what "regression to the mean" actually means, repeat (a) but with 50 000 sequences of Nflips flips. Consider Nflips = 10, 100, 500, and 2500. [*Hint:* Instead of writing similar code four times, write it once, but put it inside a loop that gives Nflips a new value each time it runs.]

d. As you get longer and longer sequences (larger values of Nflips), your answer in (c) will become insignificant compared with the *spread* in the results among trials. Confirm this as follows. Again, start with Nflips = 10. For each of your sequences, save the value of X, creating a list with 50 000 entries. Then find the spread (standard deviation) of all the values you found. Repeat with Nflips = 100, 500, and 2500. Discuss whether the proposition

> The effect of unusual past behavior doesn't disappear; it just gets **diluted** as time goes on.

is more appropriate than the idea in quotes above.

3.6 Virus evolution

The genome of the HIV virus, like any genome, is a string of "letters" (base pairs) in an "alphabet" containing only four letters. The message for HIV is rather short, just $n \approx 10^4$ letters in all.

The probability of errors in reverse transcribing the HIV genome is about one error for every $3 \cdot 10^4$ "letters" copied. Suppose that each error replaces a DNA base by one of the three other bases, chosen at random. Each time a virion infects a T cell, the reverse transcription step creates opportunities for such errors, which will then be passed on to the offspring virions. The total population of infected T cells in a patient's blood, in the quasi-steady state, is roughly 10^7 (see Problem 1.7).

a. Find the probability that a T cell infection event will generate one *particular* error, for example, the one lucky spontaneous mutation that could confer resistance to a drug. Multiply by the population to estimate the number of T cells already present with a specified mutation, prior to administering any drug. Those cells will later release resistant virions.

b. Repeat (a), but this time for the probability of spontaneously finding *two* or *three* specific errors, and comment.

[*Note:* Make the conservative approximation that each infected T cell was infected by a wild-type virion, so that mutations do not accumulate. For example, the wild-type may reproduce faster than the mutant, crowding it out in the quasi-steady state.]

3.7 Weather

Figure 3.7a is a graphical depiction of the probability distribution for the weather on consecutive days in an imagined place and season. The outcomes are labeled $X_1 X_2$, where $X = r$ or s indicates the mutually exclusive options "rain" or "sunny," and the subscripts 1 and 2 denote today and tomorrow, respectively.

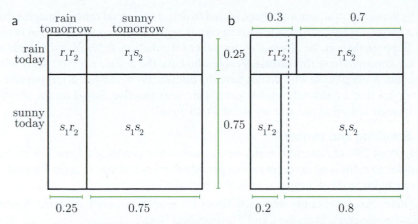

Figure 3.7 [Box diagrams.] **Probabilities for outcomes on consecutive days.** (a) A case where the two outcomes are independent. (b) A modified set of probabilities. The *dashed line* indicates the situation in (a) for comparison.

Panel (b) shows a more realistic situation. Compute $\mathcal{P}(\text{rain tomorrow})$, $\mathcal{P}(\text{rain tomorrow} \mid \text{rain today})$, and $\mathcal{P}(\text{rain tomorrow} \mid \text{sunny today})$ and comment. Repeat for the situation in panel (a).

3.8 Family history
Review Section 3.4.3. How does the probability of being sick given a positive test result change if you also know that you have some family history predisposing you to the disease? Discuss how to account for this information by using the Bayes formula.

3.9 Doping in sports
Background: A laboratory flagged a cyclist based on a urine sample taken following stage 17 of the 2006 Tour de France. The lab claimed that the test was highly unlikely to turn out positive unless the subject had taken illegal steroid drugs. Based on this determination, the International Court of Arbitration for Sport upheld doping charges against the cyclist.

In fact, the cyclist was tested eight times during the race, and a total of 126 tests were made on all contestants.

a. Suppose that the cyclist was innocent, but the false-positive rate of the test was 2%. What is the probability that at least one of the 8 tests would come out positive?

b. Suppose that the false-positive rate was just 1% and that *all* contestants in the race were innocent. What is the chance that *some* contestant (that is, one or more) would test positive at least once?

c. Actually, it's not enough to know the false-positive rate. If we wish to know the probability of guilt given the test results, we need one *additional* piece of quantitative information (which the court did not have). What is that needed quantity? [*Hint:* You may assume that the false-negative rate is small. It is not the quantity being requested.]

3.10 Hemoccult test
Figure 3.5 (page 51) represents a situation in which the sensitivity of a medical test is equal to its selectivity. This is actually not a very common situation.

The hemoccult test, among others, is used to detect colorectal cancer. Imagine that you conduct mass screening with this test over a certain region of the country, in a particular age group. Suppose that, in the absence of any other information, 0.3% of individuals in this group are known to have this disease. People who have the disease are 50% likely to have a positive test. Among those who do not have the disease, 3% nevertheless test positive.

Suppose that a randomly chosen participant tests positive. Based on the above data, and that single test, what can you say about $\mathcal{P}(\text{sick} \mid \text{pos})$?

3.11   Smoking and cancer

In 1993, about 28% of American males were classified as cigarette smokers. The probability for a smoker to die of lung cancer in a given period of time was about 11 times the probability for a nonsmoker to die of lung cancer in that period.

a. Translate these statements into facts about $\mathcal{P}(\text{die of lung cancer} \mid \text{smoker})$, $\mathcal{P}(\text{die of lung cancer} \mid \text{nonsmoker})$, $\mathcal{P}(\text{smoker})$, and $\mathcal{P}(\text{nonsmoker})$.

b. From these data, compute the probability that an American male who died of lung cancer in the specified period was a smoker.

3.12   Effect of new information

The "Monty Hall" puzzle is a classic problem that can be stated and analyzed in the language we are developing.

A valuable prize is known to lie behind one of three closed doors. All three options are equally probable. The director of the game ("Monty") knows which door conceals the prize, but you don't. The rules state that after you make a preliminary choice, Monty will choose one of the *other* two doors, open it, and reveal that the prize is *not* there. He then gives you the option of changing your preliminary choice, or sticking with it. After you make this decision, your final choice of door is opened. The puzzle is to find the best strategy for playing this game.

Let's suppose that you initially choose door #1.[28] Certainly, either #2 or #3, or both, has no prize. After Monty opens one of these doors, have you now gained any relevant additional information? If not, there's no point in changing your choice (analogous to scenario **a** in Section 3.4.1). If so, then maybe you should change (analogous to scenario **b**). To analyze the game, make a grid with six cells:

		It's actually behind door #		
		1	2	3
Monty reveals it's not behind door #	2	A	B	C
	3	D	E	F

In this table,

$$A = \mathcal{P}\big(\text{it's behind door \#1 and Monty shows you it's not behind \#2}\big),$$

$$D = \mathcal{P}\big(\text{it's behind door \#1 and Monty shows you it's not behind \#3}\big),$$

----
[28] By symmetry, it's enough to analyze only this case.

and so on. Convince yourself that

$$A = 1/6, \ D = 1/6, \text{ but}$$

$$B = 0, \ C = 1/3 \ (\text{Monty has no choice if it's not behind the door you chose}), \text{ and}$$

$$E = 1/3, \ F = 0.$$

a. Now compute $\mathcal{P}(\text{it's behind #1}|\text{Monty showed you #2})$ by using the definition of conditional probability.

b. Compute $\mathcal{P}(\text{it's behind #3} \mid \text{Monty showed you #2})$, and compare it with your answer in (a). Also compute $\mathcal{P}(\text{it's behind #2} \mid \text{Monty showed you #2})$. (The second quantity is zero, because Monty wouldn't do that.)

c. Now answer this question: If you initially chose #1, and then Monty showed you #2, should you switch your initial choice to #3 or remain with #1?

3.13   Negation rule

a. Suppose that you are looking for a special type of cell, perhaps those tagged by expressing a fluorescent protein. You spread a drop of blood on a slide marked with a grid containing N boxes, and examine each box for the cell type of interest. Suppose that a particular sample has a total of M tagged cells. What is the probability that at least one box on the grid contains more than one of these M cells? [*Hint:* Each tagged cell independently "chooses" a box, so each has probability $1/N$ to be in any particular box. Use the product rule to compute the probability that *no* box on the grid has more than one tagged cell, and then use the negation rule.]

b. Evaluate your answer for $N = 400, M = 20$.

3.14   Modified Bernoulli trial

The Example on page 56 found the expectation and variance of the Bernoulli trial distribution as functions of its parameter ξ, if <u>heads</u> is assigned the numerical value 1 and <u>tails</u> 0. Repeat, but this time, <u>heads</u> counts as $1/2$ and <u>tails</u> as $-1/2$.

3.15   Perfectly random?

Let ℓ be an integer random variable with the Uniform distribution on the range $3 \leq \ell \leq 6$. Find the variance of ℓ.

3.16 $\boxed{T_2}$ Variance of a general sum

In Your Turn 3M(b) (page 57), you found the variance of the sum of two random variables, assuming that they were independent. Generalize your result to find the variance of $f + g$ in terms of var f, var g, and the covariance $\mathrm{cov}(f, g)$, *without* assuming independence.

3.17 $\boxed{T_2}$ Multiple tests

Suppose that you are a physician. You examine a patient, and you think it's quite likely that she has strep throat. Specifically, you believe this patient's symptoms put her in a group of people with similar symptoms, of whom 90% are sick. But now you refine your estimate by taking throat swabs and sending them to a lab for testing.

The throat swab is not a perfect test. Suppose that if a patient is sick with strep, then in 70% of cases, the test comes back positive; the rest of the time, it's a false negative. Suppose that, if a patient is not sick, then in 90% of cases, the test comes back negative; the rest of the time, it's a false positive.

You run five successive swabs from the same patient and send them to the lab, where they are all tested independently. The results come back $(+ - + - +)$, apparently a total muddle. You'd like to know whether any conclusion can be drawn from such data. Specifically, do they revise your estimate of the probability that the patient is sick?

a. Based on this information, what is your new estimate of the probability that the patient is sick? [*Hint:* Prove, then use, the result about independence stated in Equations 3.25–3.26 on page 60.]

b. Work the problem again, but this time from the viewpoint of a worker at the lab, who has no information about the patient other than the five test results. This worker interprets the information in the light of a prior assumption that the patient's chance of being sick is 50% (not 90%).

3.18 $\boxed{T_2}$ **Binary fractions**

Figure 3.1b (page 37)

Find the expectation and variance of the random, m-bit binary fractions discussed in Section 3.2.1 on page 36 (see Figure 3.1). Use an analytic (exact) argument, not a computer simulation.

Some Useful Discrete Distributions

It may be that universal history is the history of the different intonations given a handful of metaphors.
—Jorge Luis Borges

4.1 Signpost

Much of the everyday business of science involves proposing a model for some phenomenon of interest, poking the model until it yields some quantitative prediction, and then testing the prediction. A theme of this book is that often what is predicted is a probability distribution. This chapter begins our discussion of how to make such predictions, starting from a proposed physical model of a living system.

Chapter 3 may have given the impression that a probability distribution is a purely empirical construction, to be deduced from repeated measurements (via Equation 3.3, page 42). In practice, however, we generally work with distributions that embody simplifying hypotheses about the system (the physical model). For example, we may have reason to believe that a variable is Uniformly distributed on some range. Generally we need more complicated distributions than that, but perhaps surprisingly, just *three* additional discrete distributions describe many problems that arise in biology and physics: the Binomial, Poisson, and Geometric distributions. We'll see that, remarkably, all three are descendants of the humble Bernoulli trial.[1] Moreover, each has rather simple mathematical properties. Knowing some general facts about a distribution at once gives useful information about all the systems to which it applies.

---

[1]Later chapters will show that the Gaussian and Exponential distributions, and the Poisson process, are also offshoots of Bernoulli.

Our Focus Question is

Biological question: How do bacteria become resistant to a drug or virus that they've never encountered?

Physical idea: The Luria-Delbrück experiment tested a model by checking a statistical prediction.

4.2 Binomial Distribution

4.2.1 Drawing a sample from solution can be modeled in terms of Bernoulli trials

Here is a question that arises in the lab: Suppose that you have 10 mL of solution containing just *four molecules* of a particular type, each of which is tagged with a fluorescent dye. You mix well and withdraw a 1 mL sample (an "aliquot"). How many of those four molecules will be in your sample?[2] One reply is, "I can't predict that; it's random," and of course that is true. But the preceding chapter suggested some more informative questions we can ask about this system.

What we really want to know is a *probability distribution* for the various values for ℓ, the number of molecules in the sample. To determine that distribution, we imagine preparing many identical solutions, extracting a 1 mL sample from each one, and counting how many labeled molecules are in each such sample. Prior to sampling, each labeled molecule wanders at random through the solution, independently of the others. At the moment of sampling, each molecule is captured or not, in a Bernoulli trial with probability ξ. Assigning the value $s = 1$ to capture and 0 to noncapture, we have that $\ell = s_1 + \cdots + s_M$, where M is the total number of tagged molecules in the original solution.

The Bernoulli trial is easy to characterize. Its probability distribution is just a graph with two bars, of heights ξ and $\xi' = 1 - \xi$. If either ξ or ξ' equals 1, then there's no randomness; the "spread" is zero. If $\xi = \xi' = \frac{1}{2}$, the "spread" is maximal (see the Example on page 56). For the problem at hand, however, we have batches of *several* Bernoulli trials (M of them in a batch). We are interested only in a reduced description of the outcomes, not the details of every individual draw in a batch. Specifically, we want the distribution, across batches, for the discrete random variable ℓ.

Before proceeding, we should first try to frame some expectations. The capture of each labeled molecule is like a coin flip. If we flip a fair coin 50 times, we'd expect to get "about" 25 <u>heads</u>, though we wouldn't be surprised to get 24 or 26.[3] In other words, we expect for a fair coin that the most probable value of ℓ is $M/2$; but we also expect to find a spread about that value. Similarly, when we draw an aliquot from solution, we expect to get about ξM tagged molecules in each sample, with some spread.

For *10 000* coin flips, we expect the fraction coming up <u>heads</u> to equal 1/2 to high accuracy, whereas for just a few flips we're not surprised at all to find some extreme results, even $\ell = 0$ or $\ell = M$. For a general Bernoulli trial, we expect the actual number not to deviate much from ξM, if that number is large. Let's make these qualitative hunches more precise.

---

[2]Modern biophysical methods really can give exact counts of individual fluorescent dye molecules in small volumes, so this is not an academic example.

[3]In fact, if we got exactly 25 <u>heads</u>, and redid the whole experiment many times and *always* got exactly 25, *that* would be surprising.

4.2.2 The sum of several Bernoulli trials follows a Binomial distribution

Sampling from solution is like flipping M coins, but recording only the *total* number ℓ of <u>heads</u> that come up. Thus, an "outcome" is one of the aggregate values $\ell = 0, \dots, M$ that may arise. We'd like to know the probability of each outcome.

The problem discussed in Section 4.2.1 had $M = 4$, and

$$\xi = \text{(sample volume)/(total volume)} = 0.1.$$

If we define $\xi' = 1 - \xi$, then certainly $(\xi + \xi')^4 = 1$. To see why this fact is useful, expand it, to get 16 terms that are guaranteed to add up to 1. Collecting the terms according to powers of ξ and ξ', we find one term containing ξ^4, four terms containing $\xi^3\xi'$, and so on. Generally, the term $\xi^\ell(\xi')^{M-\ell}$ corresponds to flipping <u>heads</u> exactly ℓ times, and by the binomial theorem it contributes

$$\mathcal{P}_{\text{binom}}(\ell; \xi, M) = \frac{M!}{\ell!(M - \ell)!}\, \xi^\ell(1 - \xi)^{M-\ell} \text{ for } \ell = 0, \dots, M \qquad \textbf{Binomial distribution}$$

(4.1)

to the total probability (see Figure 4.1). This probability distribution is really a *family* of discrete distributions of ℓ, with two parameters M and ξ. By its construction, it has the normalization property: We get 1 when we sum it over ℓ, holding the two parameters fixed.

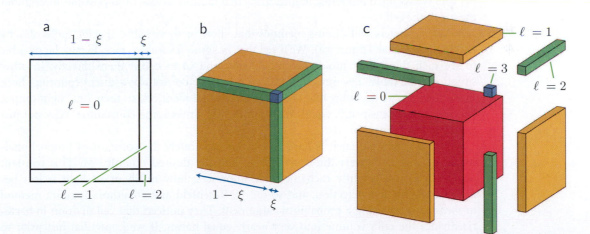

Figure 4.1 [Sketches.] **Graphical representation of the binomial theorem.** (a) For $M = 2$ and $\xi = 1/10$, the small block representing two <u>heads</u> has area ξ^2; the two blocks representing one <u>heads</u>/one <u>tails</u> have combined area $2\xi(1 - \xi)$, and the remaining block has area $(1-\xi)^2$. Thus, the three classes of outcomes have areas corresponding to the expressions in Equation 4.1 with $M = 2$ and $\ell = 0, 1$, and 2. The sum of these expressions equals the area of the complete unit square, so the distribution is properly normalized. (b) For $M = 3$, the small cube in the front represents all three flips coming up <u>heads</u>, and so on. (The large cube representing $\ell = 0$ is hidden in the back of the picture.) This time there are four classes of outcomes, again with volumes that correspond to terms of Equation 4.1. (c) Exploded view of panel (b).

Your Turn 4A

a. Evaluate the Binomial distribution for $M = 4$ and $\xi = 0.1$. Is there any significant chance of capturing more than one tagged molecule?
b. Expand the $M = 5$ case, find all six terms, and compare them with the values of $\mathcal{P}_{\text{binom}}(\ell; \xi, M)$ in the general formula above.

4.2.3 Expectation and variance

Example a. What are the expectation and variance of ℓ in the Binomial distribution? [*Hint:* Use the Example on page 56.]
b. Use your answer to (a) to confirm and make precise our earlier intuition that we should get about $M\xi$ <u>heads</u>, and that for large M we should get very little spread about that value.

Solution a. The expectation is $M\xi$, and the variance is $M\xi(1 - \xi)$. These are very easy when we recall the general formulas for expectation and variance for the sum of independent random variables.[4]
b. More precisely, we'd like to see whether the standard deviation is small relative to the expectation. Indeed, their ratio is $\sqrt{M\xi(1 - \xi)}/(M\xi)$, which gets small for large enough M.

4.2.4 How to count the number of fluorescent molecules in a cell

Some key molecular actors in cells are present in small numbers, perhaps a few dozen copies per cell. We are often interested in measuring that number as exactly as possible, throughout the life of the cell.

Later chapters will discuss methods that allow us to visualize specific molecules, by making them glow (fluoresce). We'll see that in some favorable cases, it may be possible to see such fluorescent molecules individually, and so to count them directly. In other situations, the molecules move too fast, or otherwise do not allow direct counting. Even then, however, we do know that the molecules are all identical, so their total light output (fluorescence intensity), y, equals their number M times some constant α. Why not just measure y as a proxy for M?

The problem is that it is hard to estimate accurately the constant of proportionality, α, needed to convert the observable y into the desired quantity M. This constant depends on how brightly each molecule fluoresces, how much of its light is lost between emission and detection, and so on. N. Rosenfeld and coauthors found a method to *measure* α, by using a probabilistic argument. They noticed that cell division in bacteria divides the cell's volume into very nearly equal halves. If we know that just prior to division there are M_0 fluorescent molecules, then after division one daughter cell gets M_1 and the other gets $M_2 = M_0 - M_1$. If, moreover, the molecules wander at random inside the cell, then for given M_0 the quantity M_1 will be distributed according to $\mathcal{P}_{\text{binom}}(M_1; M_0, 1/2)$. Hence, the variance of M_1 equals $\frac{1}{2}(1 - \frac{1}{2})M_0$. Defining the "error of partitioning" $\Delta M = M_1 - M_2$ then gives $\Delta M = M_1 - (M_0 - M_1) = 2M_1 - M_0$.

---

[4]See Your Turn 3M (page 57).

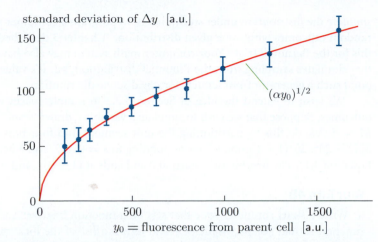

Figure 4.2 [Experimental data with fit.] **Calibration of a single-molecule fluorescence measurement.** *Horizontal axis:* Measured fluorescence intensity of cells prior to division. *Vertical axis:* Sample standard deviation of the partitioning error of cell fluorescence after division. *Error bars* indicate that this quantity is uncertain due in part to the finite number of cells observed. *Red curve:* The predicted function from Idea 4.2. The best-fit value of the parameter α is 15 fluorescence units per tagged molecule. [Data from Rosenfeld et al., 2005.]

Thus,[5]

$$\mathrm{var}(\Delta M) = 4\,\mathrm{var}(M_1) = M_0.$$

We wish to re-express this result in terms of the observed fluorescence, so let $y = \alpha M$, where α is the constant we are seeking:

$$\mathrm{var}(\Delta y) = \alpha^2\,\mathrm{var}(\Delta M) = \alpha^2 M_0 = \alpha y_0.$$

That is, we have predicted that

> *The standard deviation of Δy, among a population of cells all with the same initial fluorescence y_0, is $(\alpha y_0)^{1/2}$.* (4.2)

Idea 4.2 involves some experimentally measurable quantities (y_0 and Δy), as well as the unknown constant α. Fitting this model to data thus yields the desired value of α. The experimenters observed a large number of cells just prior to and just after division;[6] thus, for each value of y_0 they found many values of Δy. Computing the variance gave them a dataset to fit to the prediction in Idea 4.2. Figure 4.2 shows that the data do give a good fit.

4.2.5 Computer simulation

It is nice to have an exact formula like Equation 4.1 for a probability distribution; sometimes important results can be proved directly from such a formula. Other times, however, a known distribution is merely the starting point for constructing something more elaborate, for which exact results are not so readily available. In such a case, it can be important to

---

[5]See Your Turn 3L(a) (page 56).

[6] $\boxed{T_2}$ Rosenfeld and coauthors arranged to have a wide range of y_0 values, and they ensured that the fluorescent molecule under study was neither created nor significantly cleared during the observed period of cell division.

simulate the distribution under study, that is, to program a computer to emit sequences of random outcomes with some given distribution.[7] Chapter 3 described how to accomplish this for the Bernoulli trial.[8] Your computer math system may also have a built-in function that simulates sampling from the Binomial distribution, but it's valuable to know how to build such a generator from scratch, for *any* discrete distribution.

We wish to extend the idea of Section 3.2.2 to sample spaces with more than two outcomes. Suppose that we wish to simulate a variable ℓ drawn from $\mathcal{P}_{\text{binom}}(\ell; M, \xi)$ with $M = 3$. We do this by partitioning the unit segment into four bins of widths $(1 - \xi)^3$, $3\xi(1 - \xi)^2$, $3\xi^2(1 - \xi)$, and ξ^3, corresponding to $\ell = 0, 1, 2,$ and 3 <u>heads</u>, respectively (see Equation 4.1). The first bin thus starts at 0 and ends at $(1 - \xi)^3$, and so on.

Your Turn 4B

a. Write a short computer code that sets up a function `binomSimSetup(xi)`. This function should accept a value of ξ and return a *list* of the locations of the bin edges appropriate for computing $\mathcal{P}_{\text{binom}}(\ell; M = 3, \xi)$ for three-flip sequences.

b. Write a short "wrapper" program that calls `binomSimSetup`. The program should then use the list of bin edges to generate 100 Binomial-distributed values of ℓ and histogram them. Show the histogram for a few different values of ξ, including $\xi = 0.6$.

c. Find the sample mean and the variance of your 100 samples, and compare your answers with the results found in the preceding Example. Repeat with 10 000 samples.

4.3 Poisson Distribution

The formula for the Binomial distribution, Equation 4.1, is complicated. For example, it has two parameters, M and ξ. Two may not sound like a large number, but fitting data to a model rapidly becomes complicated and unconvincing when there are too many parameters. Fortunately, often a simpler, approximate form of this distribution can be used instead. The simplified distribution to be derived in this section has just *one* parameter, so using it can improve the predictive power of a model.

The derivation that follows is so fundamental that it's worth following in detail. It's important to understand the approximation we will use, in order to say whether it is justified for a particular problem.

4.3.1 The Binomial distribution becomes simpler in the limit of sampling from an infinite reservoir

Here is a physical question similar to the one that introduced Section 4.2.1, but with more realistic numbers: Suppose that you take a liter of pure water (10^6 mm^3) and add five million fluorescently tagged molecules. You mix well, then withdraw one cubic millimeter. How many tagged molecules, ℓ, will you get in your sample?

Section 4.2 sharpened this question to one involving a Binomial probability distribution. For the case under study now, the expectation of that distribution is $\langle \ell \rangle = M\xi = (5 \cdot 10^6)(1 \text{ mm}^3)/(10^6 \text{ mm}^3) = 5$. Suppose next that we instead take a cubic *meter* of water and add five *billion* tagged molecules: That's the same concentration, so we again expect

---

[7] For example, you'll use this skill to simulate bacterial genetics later in this chapter, and cellular mRNA populations in Chapter 8.

[8] See Section 3.2.2 (page 40).

$\langle \ell \rangle = 5$ for a sample of the same volume $V = 1\,\text{mm}^3$. Moreover, it seems reasonable that the entire distribution $\mathcal{P}(\ell)$ is essentially the same in this case as it was before. After all, each liter of that big thousand-liter bathtub has about five million tagged molecules, just as in the original situation. And in a $100\,\text{m}^3$ swimming pool, with $5 \cdot 10^{11}$ tagged molecules, the situation should be essentially the same. In short, it's reasonable to expect that there should be some *limiting distribution*, and that any large enough reservoir with concentration $c = 5 \cdot 10^6$ molecules per liter will give the same result for that distribution as any other. But "reasonable" is not enough. We need a proof. And anyway, we'd like to find an explicit formula for that limiting distribution.

4.3.2 The sum of many Bernoulli trials, each with low probability, follows a Poisson distribution

Translating the words of Section 4.3.1 into math, we are given values for the concentration c of tagged molecules and the sample volume V. We wish to find the distribution of the number ℓ of tagged molecules found in a sample, in the limit where the reservoir is huge but $\langle \ell \rangle$ is kept fixed. The discussion will involve several named quantities, so we summarize them here for reference:

V	sample volume, held fixed
V_*	reservoir volume, $\to \infty$ in the limit
$\xi = V/V_*$	probability that any one molecule is captured, $\to 0$ in the limit
c	concentration (number density), held fixed
$M_* = cV_*$	total number of molecules in the reservoir, $\to \infty$ in the limit
$\mu = cV = M_*\xi$	a constant as we take the limit
ℓ	number of tagged molecules in a particular sample, a random variable

Suppose that M_* molecules each wander through a reservoir of volume V_*, so $c = M_*/V_*$. We are considering a series of experiments all with the same concentration, so any chosen value of V_* also implies the value $M_* = cV_*$. Each molecule wanders independently of the others, so each has probability $\xi = V/V_* = Vc/M_*$ to be caught in the sample.

The total number caught thus reflects the sum of M_* identical, independent Bernoulli trials, whose distribution we have already worked out. Thus, we wish to compute

$$\lim_{M_* \to \infty} \mathcal{P}_{\text{binom}}(\ell; \xi, M_*), \text{ where } \xi = Vc/M_*. \tag{4.3}$$

The parameters V, c, and ℓ are to be held fixed when taking the limit.

Your Turn 4C

Think about how this limit implements the physical situation discussed in Section 4.3.1.

Notice that V and c enter our problem only via their product, so we will have one fewer symbol in our formulas if we eliminate them by introducing a new abbreviation $\mu = Vc$. The parameter μ is dimensionless, because the concentration c has dimensions of inverse volume (for example, "molecules per liter").

Substituting the Binomial distribution (Equation 4.1) into the expression above and rearranging gives

$$\lim_{M_* \to \infty} \left(\frac{\mu^\ell}{\ell!} \right) \left(1 - \frac{\mu}{M_*} \right)^{M_*} \left(1 - \frac{\mu}{M_*} \right)^{-\ell} \frac{M_*(M_* - 1) \cdots (M_* - (\ell - 1))}{M_*^{\ell}}. \tag{4.4}$$

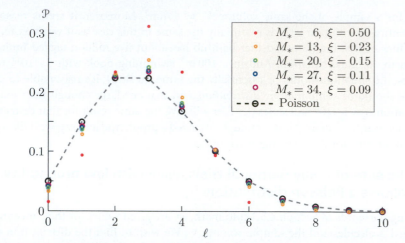

Figure 4.3 [Mathematical functions.] **Poisson distribution as a limit.** *Black circles* show the Poisson distribution for $\mu = 3$. The *dashed line* just joins successive points; the distribution is defined only at integer values of ℓ. The *colored circles* show how the Binomial distribution (Equation 4.3) converges to the Poisson distribution for large M_*, holding fixed $M_*\xi = 3$.

The first factor of expression 4.4 doesn't depend on M_*, so it may be taken outside of the limit. The third factor just equals 1 in the large-M_* limit, and the last one is

$$(1 - M_*^{-1})(1 - 2M_*^{-1}) \cdots \left(1 - (\ell - 1)M_*^{-1}\right).$$

Each of the factors above is very nearly equal to 1, and there are only $\ell - 1 \ll M_*$ of them, so in the limit the whole thing becomes another factor of 1, and may be dropped.

The second factor in parentheses in expression 4.4 is a bit more tricky, because its exponent is becoming large in the limit. To evaluate it, we need the compound interest formula:[9]

$$\lim_{M_* \to \infty} \left(1 - \frac{\mu}{M_*}\right)^{M_*} = \exp(-\mu). \tag{4.5}$$

To convince yourself of Equation 4.5, let $X = M_*/\mu$; then we want $\left((1 - X^{-1})^X\right)^\mu$. You can just evaluate the quantity $(1 - X^{-1})^X$ for large X on a calculator, and see that it approaches $\exp(-1)$. So the left side of Equation 4.5 equals e^{-1} raised to the power μ, as claimed.

Putting everything together then gives

$$\mathcal{P}_{\text{pois}}(\ell; \mu) = \frac{1}{\ell!} \mu^\ell e^{-\mu}. \quad \textbf{Poisson distribution} \tag{4.6}$$

Figure 4.3 illustrates the limit we have found, in the case $\mu = 3$. Figure 4.4 compares two Poisson distributions that have different values of μ. These distributions are not symmetric; for example, ℓ cannot be smaller than zero, but it can be arbitrarily large (because we took

---

[9] See page 20.

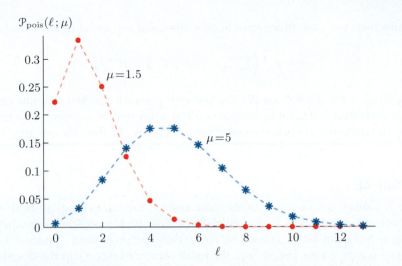

Figure 4.4 [Mathematical functions.] **Two examples of Poisson distributions.** Again, *dashed lines* just join successive points; Poisson distributions are defined only at integer values of ℓ.

the limit of large M_*). If μ is small, the distribution has a graph that is tall and narrow. For larger values of μ, the bump in the graph moves outward, and the distribution gets broader too.[10]

Your Turn 4D

Also graph the cases with $\mu = 0.1,\ 0.2$, and 1.

Example Confirm that the Poisson distribution is properly normalized for any fixed value of μ. Find its expectation and variance, as functions of the parameter μ.

Solution When we sum all the infinitely many entries in $\mathcal{P}_{\text{pois}}(\ell; \mu)$, we obtain $e^{-\mu}$ times the Taylor expansion for e^{μ} (see page 19). The product thus equals 1.

There are various ways to compute expectation and variance, but here is a method that will be useful in other contexts as well.[11] To find the expectation, we must evaluate $\sum_{\ell=0}^{\infty} \ell \mu^{\ell} e^{-\mu}/(\ell!)$. The trick is to start with the related expression $\frac{d}{d\mu}\left(\sum_{\ell=0}^{\infty} \mu^{\ell}/(\ell!)\right)$, evaluate it in two different ways, and compare the results.

On one hand, the quantity in parentheses equals e^{μ}, so its derivative is also e^{μ}. On the other hand, differentiating each term of the sum gives

$$\sum_{\ell=1}^{\infty} \ell \mu^{\ell-1}/(\ell!).$$

The derivative has pulled down a factor of ℓ from the exponential, making the expression almost the same as the quantity that we need.

---

[10] See Your Turn 4E.
[11] See Problem 7.2 and Section 5.2.4 (page 102).

Setting these two expressions equal to each other, and manipulating a bit, yields

$$1 = \mu^{-1} \left(\sum_{\ell=1}^{\infty} e^{-\mu} \ell \mu^{\ell} / (\ell!) \right) = \mu^{-1} \langle \ell \rangle.$$

Thus, $\langle \ell \rangle = \mu$ for the Poisson distribution with parameter μ. You can now invent a similar derivation and use it to compute var ℓ as a function of μ. [*Hint:* This time try taking *two* derivatives, in order to pull down two factors of ℓ from the exponent.]

Your Turn 4E

There is a much quicker route to the same answer. You have already worked out the expectation and variance of the Binomial distribution (the Example on page 72), so you can easily find them for the Poisson, by taking the appropriate limit (Equation 4.3). Do that, and compare your answer with the result computed directly in the Example just given.

To summarize,

- The Poisson distribution is useful whenever we are interested in *the sum of a lot of Bernoulli trials, each of which is individually of low probability.*
- In this limit, the two-parameter family of Binomial distributions collapses to a one-parameter family, a useful simplification in many cases where we know that M_* is large, but don't know its specific value.
- The expectation and variance have the key relationship

$$\text{var } \ell = \langle \ell \rangle \quad \text{for any Poisson distribution.} \tag{4.7}$$

4.3.3 Computer simulation

The method in Your Turn 4B can be used to simulate a Poisson-distributed random variable.[12] Although we cannot partition the unit interval into infinitely many bins, nevertheless in practice the Poisson distribution is very small for large ℓ, and so only a finite number of bins actually need to be set up.

4.3.4 Determination of single ion-channel conductance

Again, *the Poisson distribution is nothing new.* We got it as an approximation, a particular limiting case of the Binomial distribution. It's far more broadly applicable than it may seem from the motivating story in Section 4.3.1:

> *Whenever a large number of independent yes/no events each have low probability, but there are enough of them to ensure that the total "yes" count is nonnegligible,* (4.8)
> *then that total will follow a Poisson distribution.*

---

[12]See Problem 4.6.

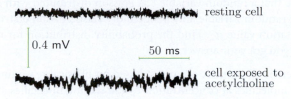

Figure 4.5 [Experimental data.] **Membrane electric potential in frog sartorius muscle.** The traces have been shifted vertically by arbitrary amounts; what the figure shows is the amplitude of the noise (randomness) in each signal. [From Katz & Miledi, 1972. ©Reproduced with permission of John Wiley & Sons, Inc.]

In the first half of the 20th century, it slowly became clear that cell membranes somehow could control their electrical conductance, and that this control lay at the heart of the ability of nerve and muscle cells to transmit information. One hypothesis for the mechanism of control was that the cell membrane is impermeable to the passage of ions (it is an insulator) but it is studded with tiny, discrete gateways. Each such gateway (or **ion channel**) can be open, allowing a particular class of ions to pass, or it can be shut. This switching, in turn, affects the electric potential across the membrane: Separating charges creates a potential difference, so allowing positive ions to reunite with negative ions reduces membrane potential.

The ion channel hypothesis was hotly debated, in part because at first, no component of the cell membrane was known that could play this role. The hypothesis made a prediction of the general magnitude of single-channel currents, but the prediction could not be tested: The electronic instrumentation of the day was not sensitive enough to detect the tiny postulated discrete electrical events.

B. Katz and R. Miledi broke this impasse, inferring the conductance of a single ion channel from a statistical analysis of the conductance of many such channels. They studied muscle cells, whose membrane conductance was known to be sensitive to the concentration of the neurotransmitter acetylcholine. Figure 4.5 shows two time series of the electric potential drop across the membrane of a muscle cell. The top trace is from a resting cell; the lower trace is from a muscle cell exposed to acetylcholine from a micropipette. Katz and Miledi noticed that the acetylcholine not only changed the resting potential but also increased the *noise* seen in the potential.[13] They interpreted this phenomenon by suggesting that the extra noise reflects independent openings and closings of a collection of many ion channels, as neurotransmitter molecules bind to and unbind from them.

In Problem 4.13, you'll follow Katz and Miledi's logic and estimate the effect of a single channel opening, from data similar to those in Figure 4.5. The experimenters converted this result into an inferred value of the channel conductance, which agreed roughly with the value expected for a nanometer-scale gateway, strengthening the ion channel hypothesis.

4.3.5 The Poisson distribution behaves simply under convolution

We have seen that the Poisson distribution has a simple relation between its expectation and variance. Now we'll find another nice property of this family of distributions, which also illustrates a new operation called "convolution."

---

[13] Other means of changing the resting potential, such as direct electrical stimulation, did not change the noisiness of the signal.

Example Suppose that a random variable ℓ is Poisson distributed with expectation μ_1, and m is another random variable, independent of ℓ, also Poisson distributed, but with a different expectation value μ_2. Find the probability distribution for the sum $\ell + m$, and explain how you got your answer.

Solution First, here is an intuitive argument based on physical reasoning: Suppose that we have blue ink molecules at concentration c_1 and red ink molecules at concentration c_2. A large chamber, of volume V_*, will therefore contain a total of $(c_1 + c_2)V_*$ molecules of either color. The logic of Section 4.3.2 then implies that the combined distribution is Poisson with $\mu = \mu_1 + \mu_2$.

Alternatively, here is a symbolic proof: First use the product rule for the independent variables ℓ and m to get the joint distribution $\mathcal{P}(\ell, m) = \mathcal{P}_{\text{pois}}(\ell; \mu_1)\mathcal{P}_{\text{pois}}(m; \mu_2)$. Next let $n = \ell + m$, and use the addition rule to find the probability that n has a particular value (regardless of the value of ℓ):

$$\mathcal{P}_n(n) = \sum_{\ell=0}^{n} \mathcal{P}_{\text{pois}}(\ell; \mu_1)\mathcal{P}_{\text{pois}}(n - \ell; \mu_2). \tag{4.9}$$

Then use the binomial theorem to recognize that this sum involves $(\mu_1 + \mu_2)^n$. The other factors also combine to give $\mathcal{P}_n(n) = \mathcal{P}_{\text{pois}}(n; \mu_1 + \mu_2)$.

Your Turn 4F

Again let $n = \ell + m$.
a. Use facts that you know about the expectation and variance of the Poisson distribution, and about the expectation and variance of a sum of independent random variables, to compute $\langle n \rangle$ and var n in terms of μ_1 and μ_2.
b. Now use the result in the Example above to compute the same two quantities and compare them with what you found in (a).

The right side of Equation 4.9 has a structure that arises in many situations,[14] so we give it a name: If f and g are any two functions of an integer, their **convolution** $f \star g$ is a new function, whose value at a particular n is

$$(f \star g)(n) = \sum_{\ell} f(\ell)g(n - \ell). \tag{4.10}$$

In this expression, the sum runs over all values of ℓ for which $f(\ell)$ and $g(n - \ell)$ are both nonzero. Applying the reasoning of the Example above to arbitrary distributions shows the significance of the convolution:

The distribution for the sum of two independent random variables is the convolution of their respective distributions. (4.11)

For the special case of Poisson distributions, the Example also showed that

The Poisson distributions have the special feature that the convolution of any two is again a Poisson distribution. (4.12)

---

[14]For example, see Sections 5.3.2 (page 108) and 7.5 (page 165). Convolutions also arise in image processing.

Your Turn 4G

Go back to Your Turn 3E (page 48). Represent the 36 outcomes of rolling two (distinct) dice as a 6 × 6 array, and circle all the outcomes for which the sum of the dice equals a particular value (for example, 6). Now reinterpret this construction as a convolution problem.

4.4 The Jackpot Distribution and Bacterial Genetics

4.4.1 It matters

Some scientific theories are pretty abstract. The quest to verify or falsify such theories may seem like a game, and indeed many scientists describe their work in those terms. But in other cases, it's clear right from the start that it matters a lot if a theory is right.

There was still active debate about the nature of inheritance at the turn of the 20th century, with a variety of opinions that we now caricature with two extremes. One pole, now associated with Charles Darwin, held that heritable changes in an organism arise spontaneously, and that evolution in the face of new environmental challenges is the result of selection applied to such mutation. The other extreme, now associated with J.-B. Lamarck, held that organisms actively create heritable changes in response to environmental challenges. The practical stakes could not have been higher. Josef Stalin imposed an agricultural policy based on the latter view that resulted in millions of deaths by starvation, and the near-criminalization of Darwinian theory in his country. The mechanism of inheritance is also critically important at the level of microorganisms, because the emergence of drug-resistant bacteria is a serious health threat today.

S. Luria and M. Delbrück set out to explore inheritance in bacteria in 1943. Besides addressing a basic biological problem, this work developed a key mode of scientific thought. The authors laid out two competing hypotheses, and sought to generate testable quantitative predictions from them. But unusually for the time, the predictions were *probabilistic* in character. No conclusion can be drawn from any single bacterium—sometimes it gains resistance; usually it doesn't. But the pattern of *large numbers* of bacteria has bearing on the mechanism. We will see how randomness, often dismissed as an unwelcome inadequacy of an experiment, turned out to be the most interesting feature of the data.

4.4.2 Unreproducible experimental data may nevertheless contain an important message

Bacteria can be killed by exposure to a chemical (for example, an antibiotic) or to a class of viruses called **bacteriophage** (abbreviated "phage"). In each case, however, some bacteria from a colony typically survive and transmit their resistance to their descendants. Even a colony founded from a *single* nonresistant individual will be found to have some resistant survivors. How is this possible?

Luria and Delbrück were aware that previous researchers had proposed both "Darwinian" and "Lamarckian" explanations for the acquisition of resistance, but that no fully convincing answer had been reached. They began their investigation by making the two alternatives more precise, and then drew predictions from them and

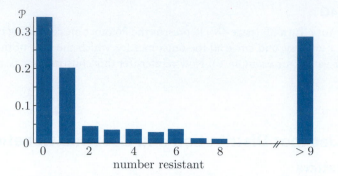

Figure 4.6 [Experimental data.] **Data from Luria and Delbrück's historic article.** This histogram represents one of their trials, consisting of 87 cultures. Figure 4.8 gives a more detailed representation of their experimental data and a fit to their model. [Data from Luria & Delbrück, 1943.]

designed an experiment intended to test the predictions. The Lamarckian hypothesis amounted to

> **H1:** *A colony descended from a single ancestor consists of identical individuals until a challenge to the population arises. When faced with the challenge, each individual struggles with it independently of the others, and most die. However, a small, randomly chosen subset of bacteria succeed in finding the change needed to survive the challenge, and are permanently modified in a way that they can transmit to their offspring.*

The Darwinian hypothesis amounted to

> **H2:** *No mutation occurs in* **response** *to the challenge. Instead, the entire colony is always spontaneously mutating, whether or not a challenge is presented. Once a mutation occurs, it is heritable. The challenge wipes out the majority, leaving behind only those individuals that had previously mutated to acquire resistance, and their descendants.*

In 1943, prior to the discovery of DNA's role in heredity, there was little convincing molecular basis for *either* of these hypotheses. An empirical test was needed.

Luria and Delbrück created a large collection of separate cultures of a particular strain of *Escherichia coli*. Each culture was given ample nutrients and allowed to grow for a time t_f, then challenged with a virus now called phage T1. To count the survivors, Luria and Delbrück spread each culture on a plate and continued to let them grow. Each surviving individual founded a colony, which eventually grew to a visible size. The survivors were few enough in number that these colonies were well separated, and so could be counted visually. Each culture had a different number m of survivors, so the experimenters reported not a single number but rather a histogram of the frequencies with which each particular value of m was observed (Figure 4.6).

Luria at once realized that the results were qualitatively unlike anything he had been trained to consider good science. In some ways, his data looked reasonable—the distribution had a peak near $m = 0$, then fell rapidly for increasing m. But there were also **outliers**, unexpected data points far from the main group.[15] Worse, when he performed the same

---

[15] Had Luria been content with two or three cultures, he might have missed the low-probability outliers altogether.

experiment a second and third time, the outliers, while always present, were quite different each time. It was tempting to conclude that this was just a bad, unreproducible experiment! In that case, the appropriate next step would be to work hard to find what was messing up the results (contamination?), or perhaps abandon the whole thing. Instead, Luria and Delbrück realized that hypothesis **H2** could explain their odd results.

The distributions we have encountered so far have either been exactly zero outside some range (like the Uniform and Binomial distributions), or at least have fallen off very rapidly outside a finite range (like Poisson or Geometric). In contrast, the empirical distribution in the Luria-Delbrück experiment is said to have a **long tail**; that is, the range of values at which it's nonnegligible extends out to very large m.[16] The more colorful phrase "jackpot distribution" is also used, by analogy to a gambling machine that generally gives a small payoff (or none), but occasionally gives a large one.

$\boxed{T_2}$ *Section 4.4.2′ (page 89) mentions one of the many additional tests that Luria and Delbrück made.*

4.4.3 Two models for the emergence of resistance

Luria and Delbrück reasoned as follows. At the start of each trial ("time zero"), a few nonresistant individuals are introduced into each culture. At the final time t_f, the population has grown to some large number $n(t_f)$; then it is subjected to a challenge, for example an attack by phage.

- **H1** states that each individual either mutates, with low probability ξ, or does not, with high probability $1 - \xi$, and that this random event is independent of every other individual. We have seen that in this situation, the total number m of individuals that succeed is distributed as a Poisson random variable. The data in Figure 4.6 don't seem to be distributed in this way.

- **H2** states that every time an individual divides, during the entire period from time zero to t_f, there is a small probability that it will spontaneously acquire the heritable mutation that confers resistance. So although the mutation event is once again a Bernoulli trial, according to **H2** it matters *when* that mutation occurred: Early mutants generate many resistant progeny, whereas mutants arising close to t_f don't have a chance to do so. Thus, in this situation there is an amplification of randomness.

Qualitatively, **H2** seems able to explain the observed jackpot distribution as a result of the occasional trial where the lucky mutant appeared early in the experiment (see Figure 4.7). A quantitative test is also required, however.

Note that both hypotheses contain a *single* unknown fitting parameter: in each case, a mutation probability. Thus, if we can adjust this one parameter to get a good fit under one hypothesis, but *no* value gives a good fit under the other hypothesis, then we will have made a fair comparison supporting the former over the latter. Note, too, that neither hypothesis requires us to understand the biochemical details of mutation, resistance, or inheritance. Both distill all of that detail into a single number, which is to be determined from data. If the winning model then makes *more* than one successful quantitative prediction (for example, if it predicts the entire shape of the distribution), then we may say that the data support it in a nontrivial way—they overconstrain the model.

---

[16]Some authors use the phrase "fat tail" to mean the same thing, because the tail of the distribution is larger numerically than we might have expected—it's "fat." Chapter 5 will give more examples illustrating the ubiquity of such distributions in Nature.

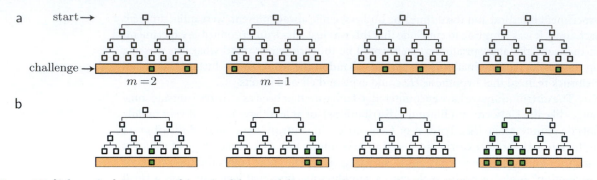

Figure 4.7 [Schematics.] **Two sets of imagined bacterial lineages relevant to the Luria-Delbrück experiment.** (a) The "Lamarckian" hypothesis states that bacterial resistance is created at the time of the challenge (*orange*). The number of resistant individuals (*green*) is then Poisson distributed. (b) The "Darwinian" hypothesis states that bacterial resistance can arise at any time. If it arises early (*second diagram*), the result can be very many resistant individuals.

T_2 *Section 4.4.3' (page 89) gives more details about Luria and Delbrück's experiment.*

4.4.4 The Luria-Delbrück hypothesis makes testable predictions for the distribution of survivor counts

Hypothesis **H1** predicts that the probability distribution of the number of resistant bacteria is of the form $\mathcal{P}_{\text{poiss}}(m; \mu)$, where μ is an unknown constant. We need to find an equally specific prediction from **H2**, in order to compare the two hypotheses. The discussion will involve several named quantities, so we summarize them here for reference:

n	cell population
g	number of doubling times (generations)
α_g	mutation probability per individual per doubling
t_f	final time
m	number of resistant mutant bacteria at time t_f
μ_{step}	expectation of number of new mutants in one doubling step
ℓ	number of new mutants actually arising in a particular doubling step, in a particular culture

Growth

Each culture starts at time zero with a known initial population n_0. (It's straightforward to estimate this quantity by sampling the bacterial suspension used to inoculate the cultures.) The growth of bacteria with plenty of food and no viral challenge can also be measured; it is exponential, doubling about every 25 minutes. Luria and Delbrück estimated $n_0 \approx 150$, and the final population to be $n(t_f) \approx 2.4 \cdot 10^8$. Thus, their growth phase consisted of $\log_2(2.4 \cdot 10^8/150) \approx 21$ doublings, a number we'll call g. We'll make the simplifying assumption that all individuals divide in synchrony, g times.

Mutation

Hypothesis **H2** states that, on every division, every individual makes an independent "decision" whether to make a daughter cell with the resistance mutation. Thus, the number of resistant individuals *newly arising on that division* is a Poisson-distributed random variable whose expectation is proportional to the total population prior to that division. The constant of proportionality is the mutation probability per cell per doubling step, α_g, which is the one free parameter of the model. After mutation, the mutant cells continue to divide; we will assume that their doubling time is the same as that for the original-type cells.[17]

Computer simulation

In principle, we have now given enough information to allow a calculation of the expected Luria-Delbrück distribution $\mathcal{P}_{LD}(m; \alpha_g, n_0, g)$. In practice, however, it's difficult to do this calculation exactly; the answer is not one of the well-known, standard distributions. Luria and Delbrück had to resort to making a rather ad hoc mathematical simplification in order to obtain the prediction shown in Figure 4.6, and even then, the analysis was very involved. However, *simulating* the physical model described above with a computer is rather easy. Every time we run the computer code, we get a history of one simulated culture, and in particular a value for the final number m of resistant individuals. Running the code many times lets us build up a histogram of the resulting m values, which we can use either for direct comparison with experiment or for a calculation of reduced statistics like $\langle m \rangle$ or var m.

Figure 4.6 (page 82)

   Such a simulation could work as follows. We maintain two population variables `Nwild` and `Nmutant`, with initial values n_0 and 0, respectively, and update them g times as follows. With each step, each population doubles. In addition, we draw a random number ℓ, representing the number of new mutants in that step, from a Poisson distribution with expectation $\mu_{\text{step}} = (\texttt{Nwild})\alpha_g$, then add ℓ to `Nmutant` and subtract it from `Nwild`. The final value of `Nmutant` after g doubling steps gives m for that simulated culture. We repeat many times for one value of the parameter α_g, compare the resulting probability distribution with experimental data, then adjust α_g and try again until we are satisfied with the fit (or convinced that no value of α_g is satisfactory).

   The strategy just outlined points out a payoff for our hard work in Section 4.3. One could imagine simply simulating `Nwild` Bernoulli trials in each doubling step. But with hundreds of millions of individuals to be polled in the later steps, we'd run out of computing resources! Because all we really need is the *number* of mutants, we can instead make a *single* draw from a Poisson distribution for each doubling step.

Results

Problem 4.14 gives more details on how to carry out these steps. Figure 4.8a shows data from the experiment, together with best-fit distributions for each of the two hypotheses. It may not be immediately apparent from this presentation just how badly *H1* fails. One way to see the failure is to note that the experimental data have sample mean $\bar{m} \approx 30$ but variance ≈ 6000, inconsistent with any Poisson distribution (and hence with hypothesis *H1*).[18]

   The figure also shows that *H2* does give a reasonable account of the entire distribution, with only one free fit parameter, whereas *H1* is unable to explain the existence of *any* cultures having more than about five mutants. To bring this out, Figure 4.8b shows the same information as panel (a) but on a logarithmic scale. This version also shows that the deviation of *H1* at $m = 20$ from the experimental observation is far more significant than that of *H2* at $m = 4$.

---

[17] $\boxed{T_2}$ See Section 4.4.3′ (page 89).
[18] See Problem 4.15.

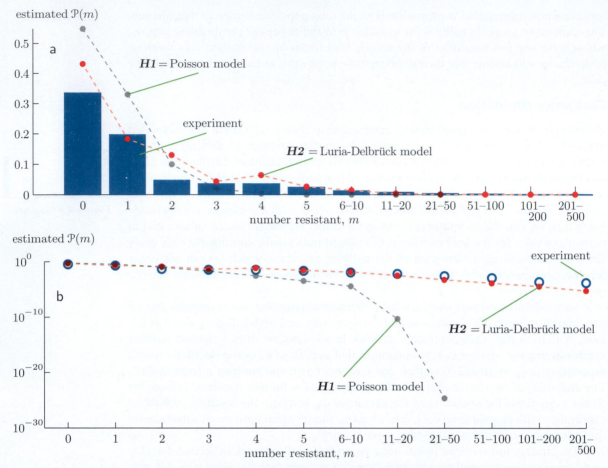

Figure 4.8 [Experimental data with fits.] **Two models compared to data on acquired resistance.** (a) *Bars:* Data from the same experiment as in Figure 4.6. The *gray dots* show a fit to data under the "Lamarckian" hypothesis *H1*. The *red dots* show a fit under the Luria-Delbrück hypothesis *H2*. (b) The same as (a), plotted in semilog form to highlight the inability of *H1* to account for the outliers in the data. Luria and Delbrück combined the data for high mutant number m by lumping several values together, as indicated in the horizontal axis labels. Both panels correct for this: When a bin contains K different values of m all lumped together, its count has been divided by K, so that the bar heights approximate the probabilities for individual values of m. That is, each bar represents the estimated $\mathcal{P}(m)$ for single values of m.

Your Turn 4H

Figure 4.6 appears to show two bumps in the probability, whereas Figure 4.8a does not. Explain this apparent discrepancy.

4.4.5 Perspective

Luria and Delbrück's experiment and analysis showed dramatically that bacteriophage resistance was the result of spontaneous mutation, not the survival challenge itself. Similar mechanisms underlie other evolutionary phenomena, including viral evolution in a single HIV patient, discussed in the Prolog to this book.[19]

_____

[19] Also see Problem 3.6.

This work also provided a framework for the quantitative measurement of extremely low mutation probabilities. Clearly α_g must be on the order of 10^{-8}, because hundreds of millions of bacteria contain only a handful of resistant mutants. It may be mind-boggling to imagine checking all the population and somehow counting the resistant members, but the authors' clever experimental method accomplished just that. At first, this counting seemed to give contradictory results, due to the large spread in the result m. Then, however, Luria and Delbrück had the insight of making a *probabilistic* prediction, comparing it to *many* trials, and finding the *distribution* of outcomes. Fitting that distribution did lead to a good measurement of α_g. As Luria and Delbrück wrote, "The quantitative study of bacterial variation has [until now] been hampered by the apparent lack of reproducibility of results, which, as we show, lies in the very nature of the problem and is an essential element for its analysis."

Your dull-witted but extremely fast assistant was a big help in this analysis. Not every problem has such a satisfactory numerical solution, just as not every problem has an elegant analytic (pencil-and-paper) solution. But the set of problems that are easy analytically, and that of problems that are easy numerically, are two different domains. Scientists with both kinds of toolkit can solve a broader range of problems.

$\boxed{T_2}$ *Section 4.4.5′ (page 89) discusses some qualifications to the Darwinian hypothesis discussed in this chapter, in the light of more recent discoveries in bacterial genetics, as well as an experiment that further confirmed Luria and Delbrück's interpretation.*

THE BIG PICTURE

Many physical systems generate partially random behavior. If we treat the distribution of outcomes as completely unknown, then we may find it unmanageable, and uninformative, to determine that distribution empirically. In many cases, however, we can formulate some well-grounded expectations that narrow the field considerably. From such "insider information"—a model—we can sometimes predict most of the behavior of a system, leaving only one or a few parameter values unknown. Doing so not only lightens our mathematical burden; it can also make our predictions specific, to the point where we may be able to falsify a hypothesis by embodying it in a model, and showing that *no* assumed values of the parameters make successful predictions.

For example, it was reasonable to suppose that a culture of bacteria suspended in liquid will all respond independently of each other to attack by phage or antibiotic. From this assumption, Luria and Delbrück got falsifiable predictions from two hypotheses, and eliminated one of them.

Chapter 6 will start to systematize the procedure for simultaneously testing a model and determining the parameter values that best represent the available experimental data. First, however, we must extend our notions of probability to include continuously varying quantities (Chapter 5).

KEY FORMULAS

- *Binomial distribution:* $\mathcal{P}_{\text{binom}}(\ell; \xi, M) = \frac{M!}{\ell!(M-\ell)!} \xi^\ell (1-\xi)^{M-\ell}$. The random variable ℓ is drawn from the sample space $\{0, 1, \ldots, M\}$. The parameters ξ and M, and $\mathcal{P}$ itself, are all dimensionless. The expectation is $\langle \ell \rangle = M\xi$, and the variance is $\text{var}\, \ell = M\xi(1-\xi)$.
- *Simulation:* To simulate a given discrete probability distribution $\mathcal{P}$ on a computer, divide the unit interval into bins of widths $\mathcal{P}(\ell)$ for each allowed value of ℓ. Then choose Uniform random numbers on that interval and assign each one to its appropriate bin. The resulting bin assignments are draws from the desired distribution.

- *Compound interest:* $\lim_{M \to \infty} \left(1 \pm (a/M)\right)^{M} = \exp(\pm a)$.
- *Poisson distribution:* $\mathcal{P}_{\mathrm{pois}}(\ell; \mu) = \mathrm{e}^{-\mu} \mu^{\ell}/(\ell!)$. The random variable ℓ is drawn from the sample space $\{0, 1, \ldots\}$. The parameter μ, and $\mathcal{P}$ itself, are both dimensionless. The expectation and variance are $\langle \ell \rangle = \mathrm{var}\, \ell = \mu$.
- *Convolution:* $(f \star g)(m) = \sum_{\ell} f(\ell) g(m - \ell)$. Then $\mathcal{P}_{\mathrm{pois}}(\bullet; \mu_1) \star \mathcal{P}_{\mathrm{pois}}(\bullet; \mu_2) = \mathcal{P}_{\mathrm{pois}}(\bullet; \mu_{\mathrm{tot}})$, where $\mu_{\mathrm{tot}} = \mu_1 + \mu_2$.

FURTHER READING

Semipopular:
Discovery of phage viruses: Zimmer, 2011.
On Delbrück and Luria: Luria, 1984; Segrè, 2011. Long-tail distributions: Strogatz, 2012.

Intermediate:
Luria-Delbrück: Benedek & Villars, 2000, §3.5; Phillips et al., 2012, chapt. 21.

Technical:
Luria & Delbrück, 1943.
Estimate of ion channel conductance: Bialek, 2012, §2.3.
Calibration of fluorescence by Binomial partitioning: Rosenfeld et al., 2005, supporting online material.

T_2 **Track 2**

4.4.2′ On resistance

Our model of the Luria-Delbrück experiment assumed that the resistant cells were like the wild type, except for the single mutation that conferred resistance to phage infection. Before concluding this, Luria and Delbrück had to rule out an alternative possibility to be discussed in Chapter 10, that their supposedly resistant cells had been transformed to a "lysogenic" state. They wrote, "The resistant cells breed true . . . . No trace of virus could be found in any pure culture of the resistant bacteria. The resistant strains are therefore to be considered as non-lysogenic."

T_2 **Track 2**

4.4.3′ More about the Luria-Delbrück experiment

The discussion in the main text hinged on the assumption that initially the cultures of bacteria contained no resistant individuals. In fact, any colony could contain such individuals, but only at a very low level, because the resistance mutation also slows bacterial growth. Luria and Delbrück estimated that fewer than one individual in 10^5 were resistant. They concluded that inoculating a few dozen cultures, each with a few dozen individuals, was unlikely to yield even one culture with one resistant individual initially.

The analysis in Section 4.4.4 neglected the reproduction penalty for having the resistance mutation. However, the penalty needed to suppress the population of initially resistant individuals is small enough not to affect our results much. If we wish to do better, it is straightforward to introduce two reproduction rates into the simulation.

T_2 **Track 2**

4.4.5′a Analytical approaches to the Luria-Delbrück calculation

The main text emphasized the power of computer simulation to extract probabilistic predictions from models such as Luria and Delbrück's. However, analytic methods have also been developed for this model as well as for more realistic variants (Lea & Coulson, 1949; Rosche & Foster, 2000).

4.4.5′b Other genetic mechanisms

The main text outlined a turning point in our understanding of genetics. But our understanding continues to evolve; no one experiment settles everything forever. Thus, the main text didn't say "inheritance of acquired characteristics is wrong"; instead, we outlined how one specific implementation of that idea led to quantitatively testable predictions about one particular system, which were falsified.

Other mechanisms of heritable change have later been found that are different from the random mutations discussed in the main text. For example,

- A virus can integrate its genetic material into a bacterium and lie dormant for many generations ("lysogeny"; see Chapter 10).
- A virus can add a "plasmid," a small autonomous loop of DNA that immediately confers new abilities on its host bacterium without any classical mutation, and that is copied and passed on to offspring.

- Bacteria can also exchange genetic material among themselves, with or without help from viruses ("horizontal gene transfer"; see Thomas & Nielsen, 2005).
- Genetic mutations themselves may not be uniform, as assumed in neo-Darwinian models: Regulation of mutation rates can itself be an adaptive response to stress, and different loci on the genome have different mutation rates.

None of these mechanisms should be construed as a failure of Darwin's insight, however. Darwin's framework was quite general; he did not assume Mendelian genetics, and in fact was unaware of it. Instead, we may point out that the mechanisms listed above that lie outside of classical genetics reflect competencies that cells possess *by virtue of their genetic makeup*, which itself evolves under natural selection.

4.4.5′c Non-genetic mechanisms

An even broader class of heritable but non-genetic changes has been found, some of which are implicated in resistance to drug or virus attack:

- The available supply of nutrients can "switch" a bacterium into a new state, which persists into its progeny, even though no change has occurred to the genome (again see Chapter 10). Bacteria can also switch spontaneously, for example, creating a subpopulation of slowly growing "persistors" that are resistant to antibiotic attack. Such "epigenetic" mechanisms (for example, involving covalent modifications of DNA without change to its sequence) have also been documented in eukaryotes.
- Clustered regularly interspaced short palindromic repeats (CRISPR) have been found to give a nearly "Lamarckian" mechanism of resistance (Barrangou et al., 2007; Koonin & Wolf, 2009).
- Drug and virus resistance have also been documented via gene silencing by RNA interference (Calo et al., 2014 and Rechavi, Minevich, and Hobert 2011; see also Section 9.3.3′, page 234).

4.4.5′d Direct confirmation of the Luria-Delbrück hypothesis

The main text emphasized testing a hypothesis based on its probabilistic predictions, but eight years after Luria and Delbrück's work it became possible to give a more direct confirmation. J. Lederberg and E. Lederberg created a bacterial culture and spread it on a plate, as usual. Prior to challenging the bacteria with antibiotic, however, they let them grow on the plate a bit longer, then *replicated* the plate by pressing an absorbent fabric onto it and transferring it to a second plate. The fabric picked up some of the bacteria in the first plate, depositing them in the same relative positions on the second. When both plates were then subjected to viral attack, they showed colonies of resistant individuals in corresponding locations, demonstrating that those subcolonies existed prior to the attack (Lederberg & Lederberg, 1952).

PROBLEMS

4.1 Risk analysis

In 1941, the mortality (death) rate for 75-year-old people in a particular region of the United States was 0.089 per year. Ten thousand people of this age were all given a vaccine, and one died within 12 hours. Should this be attributed to the vaccine? Calculate the probability that at least one would have died in 12 hours, *even without* the vaccine.

4.2 Binning jitter

Here is a more detailed version of Problem 3.3. Nora asked a computer to generate 3000 Uniformly distributed, random binary fractions, each six bits long (see Equation 3.1, page 36), and made a histogram of the outcomes, obtaining Figure 4.9. It doesn't look very Uniform. Did Nora (or her computer) make a mistake? Let's investigate.

a. Qualitatively, why isn't it surprising that the bars are not all of equal height?
   Now get more quantitative. Consider the first bar, which represents the binary fraction corresponding to 000000. The probability of that outcome is 1/64. The computer made 3000 such draws and tallied how many had this outcome. Call that number N_{000000}.

b. Compute the expectation, variance, and standard deviation of N_{000000}.

c. The height of the first bar is $N_{000000}/3000$. Compute the standard deviation of this quantity. The other bars will also have the same standard deviation, so comment on whether your calculated value appears to explain the behavior seen in the figure.

4.3 Gene frequency

Consider a gene with two possible variants (alleles), called A and a.

   Father Fish has two copies of this gene in every somatic (body) cell; suppose that each cell has one copy of allele A and one copy of a. Father fish makes a zillion sperm, each with just one copy of the gene. Mother Fish also has genotype Aa. She makes a zillion eggs, again each with just one copy of the gene.

   Four sperm and four eggs are drawn at random from these two pools and fuse, giving four fertilized eggs, which grow as usual.

a. What is the total number of copies of A in these four fertilized eggs? Re-express your answer in terms of the "frequency of allele A" in the new generation, which is the total number of copies of A in these four individuals, divided by the total number of either A or a. Your answer should be a symbolic expression.

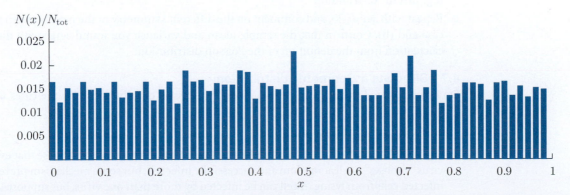

Figure 4.9 [Simulated data.] See Problem 4.2.

b. What is the probability that the frequency of allele A is exactly the same in the new generation as it was in the parent generation? Your answer should be a number. What is the probability that the frequency of allele A is *zero* in the new generation?

4.4 Partitioning error

Idealize a dividing bacterium as a well-mixed box of molecules that suddenly splits into two compartments of equal volume. Suppose that, prior to the division, there are 10 copies of a small, mobile molecule of interest to us. Will we always get exactly 5 on each side? If not, how probable is it that one side, or the other, will get 3 or fewer copies?

4.5 More about random walks

If you haven't done Problem 3.4 yet, do it before starting this problem. Again set up a simulation of a random walk, initially in a single dimension x with steps of length $d = 1 \, \mu$m. Let x_* be the location relative to the origin at time $t = 20$ s. It's a random variable, because it's different each time we make a new trajectory.

a. Compute $\langle (x_*)^2 \rangle$.

b. Let x_{**} be the location at $t = 40$ s, and again compute $\langle (x_{**})^2 \rangle$.

c. Now consider a *two*-dimensional random walk: Here the chess piece makes moves in a *plane*. The x and y components of each move are each random, and independent of each other. Again, x steps by $\pm 1 \, \mu$m in each move; y has the same step distribution. Find $\langle (r_*)^2 \rangle$ for this situation, where $r^2 = x^2 + y^2$ and again the elapsed time is 20 s.

d. Returning to the one-dimensional walker in part (a), this time suppose that it steps in the $+$ direction 51% of the time and in the $-$ direction 49% of the time. What are the expectation and variance of x_* in this situation?

4.6 Simulate a Poisson distribution

a. Write a function for your computer called `poissonSetup(mu)`, similar to the one described in Your Turn 4B, but which prepares a set of bin edges suitable for simulating a Poisson distribution with expectation mu. In principle, this distribution has infinitely many bins, but in practice you can cut it off; that is, use either 10 or 10mu bins (rounded to an integer), whichever is larger. (Or you may invent a more clever way to find a suitable finite cutoff.)

b. Write a little "wrapper" program that calls `poissonSetup(2)`, and then generates 10 000 numbers from the distribution, finds the sample mean and variance, and histograms the distribution.

c. Repeat with mu = 20, and comment on the different symmetry of the peak between this case and (b). Confirm that the sample mean and variance you found agree with direct calculation from the definition of the Poisson distribution.

4.7 Simulate a Geometric distribution

Do Problem 4.6, but with Geometric instead of Poisson distributions. Try the cases with $\xi = 1/2$ and $1/20$.

4.8 Cultures and colonies

a. Suppose that you add $2 \cdot 10^8$ virions to a culture containing 10^8 cells. Suppose that every virus "chooses" a cell at random and successfully infects it, but some mechanism prevents infected cells from lysing. A cell can be infected by more than one virus, but suppose that a prior infection doesn't alter the probability of another one. What fraction of the cells will remain uninfected? How many virions would have been required had you wished for over 99% of the cells in the culture to be infected?

b. Suppose that you take a bacterial culture and dilute it by a factor of one million. Then you spread 0.10 mL of this well-mixed, diluted culture on a nutrient plate, incubate, and find 110 well-separated colonies the next day. What was the concentration of live bacteria (colony forming units, or CFU) in the original culture? Express your answer as CFU/mL and also give the standard deviation of your estimate.

4.9 Poisson limit

The text argued analytically that the Poisson distribution becomes a "good" approximation to the Binomial distribution in a certain limiting case. Explore the validity of the argument:

a. Compute the natural log of the Binomial distribution with $M = 100$ and $\xi = 0.03$, at all values of ℓ. Compare the log of the corresponding Poisson distribution by graphing both. Make another graph showing the actual value (not the log) of each distribution for a range of ℓ values close to $M\xi$.

b. Repeat, but this time use $\xi = 0.5$.

c. Repeat, but this time use $\xi = 0.97$.

d. Comment on your results in the light of the derivation in Section 4.3.2 (page 75).

[*Hint:* Your computer math package may be unable to compute quantities like 100! directly. But it will have no difficulty computing $\ln(1) + \cdots + \ln(100)$. It may be particularly efficient to start with $\ell = 1$, where $\mathcal{P}_{\text{binom}}$ and $\mathcal{P}_{\text{pois}}$ are both simple, then obtain each succeeding $\mathcal{P}(\ell)$ value from its predecessor.]

4.10 Cancer clusters

Obtain Dataset 6. The variable `incidents` contains a list of (x, y) coordinate pairs, which we imagine to be the geographic coordinates of the homes of persons with some illness.

a. First create a graphical representation of these points. The variable `referencepoints` contains coordinates of four landmarks; add them to your plot in a different color.

Suppose that someone asks you to investigate the cause of that scary cluster near reference point #3, and the relative lack of cases in some other regions. Before you start looking for nuclear reactors or cell-phone towers, however, the first thing to check is the "null hypothesis": Maybe these are just points randomly drawn from a Uniform distribution. There's no way to prove that a single instance of dots "is random." But we can try to make a quantitative prediction from the hypothesis and then check whether the data roughly obey it.[20]

b. Add vertical lines to your plot dividing it into N equal strips, either with your computer or by drawing on a hard copy of your plot. Choose a value of N somewhere between 10 and 20. Also add the same number of horizontal lines dividing it into N equal strips. Thus, you have divided your graph into a grid of N^2 blocks. (What's wrong with setting up fewer than 100 blocks? What's wrong with more than 400?)

c. Count how many dots lie in each of the blocks. Tally up how many blocks have 0, 1, ... dots in them. That gives you the frequency $F(\ell)$ to find ℓ dots in a block, and hence an estimate for the probability $\mathcal{P}_{\text{est}}(\ell) = F(\ell)/N^2$ that a block will have ℓ dots. The dataset contains a total of 831 points, so the average number of dots per block is $\mu = 831/N^2$.

---

[20]The next step would be to obtain new data and see if the same hypothesis, *with no further tweaking*, also succeeds on them, but this is not always practical.

d. If we had a huge map with lots of blocks, and dots distributed Uniformly and independently over that map with an average of μ per block, then the actual number observed in a block would follow a known distribution. Graph this probability distribution for a relevant range of ℓ values. Overlay a graph of the estimated distribution $\mathcal{P}_{est}$ that you obtained in (c). Does the resulting picture seem to support the null hypothesis?

e. For comparison, generate 831 simulated data points that really are Uniformly distributed over the region shown, and repeat the above steps.

4.11 Demand fluctuations

In a large fleet of delivery trucks, the average number inoperative on any day, due to breakdowns, is two. Some standby trucks are also available. Find numerical answers for the probability that on any given day

a. No standby trucks are needed.

b. More than one standby truck is needed.

4.12 Low probability

a. Suppose that we have an unfair "coin," for which flipping <u>heads</u> is a rather rare event, a Bernoulli trial with $\xi = 0.08$. Imagine making $N = 1000$ trials, each consisting of 100 such coin flips. Write a computer simulation of such an experiment, and for each trial compute the total number of heads that appeared. Then plot a histogram of the frequencies of various outcomes.

b. Repeat for $N = 30\,000$ and comment. What was the most frequent outcome?

c. Superimpose on the plot of (a) the function $1000\mathcal{P}_{pois}(\ell; 8)$, and compare the two graphs.

4.13 Discreteness of ion channels

Section 4.3.4 introduced Katz and Miledi's indirect determination of the conductance of a single ion channel, long before biophysical instruments had developed enough to permit direct measurement. In this problem, you'll follow their logic with some simplifying assumptions to make the math easier.

    For concreteness, suppose that each channel opening causes the membrane to depolarize slightly, increasing its potential by an amount a for a fixed duration τ; afterward the channel closes again. There are M channels; suppose that M is known to be very large. Each channel spends a small fraction ξ of its time open in the presence of acetylcholine, and all channels open and close independently of one another. Suppose also that when ℓ channels are simultaneously open, the effect is linear (the excess potential is ℓa).

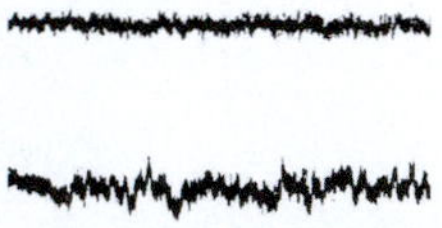

a. None of the parameters a, τ, M, or ξ is directly measurable from data like those in Figure 4.5. However, two quantities *are* measurable: the mean and the variance of the membrane potential. Explain why the Poisson distribution applies to this problem, and use it to compute these quantities in terms of the parameters of the system.

b. The top trace in the figure shows that even in the resting state, where all the channels are closed, there is still some electrical noise for reasons unrelated to the hypothesis being considered. Explain why it is legitimate to simply subtract the average and variance of this resting-state signal from that seen in the lower trace.

c. Show how Katz and Miledi's experimental measurement of the change in the average and the variance of the membrane potential upon acetylcholine application allows us to deduce the value of a. (This value is the desired quantity, a measure of the effect of a single channel opening; it can be converted to a value for the conductance of a single channel.)

Figure 4.5 (page 79)

d. In a typical case, Katz and Miledi found that the average membrane potential increased by 8.5 mV and that the variance increased by $(29.2\,\mu V)^2$ after application of acetylcholine. What then was a?

4.14 Luria-Delbrück experiment

First do Problem 4.6, and be sure that your code is working the way you expect before attempting this problem.

Imagine a set of C cultures (separate flasks) each containing n_0 bacteria initially. Assume that all the cells in a culture divide at the same time, and that every time a cell divides, there is a probability α_g that one of the daughter cells will mutate to a form that is resistant to phage attack. Assume that the initial population has no resistant mutants ("pure wild-type"), and that all progeny of resistant cells are resistant ("no reversion"). Also assume that mutant and wild-type bacteria multiply at the same rate (no "fitness penalty"), and that at most one of the two daughter cells mutate (usually neither).

a. Write a computer code to simulate the situation and find the number of resistant mutant cells in a culture after g doublings. The Poisson distribution gives a good approximation to the number of new mutants after each doubling, so use the code you wrote in Problem 4.6. Each simulated culture will end up with a different number m of resistant mutant cells, due to the random character of mutation.

b. For $C = 500$ cultures with $n = 200$ cells initially, and $\alpha_g = 2 \cdot 10^{-9}$, find the number of cultures with m resistant mutant cells after $g = 21$ doublings, as a function of m. Plot your result as an estimated probability distribution. Compare its sample mean to its variance and comment.

c. Repeat the simulation $M = 3$ times (that is, M sets of C cultures), and comment on how accurately we can expect to find the true expectation and variance of the distribution from such experiments.

d. The chapter claimed that the origin of the long tail in the distribution is that on rare occasions a resistant mutant occurs earlier than usual, and hence has lots of offspring. For each simulated culture, let i_* denote at which step (number of doublings) the *first* mutant appears (or $g + 1$ if never). Produce a plot with m on one axis and i_* on the other, and comment.

[*Hints:* (*i*) This project will require a dozen or so lines of code, more complex than what you've done so far. Outline your algorithm before you start to code. Keep a list of all the variables you plan to define, and give them names that are meaningful to you. (You don't want two unrelated variables both named n.)
(*ii*) Start with smaller numbers, like $C = 100, M = 1$, so that your code runs fast while you're debugging it. When it looks good, then substitute the requested values of those parameters.
(*iii*) One way to proceed is to use three nested loops: The outermost loop repeats the code for each simulated experiment, from 1 to M. The middle loop involves which culture in a particular experiment is being simulated, from 1 to C. The innermost loop steps through the doublings of a particular experiment, in a particular culture.[21]
(*iv*) Remember that in each doubling step the only candidates for mutation are the remaining unmutated cells.]

4.15 Luria-Delbrück data

a. Obtain Dataset 5, which contains counts of resistant bacteria in two of the Luria-Delbrück experiments. For their experiment #23, find the sample mean and variance in the number

---

[21] More efficient algorithms are possible.

of resistant mutants, and comment on the significance of the values you obtain. [*Hint:* The count data are presented in bins of nonuniform size, so you'll need to correct for that. For example, five cultures were found to have between 6 and 10 mutants, so assume that the five instances were spread uniformly across those five values (in this case, one each with 6, 7, 8, 9, and 10 mutants).]

b. Repeat for their experiment #22.

4.16 $\boxed{T_2}$ Skewed distribution

Suppose that ℓ is drawn from a Poisson distribution. Find the expectation $\langle(\ell - \langle\ell\rangle)^3\rangle$, which depends on μ. Compare your answer with the case of a symmetric distribution, and suggest an interpretation of this statistic.

Continuous Distributions

The generation of random numbers is too important to be left to chance.
—Robert R. Coveyou

5.1 Signpost

Some of the quantities that we measure are discrete, and the preceding chapters have used discrete distributions to develop many ideas about probability and its role in physics, chemistry, and biology. Most measured quantities, however, are inherently continuous, for example, lengths or times.[1] Figure 3.2b showed one attempt to represent the distribution of such a quantity (a waiting time) by artificially dividing its range into bins, but Nature does not specify any such binning. In other cases, a random quantity may indeed be discrete, but with a distribution that is roughly the same for neighboring values, as in Figure 3.1b; treating it as continuous may eliminate an irrelevant complication.

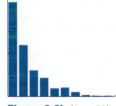

Figure 3.2b (page 38)

Figure 3.1b (page 37)

This chapter will extend our previous ideas to the continuous case. As in the discrete case, we will introduce just a few standard distributions that apply to many situations that arise when we make physical models of living systems.

This chapter's Focus Question is

Biological question: What do neural activity, protein interaction networks, and the diversity of antibodies all have in common?

Physical idea: Power-law distributions arise in many biophysical contexts.

---

[1] Some authors call a continuous random variable a "metric character," in distinction to the discrete case ("meristic characters").

5.2 Probability Density Function

5.2.1 The definition of a probability distribution must be modified for the case of a continuous random variable

In parallel with Chapter 3, consider a replicable random system whose samples are described by a continuous quantity x—a **continuous random variable**. x may have dimensions. To describe its distribution, we *temporarily* partition the range of allowed values for x into bins of width Δx, each labeled by the value of x at its center. As in the discrete case, we again make many measurements and find that ΔN of the N measurements fall in the bin centered on x_0, that is, in the range from $x_0 - \frac{1}{2}\Delta x$ to $x_0 + \frac{1}{2}\Delta x$. The integer ΔN is the frequency of the outcome.

We may be tempted now to define $\wp(x_0) \overset{?}{=} \lim_{N_{\text{tot}} \to \infty} \Delta N / N_{\text{tot}}$, as in the discrete case. The problem with this definition is that, in the limit of small Δx, it always goes to *zero*—a correct but uninformative answer. After all, the fraction of students in a class with heights between, say, 199.999 999 and 200.000 001 cm is very nearly zero, regardless of how large the class is. More generally, we'd like to invent a description of a continuous random system that doesn't depend on any extrinsic choice like a bin width.

The problem with the provisional definition just proposed is that when we cut the bin width in half, each of the resulting half-bins will contain roughly half as many observations as previously.[2] To resolve this problem, in the continuous case we modify the provisional definition of probability distribution by introducing a factor of $1/(\Delta x)$. Dividing by the bin width has the desirable effect that, if we subdivide each bin into two, then we get canceling factors of 1/2 in numerator and denominator, and no net change in the quotient. Thus, at least in principle, we can keep reducing Δx until we obtain a continuous function of x, at the value x_0:

$$\wp_{\text{x}}(x_0) = \lim_{\Delta x \to 0} \left(\lim_{N_{\text{tot}} \to \infty} \frac{\Delta N}{N_{\text{tot}} \Delta x} \right). \qquad (5.1)$$

As with discrete distributions, we may drop the subscript "x" if the value of x completely describes our sample space, or more generally if this abbreviation will not cause confusion.

Even if we have only a finite number of observations, Equation 5.1 gives us a way to make an estimate of the pdf from data:

> *Given many observations of a continuous random variable x, choose a set of bins that are narrow, yet wide enough to each contain many observations. Find the frequencies ΔN_i for each bin centered on x_i. Then the estimated pdf at x_i is $\wp_{\text{x,est}}(x_i) = \Delta N_i / (N_{\text{tot}} \Delta x)$.* (5.2)

For example, if we want to find the pdf of adult human heights, we'll get a fairly continuous distribution if we take Δx to be about 1 cm or less, and N_{tot} large enough to have many samples in each bin in the range of interest. Notice that Equation 5.1 implies that[3]

> *A probability density function for x has dimensions inverse to those of x.* (5.3)

---

[2]Similarly, in Figure 3.1 (page 37), the larger number of bins in panel (b) means that each bar is shorter than in (a).
[3]Just as mass density (kilograms per cubic meter) has different units from mass (kilograms), so the terms "probability density" here, and "probability mass" in Section 3.3.1, were chosen to emphasize the different units of these quantities. Many authors simply use "probability distribution" for either the discrete or continuous case.

We can also express a continuous pdf in the language of events:[4] Let $E_{x_0,\Delta x}$ be the event containing all outcomes for which the value of x lies within a range of width Δx around the value x_0. Then Equation 5.1 says that

$$\wp(x_0) = \lim_{\Delta x \to 0} \left(\mathcal{P}(E_{x_0,\Delta x})/(\Delta x) \right). \quad \text{probability density function} \qquad (5.4)$$

$\wp_x(x_0)$ is not the probability to observe a particular value for x; as mentioned earlier, that's always zero. But once we know $\wp(x)$, then the probability that a measurement will fall into a finite *range* is $\int_{x_1}^{x_2} \mathrm{d}x\, \wp(x)$. Thus, the normalization condition, Equation 3.4 (page 42), becomes

$$\int \mathrm{d}x\, \wp(x) = 1, \quad \textbf{normalization condition}, \text{ continuous case} \qquad (5.5)$$

where the integral runs over all allowed values of x. That is, the area under the curve defined by $\wp(x)$ must always equal 1. As in the discrete case, a pdf is always nonnegative. *Unlike* the discrete case, however, a pdf need not be everywhere smaller than 1: It can have a high, but narrow, spike and still obey Equation 5.5.

T₂ *Section 5.2.1′ (page 114) discusses an alternative definition of the pdf used in mathematical literature.*

5.2.2 Three key examples: Uniform, Gaussian, and Cauchy distributions

Uniform, continuous distribution

Consider a probability density function that is *constant* throughout the range $x_{\min}$ to $x_{\max}$:

$$\wp_{\mathrm{unif}}(x) = \begin{cases} 1/(x_{\max} - x_{\min}) & \text{if } x_{\min} \le x \le x_{\max}; \\ 0 & \text{otherwise.} \end{cases} \qquad (5.6)$$

The formula resembles the discrete case,[5] but note that now $\wp_{\mathrm{unif}}(x)$ will have dimensions, if the variable x does.

Gaussian distribution

The famous "bell curve" is actually a family of functions defined by the formula

$$f(x; \mu_x, \sigma) = A e^{-(x-\mu_x)^2/(2\sigma^2)}, \qquad (5.7)$$

where x ranges from $-\infty$ to $+\infty$. Here A and σ are positive constants; μ_x is another constant.

---

[4]See Section 3.3.1 (page 41).
[5]See Section 3.3.2 (page 43).

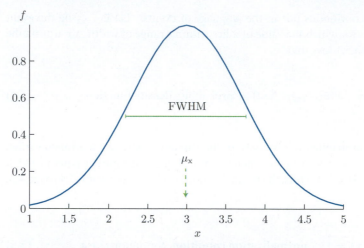

Figure 5.1 [Mathematical function.] **The function defined in Equation 5.7,** with $A = 1$, $\mu_x = 3$, and $\sigma = 1/\sqrt{2}$. Although the function is very small outside the range shown, it is nonzero for *any* x. The abbreviation FWHM refers to the full width of this curve at one half the maximum value, which in this case equals $2\sqrt{\ln 2}$. The Gaussian distribution $\wp_{\text{gauss}}(x; 3, 1/\sqrt{2})$ equals this f times $1/\sqrt{\pi}$ (see Equation 5.8).

Figure 5.1 shows an example of this function. Graphing it for yourself, and playing with the parameters, is a good way to bring home the point that the bell curve is a bump function centered at μ_x (that is, it attains its maximum there) with *width* controlled by the parameter σ. Increasing the value of σ makes the bump wider.

The function f in Equation 5.7 is everywhere nonnegative, but this is not enough: It's only a candidate for a probability density function if it also satisfies the normalization condition, Equation 5.5. Thus, the constant A appearing in it isn't free; it's determined in terms of the other parameters by

$$1/A = \int_{-\infty}^{\infty} dx \, e^{-(x-\mu_x)^2/(2\sigma^2)}.$$

Even if you don't have your computer handy, you can make some progress evaluating this integral. Changing variables to $y = (x - \mu_x)/(\sigma\sqrt{2})$ converts it to

$$1/A = \sigma\sqrt{2} \int_{-\infty}^{\infty} dy \, e^{-y^2}.$$

At this point we are essentially done: We have extracted all the dependence of A on the other parameters (that is, $A \propto \sigma^{-1}$). The remaining integral is just a universal constant, which we could compute just once, or look up. In fact, it equals $\sqrt{\pi}$. Substituting into Equation 5.7 yields "the" Gaussian distribution, or rather a family of distributions defined by the probability density functions[6]

$$\wp_{\text{gauss}}(x; \mu_x, \sigma) = \frac{1}{\sigma\sqrt{2\pi}} e^{-(x-\mu_x)^2/(2\sigma^2)}. \quad \textbf{Gaussian distribution} \qquad (5.8)$$

---

[6]The special case $\mu_x = 0, \sigma = 1$ is also called the **normal distribution**.

The appearance of $1/\sigma$ in front of the exponential has a simple interpretation. Decreasing the value of σ makes the exponential function more narrowly peaked. In order to maintain fixed area under the curve, we must therefore make the curve taller; the factor $1/\sigma$ accomplishes this. This factor also gives $\wp(x)$ the required dimensions (inverse to those of x).

The Gaussian distribution has arrived with little motivation, merely a nontrivial example of a continuous distribution on which to practice some skills. We'll understand its popularity a bit later, when we see how it emerges in a wide class of real situations. First, though, we introduce a counterpoint, another similar-looking distribution with some surprising features.

Cauchy distribution

Consider the family of probability density functions of the form[7]

$$\wp_{\text{cauchy}}(x; \mu_x, \eta) = \frac{A}{1 + \left(\frac{x - \mu_x}{\eta}\right)^2}. \quad \textbf{Cauchy distribution} \quad (5.9)$$

Here, as before, μ_x is a parameter specifying the most probable value of x (that is, it specifies the distribution's center). η is a constant a bit like σ in the Gaussian distribution; it determines how wide the bump is.

Your Turn 5A

a. Find the required value of the constant A in Equation 5.9 in terms of the other constants, using a method similar to the one that led to Equation 5.8. Graph the resulting pdf, and compare with a Gaussian having the same FWHM.
b. Your graph may seem to say that there isn't much difference between the Gaussian and Cauchy pdfs. To see the huge difference more clearly, plot them together on semilog axes (logarithmic axis for the $\wp$, linear for x), and compare them again.

Section 5.4 will discuss real situations in which Cauchy, and related, distributions arise.

5.2.3 Joint distributions of continuous random variables

Just as in the discrete case, we will often be interested in joint distributions, that is, in random systems whose outcomes are sets of two or more continuous values (see Section 3.4.2). The same reasoning that led to the definition of the pdf (Equation 5.1) then leads us to define ΔN as the number of observations for which x lies in a particular range around x_0 of width Δx, *and* y also lies in a particular range of width Δy, and so on. To get a good limit, then, we must divide ΔN by the product $(\Delta x)(\Delta y) \cdots$. Equivalently, we can imitate Equation 5.4:

$$\wp(x_0, y_0) = \lim_{\Delta x, \Delta y \to 0} \left(\mathcal{P}(\mathsf{E}_{x_0, \Delta x} \textbf{ and } \mathsf{E}_{y_0, \Delta y}) / (\Delta x \, \Delta y) \right).$$

---

[7]Some authors call them Lorentzian or Breit-Wigner distributions.

Find appropriate generalizations of the dimensions and normalization condition (Idea 5.3 and Equation 5.5) for the case of a continuous, joint distribution.

We can also extend the notion of conditional probability (see Equation 3.10, page 45):

$$\wp(x \mid y) = \wp(x, y)/\wp(y). \tag{5.10}$$

Thus,

The dimensions of the conditional pdf $\wp(x \mid y)$ are inverse to those of x, regardless of the dimensions of y.

Example Write a version of the Bayes formula for $\wp(x \mid y)$, and verify that the units work out properly.

Solution Begin with a formula similar to Equation 5.10 but with x and y quantities reversed. Comparing the two expressions and imitating Section 3.4.4 (page 52) yields

$$\wp(y \mid x) = \wp(x \mid y)\wp(y)/\wp(x). \tag{5.11}$$

On the right-hand side, $\wp(x)$ in the denominator cancels the units of $\wp(x \mid y)$ in the numerator. Then the remaining factor $\wp(y)$ gives the right-hand side the appropriate units to match the left-hand side.

The continuous form of the Bayes formula will prove useful in the next chapter, providing the starting point for localization microscopy. You can similarly work out a version of the formula for the case when one variable is discrete and the other is continuous.

5.2.4 Expectation and variance of the example distributions

Continuous distributions have descriptors similar to the discrete case. For example, the expectation is defined by [8]

$$\langle f \rangle = \int \mathrm{d}x\, f(x)\wp(x).$$

Note that $\langle f \rangle$ has the same dimensions as f, because the units of $\mathrm{d}x$ cancel those of $\wp(x)$.[9] The variance of f is defined by the same formula as before, Equation 3.20 (page 55); thus it has the same dimensions as f^2.

a. Find $\langle x \rangle$ for the Uniform continuous distribution on some range $a < x < b$. Repeat for the pdf $\wp_{\mathrm{gauss}}(x; \mu_x, \sigma)$.
b. Find var x for the Uniform continuous distribution.

---

[8] Compare the discrete version Equation 3.19 (page 53).
[9] The same remark explains how the normalization integral (Equation 5.5) can equal the pure number 1.

Your Turn 5D

The Gaussian distribution has the property that its expectation and most probable value are equal. Think: What sort of distribution could give *unequal* values?

The variance of a Gaussian distribution is a bit more tricky; let's first guess its general form. The spread of a distribution is unchanged if we just shift it.[10] Changing μ_x just shifts the Gaussian, so we don't expect μ_x to enter into the formula for the variance. The only other relevant parameter is σ. Dimensional analysis shows that the variance must be a constant times σ^2.

To be more specific than this, we must compute the expectation of x^2. We can employ a trick that we've used before:[11] Define a function $I(b)$ by

$$I(b) = \int_{-\infty}^{\infty} dx\, e^{-bx^2}.$$

Section 5.2.2 explained how to evaluate this normalization-type integral; the result is $I(b) = \sqrt{\pi/b}$. Now consider the derivative dI/db. On one hand, it's

$$dI/db = -(1/2)\sqrt{\pi/b^3}. \tag{5.12}$$

But also,

$$dI/db = \int_{-\infty}^{\infty} dx\frac{d}{db}e^{-bx^2} = -\int_{-\infty}^{\infty} dx\, x^2 e^{-bx^2}. \tag{5.13}$$

That last integral is the one we need in order to compute $\langle x^2 \rangle$. Setting the right sides of Equations 5.12 and 5.13 equal to each other and evaluating at $b = (2\sigma^2)^{-1}$, gives

$$\int_{-\infty}^{\infty} dx\, x^2 e^{-x^2/(2\sigma^2)} = \tfrac{1}{2}\pi^{1/2}(2\sigma^2)^{3/2}.$$

With this preliminary result, we can finally evaluate the variance of a Gaussian distribution centered on zero:

$$\text{var } x = \langle x^2 \rangle = \int dx\, \wp_{\text{gauss}}(x;0,\sigma)x^2 = \left[(2\pi\sigma^2)^{-1/2}\right]\left[\tfrac{1}{2}\pi^{1/2}(2\sigma^2)^{3/2}\right] = \sigma^2. \tag{5.14}$$

Because the variance doesn't depend on where the distribution is centered, we conclude more generally that

$$\text{var } x = \sigma^2 \quad \text{if } x \text{ is drawn from } \wp_{\text{gauss}}(x;\mu_x,\sigma). \tag{5.15}$$

Example Find the variance of the Cauchy distribution.

Solution Consider the Cauchy distribution centered on zero, with $\eta = 1$. This time, the integral that defines the variance is

---

[10] See Your Turn 3L (page 56).
[11] See the Example on page 77.

$$\int_{-\infty}^{\infty} dx \, \frac{x^2}{\pi} \frac{1}{1+x^2}.$$

This integral is infinite, because at large $|x|$ the integrand approaches a constant.

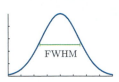

Figure 5.1 (page 100)

Despite this surprising result, the Cauchy distribution is normalizable, and hence it's a perfectly legitimate probability density function. The problem lies not with the distribution, but with the choice of variance as a descriptor: The variance is very sensitive to outliers, and a Cauchy distribution has many more of these than does a Gaussian.

Other descriptors of spread work just fine for the Cauchy distribution, however. For example, we can use <u>f</u>ull <u>w</u>idth at <u>h</u>alf <u>m</u>aximum (FWHM; see Figure 5.1[12]) instead of variance to describe its spread.

$\boxed{T_2}$ *Section 5.2.4′ (page 114) introduces another measure of spread that is useful for long-tail distributions: the interquartile range.*

5.2.5 Transformation of a probability density function

The definition of probability density function creates an important difference from the discrete case. Suppose that you have recorded many vocalizations of some animal, perhaps a species of whale. The intensity and pitch vary over time. You'd like to characterize these sounds, perhaps to see how they vary with species, season, and so on. One way to begin might be to define x as the intensity of sound emitted (in watts, abbreviated W) and create an estimated pdf $\wp_x$ from many observations of x. A colleague, however, may believe that it's more meaningful to report the related quantity $y = 10 \log_{10}(x/(1\,\text{W}))$, the sound intensity on a "decibel" scale. That colleague will then report the pdf $\wp_y$.

To compare your results, you need to *transform* your result from your choice of variable x to your colleague's choice y. To understand transformation in general, suppose that x is a continuous random variable with some known pdf $\wp_x(x)$. If we collect a large number of draws from that distribution ("measurements of x"), the fraction that lie between $x_0 - \frac{1}{2}\Delta x$ and $x_0 + \frac{1}{2}\Delta x$ will be $\wp_x(x_0)\Delta x$.[13] Now define a new random variable y to be some function applied to x, or $y = G(x)$. This y is not independent of x; it's just another description of the same random quantity reported by x. Suppose that G is a strictly increasing or decreasing function—a **monotonic** function.[14] In the acoustic example above, $G(x) = 10 \log_{10}(x/1\,\text{W})$ is a strictly increasing function (see Figure 5.2); thus, its derivative dG/dx is everywhere positive.

To find $\wp_y$ at some point y_0, we now ask, for a small interval Δy: How often does y lie within a range $\pm\frac{1}{2}\Delta y$ of y_0? Figure 5.2 shows that, if we choose the y interval to be the image of the x interval, then the *same fraction* of all the points lie in this interval in either description. We know that $y_0 = G(x_0)$. Also, because Δx is small, Taylor's theorem gives $\Delta y \approx (\Delta x)(dG/dx|_{x_0})$, and so

$$\left[\wp_y(G(x_0))\right]\left[(\Delta x)\frac{dG}{dx}\Big|_{x_0}\right] = \wp_x(x_0)(\Delta x).$$

---

[12] See also Problem 5.10.

[13] See Equation 5.1 (page 98).

[14] $\boxed{T_2}$ Thus, G associates exactly one x value to each y in its range. If G is not monotonic, the notation gets more awkward but we can still get a result analogous to Equation 5.16.

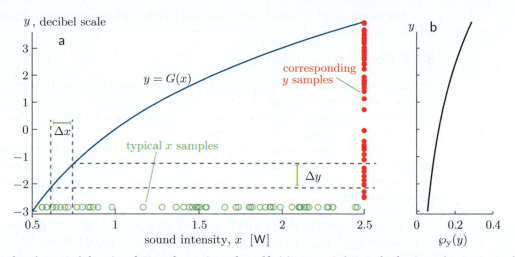

Figure 5.2 [Mathematical functions.] **Transformation of a pdf.** (a) *Open circles* on the horizontal axis give a cloud representation of a Uniform distribution $\wp_x(x)$. These representative samples are mapped to the vertical axis by the function $G(x) = 10\log_{10}(x/(1\,\text{W}))$ and are shown there as *solid circles*. They give a cloud representation of the transformed distribution $\wp_y(y)$. One particular interval of width Δx is shown, along with its transformed version on the y axis. Both representations agree that this bin contains five samples, but they assign it different widths. (b) The transformed pdf (horizontal axis), determined by using Equation 5.16, reflects the non-Uniform density of the solid circles in (a).

Dividing both sides by $(\Delta x)(\mathrm{d}G/\mathrm{d}x|_{x_0})$ gives the desired formula for $\wp_y$:

$$\wp_y(y_0) = \wp_x(x_0) \Big/ \frac{\mathrm{d}G}{\mathrm{d}x}\Big|_{x_0} \qquad \text{for monotonically increasing } G. \qquad (5.16)$$

The right side of this formula is a function of y_0, because we're evaluating it at $x_0 = G^{-1}(y_0)$, where G^{-1} is the inverse function to G.

Your Turn 5E

a. Think about how the dimensions work in Equation 5.16, for example, in the situation where x has dimensions $\mathbb{L}$ and $G(x) = x^3$. Your answer provides a useful mnemonic device for the formula.
b. Why aren't the considerations of this section needed when we study *discrete* probability distributions?

Example Go through the above logic again for the case of a function G that's monotonically *decreasing,* and make any necessary changes.

Solution In this case, the width of the y interval corresponding to Δx is $-\Delta x(\mathrm{d}G/\mathrm{d}x)$, a positive quantity. Using the absolute value covers both cases:

$$\wp_y(y_0) = \wp_x(x_0) \Big/ \left|\frac{\mathrm{d}G}{\mathrm{d}x}\Big|_{x_0}\right|. \quad \text{transformation of a pdf, where } x_0 = G^{-1}(y_0)$$

$$(5.17)$$

The transformation formula just found will have repercussions when we discuss model selection in Section 6.2.3.

5.2.6 Computer simulation

The previous section stressed the utility of transformations when we need to convert a result from one description to another. We now turn to a second practical application, simulating draws from a specified distribution by using a computer. Chapter 8 will use these ideas to create simulations of cell reaction networks.

Equation 5.17 has an important special case: If y is the *Uniformly* distributed random variable on the range $[0, 1]$, then $\wp_y(y) = 1$ and $\wp_x(x_0) = |dG/dx|_{x_0}|$. This observation is useful when we wish to simulate a random system with some arbitrarily specified probability density function:

> To simulate a random system with a specified pdf $\wp_x$, find a function G whose derivative equals $\pm\wp_x$ and that maps the desired range of x onto the interval $[0, 1]$. Then apply the inverse of G to a Uniformly distributed variable y; the resulting x values will have the desired distribution.   (5.18)

Example The probability density function $\wp(x) = e^{-x}$, where x lies between zero and infinity, will be important in later chapters. Apply Idea 5.18 to simulate draws from a random variable with this distribution.

Solution To generate x values, we need a function G that solves $|dG/dx| = e^{-x}$. Thus,

$$G(x) = \text{const} \pm e^{-x}.$$

Applying functions of this sort to the range $[0, \infty)$, we see that the choice e^{-x} works. The inverse of that function is $x = -\ln y$.
Try applying $-\ln$ to your computer's random number generator, and making a histogram of the results.

Your Turn 5F

Think about the discussion in Section 4.2.5 (page 73) of how to get a computer to draw from a specified *discrete* distribution (for example, the Poisson distribution). Make a connection to the above discussion.

Your Turn 5G

Apply Idea 5.18 to the Cauchy distribution, Equation 5.9 (page 101), with $\mu_x = 0$ and $\eta = 1$. Use a computer to generate some draws from your distribution, histogram them, and confirm that they have the desired distribution.

5.3 More About the Gaussian Distribution

5.3.1 The Gaussian distribution arises as a limit of Binomial

The Binomial distribution is very useful, but it has two unknown parameters: the number of draws M and the probability ξ to flip <u>heads</u>. Section 4.3 described a limiting case, in which

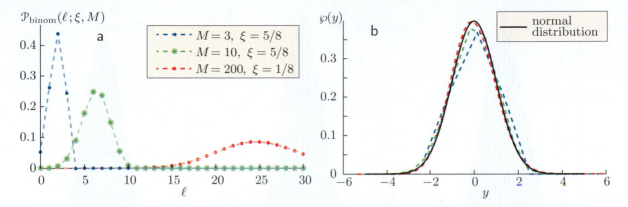

Figure 5.3 [Mathematical functions.] **The Gaussian distribution as a limit.** (a) Three examples of Binomial distributions. (b) The same three discrete distributions as in (a) have been modified as described in the text. In particular, each curve has been rescaled: Instead of all the points summing to 1, the scale factor $1/\Delta y$ has been applied to ensure that the *area* under each curve equals 1. For comparison, the *solid curve* shows the Gaussian distribution $\wp_{\text{gauss}}$ with $\mu_y = 0$ and $\sigma = 1$. The figure shows that $M = 3$ or 10 give only roughly Gaussian-shaped distributions but that distributions with large M and $M\xi$ are very nearly Gaussian. See also Problem 5.14.

the Binomial distribution "forgets" the individual values of these parameters, "remembering" only their product $\mu = M\xi$. The appropriate limit was large M at fixed μ.

You may already have noticed, however, an even greater degree of universality when μ is also large.[15] Figure 5.3a shows three examples of Binomial distributions. When both M and $M\xi$ are large, the curves become smooth and symmetric, and begin to look very similar to Gaussian distributions.

It's true that the various Binomial distributions all differ in their expectations and variances. But these superficial differences can be eliminated by changing our choice of variable, as follows: First, let

$$\mu_\ell = M\xi \quad \text{and} \quad s = \sqrt{M\xi(1-\xi)}.$$

Then define the new random variable $y = (\ell - \mu_\ell)/s$. Thus, y always has the same expectation, $\langle y \rangle = 0$, and variance, $\text{var}\, y = 1$, regardless of what values we choose for M and ξ.

We'd now like to compare other features of the y distribution for different values of M and ξ, but first we face the problem that the list of allowed values (the sample space) for y depends on the parameters. For example, the spacing between successive discrete values is $\Delta y = 1/s$. But instead of a direct comparison, we can divide the discrete distribution $\mathcal{P}(y; M, \xi)$ by Δy. If we then take the limit of large M, we obtain a family of probability density functions, each for a continuous random variable y in the range $-\infty < y < \infty$. It does make sense to compare these pdfs for different values of ξ, and remarkably[16]

> For any fixed value of ξ, the distribution of y approaches a universal form. That is, it "forgets" the values of both M and ξ, as long as M is large enough. The universal limiting pdf is Gaussian.

---

[15]See Problem 4.6.

[16] $\boxed{T_2}$ In Problem 5.14 you'll prove a more precise version of this claim.

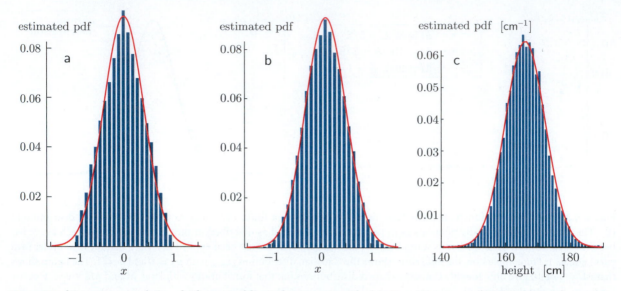

Figure 5.4 [Computer simulations.] **The central limit theorem at work.** (a) *Bars:* Histogram of 50 000 draws of a random variable defined as the sum of two independent random variables, each Uniformly distributed on the range $[-1/2, +1/2]$. The *curve* shows a Gaussian distribution with the same expectation and variance for comparison. (b) *Bars:* 50 000 draws from the sum of four such variables, scaled by $1/\sqrt{2}$ to give the same variance as in (a). The *curve* is the same as in (a). (c) [Empirical data with fit.] Distribution of the heights of a similarly large sample of 21-year-old men born in southern Spain. The curve is the best-fitting Gaussian distribution. [Data from María-Dolores & Martínez-Carrión, 2011.]

Figure 5.3b illustrates this result. For example, the figure shows that even a highly asymmetric Bernoulli trial, like $\xi = 1/8$, gives rise to the symmetric Gaussian for large enough M.

5.3.2 The central limit theorem explains the ubiquity of Gaussian distributions

The preceding subsection began to show why Gaussian distributions arise frequently: Many interesting quantities really can be regarded as sums of many independent Bernoulli trials, for example, the number of molecules of a particular type captured in a sample drawn from a well-mixed solution. And in fact, the phenomenon noted in Section 5.3.1 is just the beginning. Here is another example of the same idea at work.

Let $\wp_1(x)$ denote the continuous Uniform distribution on the range $-1/2 < x < 1/2$. Its probability density function does not look very much like any Gaussian, not even one chosen to have the same expectation and variance. Nevertheless, Figure 5.4a shows that the *sum of two* independent random variables, each drawn from $\wp_1$, has a distribution that looks a bit more like a Gaussian, although unlike a Gaussian, it equals zero outside a finite range. And the sum of *just four* such variables looks very much like a Gaussian (Figure 5.4b).[17]

This observation illustrates a key result of probability theory, the **central limit theorem**. It applies when we have M independent random variables (continuous or discrete), each

---

[17]See Problem 5.7. Incidentally, this exercise also illustrates the counterpoint between analytic and simulation methods. The distribution of a sum of random variables is the convolution of their individual distributions (Section 4.3.5). It would be tedious to work out an exact formula for the convolution of even a simple distribution with itself, say 10 times. But it's easy to make sets of 10 draws from that distribution, add them, and histogram the result.

drawn from identical distributions. The theorem states roughly that, for large enough M, the quantity $x_1 + \cdots + x_M$ is always distributed as a Gaussian. We have discussed examples where x was Bernoulli or Uniform, but actually the theorem holds *regardless* of the distribution of the original variable, as long as it has finite expectation and variance.

5.3.3 When to use/not use a Gaussian

Indeed, in Nature we often do observe quantities that reflect the additive effects of many independent random influences. For example, human height is a complex phenotypic trait, dependent on hundreds of different genes, each of which is dealt to us at least partially independently of the others. It's reasonable to suppose that these genes have at least partially additive effects on overall height,[18] and that the ones making the biggest contributions are roughly equal in importance. In such a situation, we may expect to get a Gaussian distribution, and indeed many phenotypic traits, including human height, do follow this expectation (Figure 5.4c).[19]

> **Your Turn 5H**
>
> Problem 4.5 introduced a model for molecular diffusion based on trajectories that take steps of $\pm d$ in each direction. But surely this is an oversimplification: The minute kicks suffered by a suspended particle must have a variety of strengths, and so must result in displacements by a variety of distances. Under what circumstances may we nevertheless expect the random walk to be a good model of diffusive motion?

In our lab work we sometimes measure the same quantity independently several times, then take the average of our measurements. The central limit theorem tells us why in these situations we generally see a Gaussian distribution of results. However, we should observe some caveats:

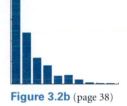

Figure 3.2b (page 38)

- Some random quantities are *not* sums of many independent, identically distributed random variables. For example, the blip waiting times shown in Figure 3.2b are far from being Gaussian distributed. Unlike a Gaussian, their distribution is very asymmetrical, reaching its maximum at its extreme low end.

- Even if a quantity does seem to be such a sum, and its distribution does appear fairly close to Gaussian near its peak, nevertheless for any finite N there may be significant discrepancies in the tail region (see Figure 5.4a), and for some applications such low-probability events may be important. Also, many kinds of experimental data, such as the *number* of blips in a time window, are Poisson distributed; for low values of μ such distributions can also be far from Gaussian.

- An observable quantity may indeed be the sum of contributions from many sources, but they may be interdependent. For example, spontaneous neural activity in the brain involves the electrical activity of many nerve cells (**neurons**), each of which is connected to many others. When a few neurons fire, they may tend to trigger others, in an "avalanche" of activity. If we add up all the electrical activity in a region of the brain, we will see a signal with peaks reflecting the total numbers of neurons firing in each event. These event magnitudes were found to have a distribution that was far from Gaussian (Figure 5.5).

---

[18] That is, we are neglecting the possibility of nonadditive gene interaction, or **epistasis**.

[19] Height also depends on nutrition and other environmental factors. Here, we are considering a roughly homogeneous population and supposing that any remaining variation in environment can also be roughly modeled as additive, independent random factors.

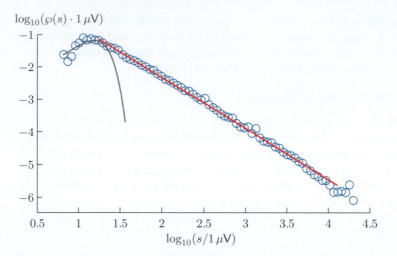

Figure 5.5 [Experimental data with fit.] **Power-law distribution of neural activity.** Slices of brain tissue from rats were cultured on arrays of extracellular electrodes that recorded the neurons' spontaneous activities. The electric potential outside the cells was found to have events consisting of bursts of activity from many cells. The total measured activity s in each burst (the "magnitude" of the event) was tabulated and used to find an estimated pdf. This log-log plot shows that the distribution has a power-law form (*red line*), with exponent -1.6. For comparison, the *gray line* shows an attempted fit to a Gaussian. Similar results were observed in intact brains of live animals. [Data from Gireesh & Plenz, 2008.]

- We have already seen that some distributions have *infinite* variance. In such cases, the central limit theorem does not apply, even if the quantity in question is a sum of independent contributions.[20]

5.4 More on Long-tail Distributions

The preceding section pointed out that not every measurement whose distribution seems to be bump-shaped will actually be Gaussian. For example, the Gaussian distribution far from its center falls as a constant times $\exp(-x^2/(2\sigma^2))$, whereas the Cauchy distribution,[21] which looks superficially similar, approaches a constant times $x^{-\alpha}$, with $\alpha = 2$. More generally, there are many random systems with **power-law distributions**; that is, they exhibit this kind of limiting behavior for some constant α. Power-law distributions are another example of the long-tail phenomenon mentioned earlier, because any power of x falls more slowly at large x than the Gaussian function.[22]

To see whether an empirically obtained distribution is of power-law form, we can make a **log-log plot**, so named because both the height of a point and its horizontal position are proportional to the logarithms of its x and y values. The logarithm of $Ax^{-\alpha}$ is $\log(A) - \alpha \log(x)$, which is a linear function of $\log x$. Thus, this function will appear on a log-log plot as a straight line,[23] with slope $-\alpha$. A power-law distribution will therefore have this straight-line form for large enough x. Figure 5.5 shows a biological example of such a distribution.

---

[20] See Problem 5.13.
[21] See Equation 5.9 (page 101).
[22] See Section 4.4.2 (page 81). You'll investigate the weight in the tails of various distributions in Problem 5.10.
[23] See Problem 5.11.

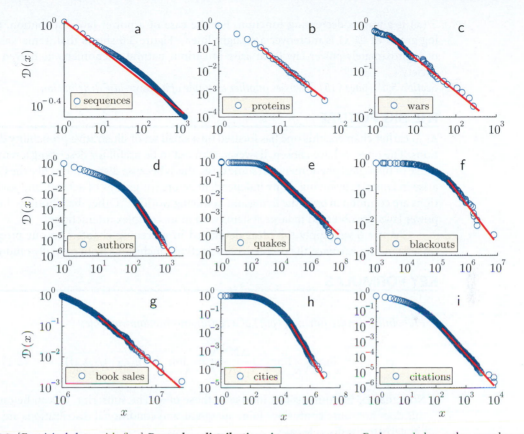

Figure 5.6 [Empirical data with fits.] **Power law distributions in many contexts.** Each panel shows the complementary cumulative distribution function $\mathcal{D}(x)$, Equation 5.19 (or its discrete analog), and a power-law fit, for a different dataset. (a) The probabilities of occurrence of various antibody sequences in the immune system of a single organism versus rank order x. The entire antibody repertoire of a zebrafish was sequenced, and a list was made of sequences of the "D region" of each antibody. Over a wide range, the probability followed the approximate form $\wp(x) \propto x^{-\alpha}$ with $\alpha \approx 1.15$. (b) The numbers of distinct interaction partners of proteins in the protein-interaction network of the yeast *S. cerevisiae* ($\alpha = 3$). (c) The relative magnitudes of wars from 1816 to 1980, that is, the number of battle deaths per 10 000 of the combined populations of the warring nations ($\alpha = 1.7$). (d) The numbers of authors on published academic papers in mathematics ($\alpha = 4.3$). (e) The intensities of earthquakes occurring in California between 1910 and 1992 ($\alpha = 1.6$). (f) The magnitudes of power failures (number of customers affected) in the United States ($\alpha = 2.3$). (g) The sales volumes of bestselling books in the United States ($\alpha = 3.7$). (h) The populations of cities in the United States ($\alpha = 2.4$). (i) The numbers of citations received by published academic papers ($\alpha = 3.2$). [Data from Clauset et al., 2009, and Mora et al., 2010.]

Equivalently, we can examine a related quantity called the **complementary cumulative distribution**,[24] the probability of drawing a value of x larger than some specified value:

$$\mathcal{D}(x) = \int_x^\infty dx' \, \wp(x'). \qquad (5.19)$$

---

[24]The qualifier "complementary" distinguishes $\mathcal{D}$ from a similar definition with the integral running from $-\infty$ to x.

$\mathcal{D}(x)$ is always a decreasing function. For the case of a power-law distribution, the log-log graph of $\mathcal{D}(x)$ is moreover a straight line.[25] Figure 5.6 shows that, remarkably, pdfs of approximately power-law form arise in various natural phenomena, and even human society.

$\boxed{T_2}$ *Section 5.4' (page 115) discusses another example of a power-law distribution.*

THE BIG PICTURE

As in earlier chapters, this one has focused on a small set of illustrative probability distributions. The ones we have chosen, however, turn out to be useful for describing a remarkable range of biological phenomena. In some cases, this is because distributions like the Gaussian arise in contexts involving many independent actors (or groups of actors), and such situations are common in both the living and nonliving worlds.[26] Other distributions, including power laws, are observed in large systems with more complex interactions.

Chapter 6 will apply the ideas developed in preceding chapters to our program of understanding *inference,* the problem of extracting conclusions from partially random data.

KEY FORMULAS

- *Probability density function (pdf) of a continuous random variable:*

$$\wp_x(x_0) = \lim_{\Delta x \to 0} \left(\lim_{N_{tot} \to \infty} \frac{\Delta N}{N_{tot} \Delta x} \right) = \lim_{\Delta x \to 0} \left(\mathcal{P}(\mathsf{E}_{x_0, \Delta x})/(\Delta x) \right). \qquad (5.1) + (5.4)$$

Note that $\wp_x$ has dimensions inverse to those of x. The subscript "x" can be omitted if this does not cause confusion. Joint, marginal, and conditional distributions are defined similarly to the discrete case.

- *Estimating a pdf:* Given some observations of x, choose a set of bins that are wide enough to each contain many observations. Find the frequencies ΔN_i for each bin centered on x_i. Then the estimated pdf at x_i is $\Delta N_i/(N_{tot} \Delta x)$.

- *Normalization and moments of continuous distribution:* $\int dx \, \wp(x) = 1$. The expectation and variance of a function of x are then defined analogously to discrete distributions, for example, $\langle f \rangle = \int dx \, \wp(x) f(x)$.

- *Continuous version of Bayes formula:*

$$\wp(y \mid x) = \wp(x \mid y)\wp(y)/\wp(x). \qquad (5.11)$$

- *Gaussian distribution:* $\wp_{gauss}(x; \mu_x, \sigma) = (\sigma\sqrt{2\pi})^{-1} \exp(-(x - \mu_x)^2/(2\sigma^2))$. The random variable x, and the parameters μ_x and σ, can have any units, but they must all match. $\langle x \rangle = \mu_x$ and $\mathrm{var}\, x = \sigma^2$.

- *Cauchy distribution:*

$$\wp_{cauchy}(x; \mu_x, \eta) = \frac{A}{1 + \left(\frac{x - \mu_x}{\eta}\right)^2}.$$

---

[25] In Problem 5.11 you'll contrast the corresponding behavior in the case of a Gaussian distribution.
[26] The Exponential distribution to be studied in Chapter 7 has this character as well, and enjoys a similarly wide range of application.

This is an example of a power-law distribution, because $\wp_{\mathrm{cauchy}} \to A\eta^2 x^{-2}$ at large $|x|$. The constant A has a specific relation to η; see Your Turn 5A (page 101). The random variable x, and the parameters μ_x and η, can have any units, but they must all match.

- *Transformation of a pdf:* Suppose that x is a continuous random variable with probability density function $\wp_x$. Let y be a new random variable, defined by drawing a sample x and applying a strictly increasing function G. Then $\wp_y(y_0) = \wp_x(x_0)/G'(x_0)$, where $y_0 = G(x_0)$ and $G' = \mathrm{d}G/\mathrm{d}x$. (One way to remember this formula is to recall that it must be valid even if x and y have different dimensions.) If G is strictly *decreasing* we get a similar formula but with $|\mathrm{d}G/\mathrm{d}x|$.

FURTHER READING

Semipopular:
Hand, 2008.

Intermediate:
Bolker, 2008; Denny & Gaines, 2000; Otto & Day, 2007, §P3; Ross, 2010.

Technical:
Power-law distributions in many contexts: Clauset et al., 2009.
Power-law distribution in neural activity: Beggs & Plenz, 2003.
Complete antibody repertoire of an animal: Weinstein et al., 2009.

$\boxed{T_2}$ **Track 2**

5.2.1′ Notation used in mathematical literature

The more complex a problem, the more elaborate our mathematical notation must be to avoid confusion and even outright error. But conversely, elaborate notation can unnecessarily obscure less complex problems; it helps to be flexible about our level of precision. This book generally uses a notation that is customary in physics literature and is adequate for many purposes. But other books use a more formal notation, and a word here may help the reader bridge to those works.

For example, we have been a bit imprecise about the distinction between a random variable and the specific values it may take. Recall that Section 3.3.2 defined a random variable to be a function on sample space. Every "measurement" generates a point of sample space, and evaluating the function at that point yields a numerical value. To make this distinction clearer, some authors reserve capital letters for random variables and lowercase for possible values.

Suppose that we have a discrete sample space, a random variable X, and a number x. Then the event $\mathsf{E}_{X=x}$ contains the outcomes for which X took the specific value x, and $\mathcal{P}_X(x) = \mathcal{P}(\mathsf{E}_{X=x})$ is a function of x—the probability mass function. The right side of this definition is a function of x.

In the continuous case, no outcomes have X exactly equal to any particular chosen value. But we can define the cumulative event $\mathsf{E}_{X\leq x}$, and from that the probability density function:

$$\wp_X(x) = \frac{\mathrm{d}}{\mathrm{d}x}\mathcal{P}(\mathsf{E}_{X\leq x}). \tag{5.20}$$

This definition makes it clear that $\wp_X(x)$ is a function of x with dimensions inverse to those of x. (Integrating both sides and using the fundamental theorem of calculus shows that the cumulative distribution is the integral of the pdf.) For a joint distribution, we generalize to

$$\wp_{X,Y}(x,y) = \frac{\partial^2}{\partial x \partial y}\mathcal{P}(\mathsf{E}_{X\leq x}\text{ and }\mathsf{E}_{Y\leq y}).$$

The formulas just outlined assume that a "probability measure" $\mathcal{P}$ has already been given that assigns a number to each of the events $\mathsf{E}_{X\leq x}$. It does not tell us how to *find* that measure in a situation of interest. For example, we may have a replicable system in which a single numerical quantity x is a complete description of what was measured; then Equation 5.1 effectively defines the probability. Or we may have a quantifiable degree of belief in the plausibility of various values of x (see Section 6.2.2).

> **Your Turn 5I**
>
> Rederive Equation 5.16 from Equation 5.20.

$\boxed{T_2}$ **Track 2**

5.2.4′ Interquartile range

The main text noted that the variance of a distribution is heavily influenced by outliers, and is not even defined for certain long-tail distributions, including Cauchy. The text also pointed out that another measure of spread, the FWHM, is usable in such cases.

Another widely used, robust measure of the spread of a one-variable distribution is the **interquartile range** (IQR). Its definition is similar to that of the median:[27] We start at the lower extreme of the values obtained in a data sample and work our way upward. Instead of stopping when we have seen half of the data points, however, we stop after 1/4; this value is the lower end of the IQR. We then continue upward until we have seen 3/4 of the data points; that value is the upper end. The range between these two limits contains half of the data points and is called the interquartile range.

A more primitive notion of spread is simply the **range of a dataset**, that is, the difference between the highest and lowest values observed. Although any finite number of observations will yield a value for the range, as we take more and more data points the range is even more sensitive to outliers than the variance; even a Gaussian distribution will yield infinite range as the number of observations gets large.

T_2

Track 2

5.4′a Terminology

Power-law distributions are also called Zipf, zeta, or Pareto distributions in various contexts. Sometimes these terms are reserved for the situation in which $\wp$ equals $Ax^{-\alpha}$ exactly for x greater than some "cutoff" value (and it equals zero otherwise). In contrast, the Cauchy distribution is everywhere nonzero but deviates from strict power-law behavior at small $|x|$; nevertheless, we will still call it an example of a power-law distribution because of its asymptotic form at large x.

5.4′b The movements of stock prices

A stock market is a biological system with countless individual actors rapidly interacting with one another, each based on partial knowledge of the others' aggregate actions. Such systems can display interesting behavior.

It may seem hopeless to model such a complex system, and yet its very complexity may allow for a simplified picture. Each actor observes events external to the market (politics, natural disasters, and so on), adjusts her optimism accordingly, and also observes other actors' responses. Predictable events have little effect on markets, because investors have already predicted them and factored them into market prices before they occur. It's the *unexpected* events that trigger big overall changes in the market. Thus, we may suspect that changes in a stock market index are, at least approximately, random in the sense of "no discernible, relevant structure" (Idea 3.2, page 39).

More precisely, let's explore the hypothesis that

> **H0:** *Successive fractional changes in a stock market index at times separated by some interval* Δt *are independent, identically distributed draws from some unknown, but*   (5.21) *fixed, distribution.*

---

[27] See Problem 5.2.

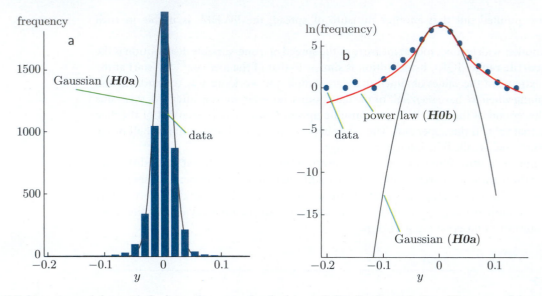

Figure 5.7 [Experimental data with fits.] **Another example of a long-tail distribution.** (a) *Bars:* Histogram of the quantity $y = \ln(x(t + \Delta t)/x(t))$, where x is the adjusted Dow Jones Industrial Average over the period 1934–2012 and $\Delta t = 1$ week. *Curve:* Gaussian distribution with center and variance adjusted to match the main part of the peak in the histogram. (b) *Dots and gray curve:* The same information as in (a), in a semilog plot. Some extreme events that were barely visible in (a) are now clearly visible. *Red curve:* Cauchy-like distribution with exponent 4. [Data from Dataset 7.]

This is not a book about finance; this hypothesis is not accurate enough to make you rich.[28] Nevertheless, the tools developed earlier in this chapter do allow us to uncover an important aspect of financial reality that was underappreciated for many years. To do this, we now address the specific question: *Under hypothesis **H0**, what distribution best reflects the available data?*

First note how the hypothesis has been phrased. Suppose that in one week a stock average changes from x to x', and that at the start of that week you owned \$100 of a fund tracking that average. If you did nothing, then at the end of that week the value of your investment would have changed to \$100$\times(x'/x)$. A convenient way to express this behavior without reference to your personal circumstance is to say that the logarithm of your stake shifted by $y = \ln x' - \ln x$. Accordingly, the bars in Figure 5.7a show the distribution of historical y values.

It may seem natural to propose the more specific hypothesis **H0a** that the distribution of y is Gaussian. After all, a stock average is the weighted sum of a very large number of individual stock prices, each in turn determined by a still larger number of transactions in the period Δt, so "surely the central limit theorem implies that its behavior is Gaussian." The figure does show a pretty good-looking fit. Still, it may be worth examining other hypotheses. For example, the Gaussian distribution predicts negligible probability of $y = -0.2$, and yet the data show that such an event did occur. It matters: In an event of that magnitude, investors lost $1 - e^{-0.2} = 18\%$ of their investment value in a single week.

Moreover, each investor is *not at all* independent of the others' behavior. In fact, large market motions are in some ways like avalanches or earthquakes (Figure 5.6e): Strains

---

[28]For example, there are significant time correlations in real data. We will minimize this effect by sampling the data at the fairly long interval of $\Delta t = 1$ week.

build up gradually, then release suddenly as many investors simultaneously change their level of confidence. Section 5.4 pointed out that such collective-dynamics systems can have power-law distributions. Indeed, Figure 5.7b shows that a distribution of the form $\wp(y) = A/\left[1 + \left((y - \mu_y)/\eta\right)^4\right]$ (hypothesis **H0b**) captures the extreme behavior of the data much better than a Gaussian.

To decide which hypothesis is *objectively* more successful, you'll evaluate the corresponding likelihoods in Problem 6.10. At the same time, you'll obtain the objectively best-fitting values of the parameters involved in each hypothesis family. For more details, see Mantegna and Stanley (2000).

PROBLEMS

5.1   Data lumping

Suppose that the body weight x of individuals in a population is known to have a pdf that is **bimodal** (two bumps; see Figure 5.8). Nora measured x on a large population. To save time on data collection, she took individuals in groups of 10, found the sample mean value of x for each group, and recorded only those numbers in her lab notebook. When she later made a histogram of those values, she was surprised to find that they didn't have the same distribution as the known distribution of x. Explain qualitatively what distribution they did have, and why.

5.2   Median mortality

The **median** of a random variable x can be defined as the value $x_{1/2}$ for which the probability $\mathcal{P}(x < x_{1/2})$ equals $\mathcal{P}(x > x_{1/2})$. That is, a draw of x is equally probable to exceed the median as it is to fall below it.

In his book *Full House*, naturalist Stephen Jay Gould describes being diagnosed with a form of cancer for which the median mortality, at that time, was eight months—in other words, of all people with this diagnosis, half would die in that time. Several years later, Gould noticed that he was alive and well. Can we conclude that the diagnosis was wrong? Sketch a possible probability distribution to illustrate your answer.

5.3   Cauchy-like distribution

Consider the family of functions

$$A \left(1 + \left| \frac{x - \mu_{\mathrm{x}}}{\eta} \right|^{\nu} \right)^{-1}.$$

Here μ_{x}, A, η, and ν are some constants; A, η, and ν are positive. For each choice of these parameters, we get a function of x. Could any of these functions be a legitimate probability density function for x in the range from $-\infty$ to ∞? If so, which ones?

5.4   Simulation via transformation

Section 5.2.6 explained how to take Uniformly distributed random numbers supplied by a computer and convert them into a modified series with a more interesting distribution.

a. As an example of this procedure, generate 10 000 Uniformly distributed real numbers x between 0 and 1, find their reciprocals $1/x$, and make a histogram of the results. (You'll

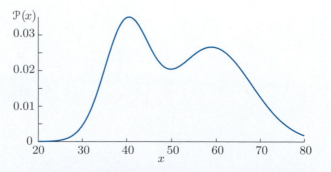

Figure 5.8 [Mathematical function.] **See Problem 5.1.**

need to make some decisions about what range of values to histogram and how many bins to use in that range.)

b. The character of this distribution is revealed most clearly if the counts in each bin (observed frequencies) are presented as a log-log plot, so make such a graph.

c. Draw a conclusion about this distribution by inspecting your graph. Explain mathematically why you got this result.

d. Comment on the high-x (lower right) end of your graph. Does it look messy? What happens if you replace 10 000 by, say, 50 000 samples?

5.5   Variance of Cauchy distribution

First work Your Turn 5G (page 106). Then consider the Cauchy distribution, Equation 5.9, with $\mu_x = 0$ and $\eta = 1$. Generate a set of 4 independent simulated draws from this distribution, and compute the sample mean of x^2. Now repeat for sets of 8, 16, ... 1024, ... draws and comment in the light of the Example on page 103.

5.6   Binomial to Gaussian (numerical)

Write a computer program to generate graphs like Figure 5.3b. That is, plot the Binomial distribution, properly rescaled to display the collapse onto the Gaussian distribution in an appropriate limit.

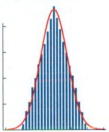

Figure 5.3b (page 107)

5.7   Central limit

a. Write a computer program to generate Figures 5.4a,b. That is, use your computer math package's Uniform random generator to simulate the two distributions shown, and histogram the result. Superimpose on your histograms the continuous probability density function that you expect in this limit.

b. Repeat for sums of 10 Uniform random numbers. Also make a semilog plot of your histogram and the corresponding Gaussian in order to check the agreement more closely in the tail regions. [*Hint:* You may need to use more than 50 000 samples in order to see the behavior out in the tails.]

Figures 5.4a (page 108)

c. Try the problem with a more exotic initial distribution: Let x take only the discrete values 1, 2, 3, 4 with probabilities 1/3, 2/9, 1/9, and 1/3, respectively, and repeat (a) and (b). This is a bimodal distribution, even more unlike the Gaussian than the one you tried in (a) and (b).

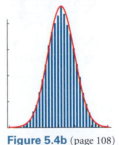

5.8   Transformation of multivariate distribution

Consider a joint probability density function for two random variables given by

Figure 5.4b (page 108)

$$\wp_{x,y}(x, y) = \wp_{\mathrm{gauss},x}(x; 0, \sigma)\wp_{\mathrm{gauss},y}(y; 0, \sigma).$$

Thus, x and y are independent, Gaussian-distributed variables.

a. Let $r = \sqrt{x^2 + y^2}$ and $\theta = \tan^{-1}(y/x)$ be the corresponding polar coordinates, and find the joint pdf of r and θ.

b. Let $u = r^2$, and find the joint pdf of u and θ.

c. Connect your result in (b) to what you found by simulation in Problem 3.4.

d. $\boxed{T_2}$ Generalize your result in (a) for the general case, where (x, y) have an arbitrary joint pdf and (u, v) are an arbitrary transformation of (x, y).

5.9   **Power-law distributions**

Suppose that some random system gives a continuous numerical quantity x in the range $1 < x < \infty$, with probability density function $\wp(x) = Ax^{-\alpha}$. Here A and α are positive constants.

a. The constant A is determined once α is specified. Find this relation.

b. Find the expectation and variance of x, and comment.

5.10   **Tail probabilities**

In this problem, you will explore the probability of obtaining extreme ("tail") values from a Gaussian versus a power-law distribution. "Typical" draws from a Gaussian distribution produce values within about one standard deviation of the expectation, whereas power-law distributions are more likely to generate large deviations. Your job is to make this intuition more precise. First work Problem 5.3 if you haven't already done so.

a. Make a graph of the Cauchy distribution with $\mu_x = 0$ and $\eta = 1$. Its variance is infinite, but we can still quantify the width of its central peak by the full width at half maximum (FWHM), which is defined as twice the value of x at which $\wp$ equals $\frac{1}{2}\wp(0)$. Find this value.

b. Calculate the FWHM for a Gaussian distribution with standard deviation σ. What value of σ gives the same FWHM as the Cauchy distribution in (a)? Add a graph of the Gaussian with this σ and expectation equal to zero to your graph in (a), and comment.

c. For the Cauchy distribution, calculate $\mathcal{P}(|x| > \text{FWHM}/2)$.

d. Repeat (c) for the Gaussian distribution you found in (b). You will need to do this calculation numerically, either by integrating the Gaussian distribution or by computing the "error function."

e. Repeat (c) and (d) for more extreme events, with $|x| > \frac{3}{2}\text{FWHM}$.

f. ⟦T_2⟧ Repeat (a)–(e) but use the interquartile range instead of FWHM (see Section 5.2.4′).

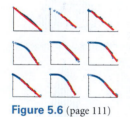

Figure 5.6 (page 111)

5.11   **Gaussian versus power law**

To understand the graphs in Figure 5.6 better, consider the following three pdfs:

a. $\wp_1(x)$: For $x > 0$, this pdf is like the Gaussian distribution centered on 0 with $\sigma = 1$, but for $x < 0$ it's zero.

b. $\wp_2(x)$: For $x > 0$, this pdf is like the Cauchy distribution centered on zero with $\eta = 1$, but for $x < 0$ it's zero.

c. $\wp_3(x)$: This pdf equals $(0.2)x^{-2}$ for $x > 0.2$ and is zero elsewhere.

For each of these distributions, find the complementary cumulative distribution, display them all on a single set of log-log axes, and compare with Figure 5.6.

5.12   ⟦T_2⟧ **Convolution of Gaussians**

Section 4.3.5 (page 79) described an unusual property of the Poisson distributions: The convolution of any two is once again Poisson. In this problem, you will establish a related result, in the domain of continuous distributions.

Consider a system with two independent random variables, x and y. Each is drawn from a Gaussian distribution, with expectations μ_x and μ_y, respectively, and variances both equal to σ^2. The new variable $z = x + y$ will have expectation $\mu_x + \mu_y$ and variance $2\sigma^2$.

a. Compute the convolution integral of the two distributions to show that the pdf $\wp_z$ is in fact precisely the Gaussian with the stated properties. [*Note:* This result is implicit in the much more difficult central limit theorem, which roughly states that a distribution becomes "more Gaussian" when convolved with itself.]

b. Try the case where σ_x and σ_y are not equal.

5.13 T_2 Convolution of Cauchy

a. Consider the Cauchy distribution with $\eta = 1$ and $\mu = 0$. Find the convolution of this distribution with itself.

b. What is the qualitative form of the convolution of this distribution with itself 2^P times? Comment in the light of the central limit theorem.

5.14 T_2 Binomial to Gaussian (analytic)

This problem pursues the idea of Section 5.3.1, that the Gaussian distribution is a particular limit of the Binomial distribution.[29] You'll need a mathematical result known as **Stirling's formula**, which states that, for large M,

$$\ln(M!) \xrightarrow[M \to \infty]{} (M + \tfrac{1}{2}) \ln M - M + \frac{1}{2} \ln(2\pi) + \cdots . \qquad (5.22)$$

The dots represent a correction that gets small as M gets large.

a. Instead of a formal proof of Equation 5.22, just try it out: Graph each side of Equation 5.22 for $M = 1$ to 30 on a single set of axes. Also graph the difference of the two sides, to compare them.

b. Following the main text, let $\mu_\ell = M\xi$ and $s = \sqrt{M\xi(1 - \xi)}$, and define $y = (\ell - \mu_\ell)/s$. Then $\langle y \rangle = 0$ and var $y = 1$, and the successive values of y differ by $\Delta y = 1/s$. Next consider the function

$$F(y; M, \xi) = (1/\Delta y) \mathcal{P}_{\text{binom}}(\ell; M, \xi), \text{ where } \ell = (sy + \mu_\ell).$$

Show that, in the limit where $M \to \infty$ holding fixed y and ξ, F defines a pdf. [*Hint:* s and ℓ also tend to infinity in this limit.]

c. Use Stirling's formula to evaluate the limit. Comment on how your answer depends on ξ.

5.15 T_2 Wild ride

Obtain Dataset 7. The dataset includes an array representing many observations of a quantity x. Let $u_n = x_{n+1}/x_n$ be the fractional change between successive observations. It fluctuates; we would like to learn something about its probability density function. In this problem, neglect the possibility of correlations between different observations of u (the "random-walk" assumption).

a. Let $y = \ln u$; compute it from the data, and put it in an array. Plot a histogram of the distribution of y.

b. Find a Gaussian distribution that looks like the one in (a). That is, consider functions of the form

$$N_{\text{tot}} \frac{\Delta y}{\sigma \sqrt{2\pi}} e^{-(y - \mu_y)^2/(2\sigma^2)},$$

---

[29] The limit studied is different from the one in Section 4.3.2 because we hold ξ, not $M\xi$, fixed as $M \to \infty$.

for some numbers μ_y and σ, where Δy is the range corresponding to a bar on your histogram. Explain why we need the factors $N_{tot}\Delta y$ in the above formula. Overlay a graph of this function on your histogram, and repeat with different values of μ_y and σ until the two graphs seem to agree.

c. It can be revealing to present your result in a different way. Plot the logarithm of the bin counts, and overlay the logarithm of the above function, using the same values of σ and μ_y that you found in (a). Is your best fit really doing a good job? (Problem 6.10 will discuss how to make such statements precise; for now just make a qualitative observation.)

d. Now explore a different family of distributions, analogous to Equation 5.9 (page 101) but with exponents larger than 2 in the denominator. Repeat steps (a)–(c) using distributions from this family. Can you do a better job modeling the data this way?

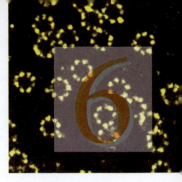

Model Selection and Parameter Estimation

Science is a way of trying not to fool yourself. The first principle is that you must not fool yourself, and you are the easiest person to fool.
—Richard Feynman

6.1 Signpost

Experimental data generally show some variation between nominally identical trials. For example, our instruments have limited precision; the system observed is subject to random thermal motions;[1] and generally each trial isn't really identical, because of some internal state that we cannot observe or control.

Earlier chapters have given examples of situations in which a physical model predicts not only the average value of some experimental observable taken over many trials, but even its full probability distribution. Such a detailed prediction is more falsifiable than just the average, so if it's confirmed, then we've found strong evidence for the model. But what exactly is required to "confirm" a model? We can just look at a graph of prediction versus experiment. But our experience with the Luria-Delbrück experiment (Figure 4.8) makes it clear that just looking is not enough: In panel (a) of that figure, it's hardly obvious which of two competing models is preferred. It is true that panel (b) seems to falsify one model, but the only difference between the panels is the style of the graph! Can't we find a more objective way to evaluate a model?

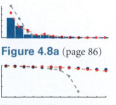

Figure 4.8a (page 86)

Figure 4.8b (page 86)

Moreover, each of the two models in Figure 4.8 was really a *family* of models, depending on a parameter. How do we find the "right" value of the parameter? So far, we have been content to adjust it until the model's predictions "look like" the data, or to observe that no

---

[1] $\boxed{T_2}$ Figure 3.2 shows randomness with a different origin: quantum physics.

suitable value exists. Again, these are subjective judgments. And suppose that one value of the parameter makes a prediction that's better on one region of the data, while another value succeeds best on a different region. Which value is better overall? This chapter will outline a generally useful method for answering questions like these.

The Focus Question is

Biological question: How can we see individual molecular motor steps when light microscopes blur everything smaller than about two hundred nanometers?

Physical idea: The *location* of a single spot can be measured to great accuracy, if we get enough photons.

6.2 Maximum Likelihood

6.2.1 How good is your model?

Up to this point, we have mainly discussed randomness from a viewpoint in which we think we know a system's underlying mechanism and wish to predict what sort of outcomes to expect. For example, we may know that a deck of cards contains 52 cards with particular markings, and model a good shuffle as one that disorganizes the cards, leaving no discernible relevant structure; then we can ask about certain specified outcomes when cards are drawn from the shuffled deck (such as being dealt a "full house"). This sort of reasoning is generally called "probability." Occasionally in science, too, we may have good reason to believe that we know probabilities a priori, as when meiosis "deals" a randomly chosen copy of a gene to a germ cell.

But often in science, we face the opposite problem: We have already measured an outcome (or many), but we don't know the underlying mechanism that generated it. Reasoning backward from the data to the mechanism is called **statistics** or **inference**. To lay out some of the issues, let's imagine two fictitious scientists discussing the problem. Their debate is a caricature of many real-world arguments that have taken place over the last two hundred years. Even today, both of their viewpoints have strong adherents in every field of natural and social science, so we need to have some feeling for each viewpoint.

Nick says: I have a candidate physical model (or several) to describe a phenomenon. Each model may also depend on an unknown parameter α. I obtained some experimental data, and now I want to choose between the models, or between different parameter values in one family. So I want to find $\wp(\text{model}_\alpha \mid \text{data})$ and find the model or parameter value that maximizes this quantity.[2]

Nora replies: But the probability of a model is meaningless, because it doesn't correspond to any replicable experiment. The mutation probability α of a particular strain of bacteria under particular conditions is a fixed number. We don't have a large collection of universes, each with a different value of α. We have *one* Universe.[3]

What's meaningful is $\wp(\text{data} \mid \text{model}_\alpha)$. It answers the question, "If we momentarily assume that we know the true model, how likely is it that the data that we did observe *would have been* observed?" If it's unacceptably low, then we should reject the model. If we can reject all but one reasonable model, then that one has the best chance of being right.

---

[2]In this section, we use $\wp$ to denote either a discrete distribution or a pdf, as appropriate to the problem being studied.

[3]Or at least we *live in* just one universe.

Similarly, although we can't nail down α precisely, nevertheless we can reject all values of α outside some range. That range is our best statement of the value of α.

Nick: When you said "reject all but one model," you didn't specify the field of "all" models. Surely, once you have got the data, you can always construct a highly contrived model that predicts *exactly those data*, and so will always win, despite having no foundation! Presumably you'd say that the contrived model is not "reasonable," but how do you make that precise?

    I'm sorry, but I really do want $\wp(\text{model}_\alpha | \text{data})$, which is different from the quantity that you proposed. If, as you claim, probability theory can't answer such practical questions, then I'll need to extend it.

$\boxed{T_2}$ *The danger that Nick is pointing out (constructing highly contrived models) is called overfitting; see Section 1.2.6 and Section 6.2.1′ (page 142).*

6.2.2 Decisions in an uncertain world

Nick has suggested that probability theory, as presented in earlier chapters, needs to be extended. We entertain many hypotheses about the world, and we wish to assign them various levels of probability, even though they may not correspond to the outcomes of any replicable random system.

    In a fair game of chance, we know a priori the sample space and the probabilities of the outcomes, and can then use the rules listed in Chapter 3 to work out probabilities for more complex events. Even if we suspect the game is not fair, we may still be able to observe someone playing for a long enough time to estimate the underlying probabilities empirically, by using the definition of probability via frequency:

$$\mathcal{P}(\mathsf{E}) = \lim_{N_{\text{tot}} \to \infty} N_{\mathsf{E}} / N_{\text{tot}}, \tag{3.3}$$

or its continuous analog.[4]

    Much more often, however, we find ourselves in a less favorable position. Imagine walking on a street. As you approach an intersection, you see a car approaching the cross street. To decide what to do, you would like to know that car's current speed and also predict its speed throughout the time between now and when you enter the intersection. And it's not enough to know the most likely thing that the car will do—you also want to know how likely it is that the car will suddenly speed up. In short, you need the *probability distribution* of the car's current and future speed, and you need to obtain it in limited time with limited information of limited relevance. Nor have you met this particular driver in this exact situation thousands of times in the past and cataloged the outcomes. Equation 3.3 is of little use in this situation. Even the word "randomness" doesn't really apply—that word implies replicable measurements. You are simply "uncertain."

    The mouse threatened by the cat faces a similar challenge, and so do all of us as we make countless daily decisions. We do not know a priori the probabilities of hypotheses, nor even the sample space of all possible outcomes. Nevertheless, each of us constantly estimates *our degree of belief* in various propositions, assigning each one a value near 0 if we are sure it's false, near 1 if we are sure that it's true, and otherwise something in between. We also constantly *update* our degree of belief in every important proposition as new information

---

[4]See Equation 5.1 (page 98).

arrives. Can this "unconscious inference" be made systematic, even mathematical? Moreover, sometimes we even actively seek those kinds of new data that will best test a proposition, that is, data that would potentially make the biggest change in our degree of belief once we obtain them. Can we systematize this process?

Scientific research often brings up the same situation: We may strongly believe some set of propositions about the world, yet be in doubt about others. We perform some experiment, obtaining a finite amount of new data, then ask, "How does this evidence alter my degree of belief in the doubted propositions?"

6.2.3 The Bayes formula gives a consistent approach to updating our degree of belief in the light of new data

To construct an extended concept of probability along these lines, let's begin by asking what properties we want such a construction to have. Denoting a proposition by E, then we want $\mathcal{P}(\mathsf{E})$ to be given by the usual expression in the case of a replicable random system (Equation 3.3). We also know that, in this case, probability values automatically obey the negation, addition, and product rules (Equations 3.7, 3.8, and 3.11). But those three rules make no direct reference to repeated trials. We can define a probability system as *any* scheme that assigns numbers to propositions and that obeys these three rules—regardless of whether the numerical values have been obtained as frequencies. Then the Bayes formula also follows (it's a consequence of the product rule).

In this framework, we can answer Nick's original question as follows: We consider each possible model as a proposition. We wish to quantify our degree of belief in each proposition, given some data. If we start with an initial estimate for $\wp(\mathsf{model}_\alpha)$, then obtain some relevant experimental data, we can update the initial probability estimate by using the Bayes formula,[5] which in this case says

$$\wp(\mathsf{model}_\alpha \mid \mathsf{data}) = \wp(\mathsf{data} \mid \mathsf{model}_\alpha)\wp(\mathsf{model}_\alpha)/\wp(\mathsf{data}). \tag{6.1}$$

This formula's usefulness lies in the fact that often it's easy to find the consequences of a known model (that is, to compute $\wp(\mathsf{data} \mid \mathsf{model}_\alpha)$), as we have done throughout Chapters 3–5.

As in Chapter 3, we'll call the updated probability estimate the posterior, the first factor on the right the likelihood, and the next factor in the numerator the prior.[6]

Nick continues: The extended notion of probability just proposed is at least mathematically consistent. For example, suppose that I get some data, I revise my probability estimate, then you tell me additional data′, and I revise again. My latest revised estimate should then be the *same* as if you had told me your data′ first and then I accounted for mine—and it is.[7]

Nora: But science is supposed to be objective! It's unacceptable that your formula should depend on your initial estimate of the probability that model is true. Why should *I* care about *your* subjective estimates?

---

[5] See Equation 3.17 (page 52).
[6] See Section 3.4.4. These traditional terms may give the mistaken impression that *time* is involved in some essential way. On the contrary, "prior" and "posterior" refer only to our own state of knowledge with and without taking account of some data.
[7] You'll confirm Nick's claim in Problem 6.8.

Nick: Well, everyone has prior estimates, whether they admit them or not. In fact, your use of $\wp(\text{data} \mid \text{model}_\alpha)$ as a stand-in for the full Equation 6.1 tacitly assumes a particular prior distribution, namely, the **uniform prior**, or $\wp(\text{model}_\alpha) = \text{constant}$. This sounds nice and unbiased, but really it isn't: If we re-express the model in terms of a different parameter (for example, $\beta = 1/\alpha$), then the probability density function for α must transform.[8] In terms of the new parameter β, it generally will no longer appear uniform!

   Also, if a parameter α has dimensions, then my factors $\wp(\text{model}_\alpha)/\wp(\text{data})$, which you omit, are needed to give my formula Equation 6.1 the required dimensions (inverse to those of α).

Nora: Actually, I use a cumulative quantity, the likelihood of obtaining the observed data *or anything more extreme than that* in the candidate model. If this dimensionless quantity is too small, then I reject the model.

Nick: But what does "more extreme" mean? It sounds as though more tacit assumptions are slipping in. And why are you even discussing values of the data that were *not* observed? The observed data are the only thing we *do* know.

6.2.4 A pragmatic approach to likelihood

Clearly, scientific inference is a subtle business. The best approach may depend on the situation. But often we can adopt a pragmatic stance that acknowledges both Nick's and Nora's points. First, we have a lot of general knowledge about the world, including, for example, the values of physical constants (viscosity of water, free energy in an ATP molecule, and so on), as well as specific knowledge that we think is relevant to the system under study (kinds of molecular actors known to be present, any known interactions between those molecules, and so on). From this knowledge, we can put together one or more physical models and attribute some prior belief to each of them. This is the step that Nora calls restricting to "reasonable" models; it is also the step that eliminates the "contrived" models that Nick worries about. Generally, the models contain some unknown parameters; again, we may have prior knowledge specifying ranges of parameter values that seem reasonable.

   Second, instead of attempting an absolute statement that any model is "confirmed," we can limit our ambition to *comparing* the posterior probabilities of the set of models identified in the first step to each other. Because they all share the common factor $1/\wp(\text{data})$ (see Equation 6.1), we needn't evaluate that factor when deciding which model is the most probable. All we need for comparison are the **posterior ratios** for all the models under consideration, that is,

$$\frac{\wp(\text{model} \mid \text{data})}{\wp(\text{model}' \mid \text{data})} = \frac{\wp(\text{data} \mid \text{model})}{\wp(\text{data} \mid \text{model}')} \times \frac{\wp(\text{model})}{\wp(\text{model}')}. \tag{6.2}$$

The units of data, which worried Nick, cancel out of this dimensionless expression. We say, "The posterior ratio is the product of the likelihood ratio and the prior ratio."

   If we have good estimates for the priors, we can now evaluate Equation 6.2 and compare the models. Even if we don't have precise priors, however, we may still be able

---

[8]See Section 5.2.5 (page 104).

to proceed, because[9]

> *If the likelihood function $\wp$(data | model) strongly favors one model, or is very* (6.3)
> *sharply peaked near one value of a model's parameter(s), then our choice of prior*
> *doesn't matter much when we compare posterior probabilities.*

Put differently,

- If your data are so compelling, and your underlying model so firmly rooted in independently known facts, that any reader is forced to your conclusion regardless of his own prior estimates, then you have a result.
- If not, then you may need more or better data, because this can lead to a more sharply peaked likelihood function.
- If more and better data still don't give a convincing fit, then it may be time to widen the class of models under consideration. That is, there may be another model, to which you assigned lower prior probability, but whose likelihood turns out to be far better than the ones considered so far. Then this model can have the largest posterior (Equation 6.2), in spite of your initial skepticism.

The last step above can be crucial—it's how we can escape a mistaken prejudice.

Sometimes we find ourselves in a situation where more than one model seems to work about equally well, or where a wide range of parameter values all seem to give good fits to our data. The reasoning in this section then codifies what common sense would also tell us: In such a situation, we need to do an *additional* experiment whose likelihood function *will* discriminate between the models. Because the two experiments are independent, we can combine their results simply by multiplying their respective likelihood ratios and optimizing the product.

This section has argued that, although Nick may be right in principle, often it suffices to compute likelihood ratios when choosing between models. This procedure is aptly named **maximum likelihood estimation**, or "the **MLE** approach." Using an explicit prior function, when one is available, is called "**Bayesian inference**." Most scientists are flexible about the use of priors. Just be sure to be clear and explicit about your method and its assumptions, or to extract this information from whatever article you may be reading.

T_2 *Section 6.2.4′ (page 142) discusses the role of data binning in model selection and parameter estimation, and introduces the concept of "odds."*

6.3 Parameter Estimation

This section will give a concrete example of likelihood maximization in action. Later sections will give more biophysical applications and show how likelihood also underpins other techniques that are sometimes presented piecemeal as recipes. Unifying those approaches helps us to see their interrelations and adapt them to new situations.

Here is a situation emblematic of many faced every day in research. Suppose that a certain strain of laboratory animal is susceptible to a certain cancer: 17% of individuals develop the disease. Now, a test group of 25 animals is given a suspected carcinogen, and 6 of them develop the disease. The quantity 6/25 is larger than 0.17—but is this a significant difference?

---

[9]See Problem 6.2.

A laboratory animal is an extremely complex system. Faced with little information about what every cell is doing, the best we can do is to suppose that each individual is an independent Bernoulli trial and every environmental influence can be summarized by a single number, the parameter ξ. We wish to assess our confidence that the experimental group can be regarded as drawn from the same distribution (the same value of ξ) as the control group.

6.3.1 Intuition

More generally, a model usually has one or more parameters whose values are to be extracted from data. As in Section 6.2.4, we will not attempt to "confirm" or "refute" any parameter values. Instead, we will evaluate $\wp(\text{model}_\alpha \mid \text{data})$, a probability distribution in α, and ask what range of α values *contains most of the posterior probability*. To see how this works in practice, let's study a situation equivalent to the carcinogen example, but for which we already have some intuition.

Suppose that we flip a coin M times and obtain ℓ heads and $(M - \ell)$ tails. We'd like to know what this information can tell us about whether the coin is fair. That is, we've got a model for this random system (it's a Bernoulli trial), but the model has an unknown parameter (the fairness parameter ξ), and we'd like to know whether $\xi \overset{?}{=} 1/2$. We'll consider three situations:

a. We observed $\ell = 6$ heads out of $M = 10$ flips.
b. We observed $\ell = 60$ heads out of $M = 100$ flips.
c. We observed $\ell = 600$ heads out of $M = 1000$ flips.

Intuitively, in situation **a** we could not make much of a case that the coin is unfair: Fair coins often do give this outcome. But we suspect that in the second and third cases we could make a much stronger claim that we are observing a Bernoulli trial with $\xi \neq 1/2$. We'd like to justify that intuition, using ideas from Section 6.2.4.

6.3.2 The maximally likely value for a model parameter can be computed on the basis of a finite dataset

The preceding section proposed a family of physically reasonable models for the coin flip: model_ξ is the proposition that each flip is an independent Bernoulli trial with probability ξ to get heads. If we have no other prior knowledge of ξ, then we use the Uniform distribution on the allowed range from $\xi = 0$ to 1 as our prior.

Before we do our experiment (that is, make M flips), both ξ and the actual number ℓ of heads are unknown. After the experiment, we have some data, in this case the observed value of ℓ. Because ℓ and ξ are not independent, we can learn something about ξ from the observed ℓ. To realize this program, we compute the posterior distribution and maximize it over ξ, obtaining our best estimate of the parameter from the data.

Equation 6.1 gives the posterior distribution $\wp(\text{model}_\xi \mid \ell)$ as the product of the prior, $\wp(\text{model}_\xi)$, times the likelihood, $\mathcal{P}(\ell \mid \text{model}_\xi)$, divided by $\mathcal{P}(\ell)$. We want to know the value of ξ that maximizes the posterior, or at least some range of values that are reasonably probable. When we do the maximization, we *hold the observed data fixed*. The experimental data (ℓ) are frozen there in our lab notebook while we entertain various hypotheses about the value of ξ. So the factor $\mathcal{P}(\ell)$, which depends only on ℓ, is a constant for our purposes; it doesn't affect the maximization.

We are assuming a uniform prior, so $\wp(\text{model}_\xi)$ also doesn't depend on ξ, and hence does not affect the maximization problem. Lumping together the constants into one symbol A, Equation 6.1 (page 126) becomes

$$\wp(\text{model}_\xi \mid \ell) = A\mathcal{P}(\ell \mid \text{model}_\xi). \tag{6.4}$$

The hypothesis model_ξ states that the distribution of outcomes ℓ is Binomial: $\mathcal{P}(\ell \mid \text{model}_\xi) = \mathcal{P}_{\text{binom}}(\ell; \xi, M) = \frac{M!}{\ell!(M-\ell)!}\xi^\ell(1-\xi)^{M-\ell}$. Section 6.2.3 defined the likelihood as this same formula. It's a bit messy, but the factorials are not interesting in this context, because they, too, are independent of ξ. We can just lump them together with A and call the result some other constant A':

$$\wp(\text{model}_\xi \mid \ell) = A' \times \xi^\ell(1-\xi)^{M-\ell}. \tag{6.5}$$

We wish to maximize $\wp(\text{model}_\xi \mid \ell)$, holding ℓ fixed, to find our best estimate for ξ. Equivalently, we can maximize the logarithm:

$$0 = \frac{d}{d\xi}\ln\wp(\text{model}_\xi \mid \ell) = \frac{d}{d\xi}\left(\ell\ln\xi + (M-\ell)\ln(1-\xi)\right) = \frac{\ell}{\xi} - \frac{M-\ell}{1-\xi}.$$

Solving this equation shows[10] that the maximum is at $\xi_* = \ell/M$. We conclude that, in all three of the scenarios stated above, our best estimate for the fairness parameter is $\xi_* = 6/10 = 60/100 = 600/1000 = 60\%$.

That was a lot of work for a rather obvious conclusion! But our calculation can easily be extended to cover cases where we have some more detailed prior knowledge, for example, if a person we trust tells us that the coin is fair; in that case, the likelihood framework tells us to use a prior with a maximum near $\xi = 1/2$, and we can readily obtain a different result for our best estimate of ξ, which accounts for both the prior and the experimental data.

We still haven't answered our original question, which was, "Is this coin fair?" But in the framework developed here, the next step, which will answer that question in the next section, is straightforward.

$\boxed{T_2}$ *Section 6.3.2' (page 143) gives more details about the role of idealized distribution functions in our calculations and discusses an improved estimator for ξ.*

6.3.3 The credible interval expresses a range of parameter values consistent with the available data

How *sure* are we that we found the true value of ξ? That is, what is the *range* of pretty-likely values for ξ? Section 6.2.4 suggested that we address such questions by asking, "How sharply peaked is the posterior about its maximum?"

The posterior distribution $\wp(\text{model}_\xi \mid \ell)$ is a probability density function for ξ. So we can find the prefactor A' in Equation 6.5 by requiring that $\int_0^1 d\xi\, \wp(\text{model}_\xi \mid \ell) = 1$. The integral is not hard to compute by hand for small M, or with a computer for larger values.[11] Figure 6.1 shows the result for the three scenarios. We can interpret the graphs as

---

[10]Some authors express this conclusion by saying that the sample mean, in this case ℓ/M, is a good "estimator" for ξ, if we know that the data are independent samples from a Bernoulli-trial distribution.

[11] $\boxed{T_2}$ Or you can notice that it's a "beta function."

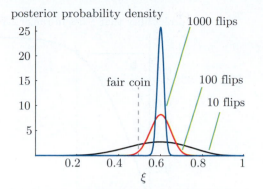

Figure 6.1 [Mathematical functions.] **Likelihood analysis of a Bernoulli random variable.** The curves show the posterior probability distributions for the coin fairness parameter ξ; see Equation 6.5. *Black* is $M = 10$ flips, of which $\ell = 6$ were heads; *red* is 100 flips, of which 60 were heads; *blue* is 1000 flips, of which 600 were heads.

follows: If you get 6 out of 10 heads, it's quite possible that the coin was actually fair (true value of ξ is $1/2$). More precisely, most of the area under the black curve is located in the wide range $0.4 < \xi < 0.8$. However, if you get 60 out of 100 heads (red curve), then most of the probability lies in the range 0.55–0.65, which does not include the value $1/2$. So in that case, it's not very probable that the coin is fair (true value of ξ is $1/2$). For 600 out of 1000, the fair-coin hypothesis is essentially ruled out.

What does "ruled out" mean quantitatively? To get specific, Figure 6.2a shows results from actually taking a bent coin and flipping it many times. In this graph, empirical frequencies of getting various values of ℓ/M, for $M = 10$ flips, are shown as gray bars. The histogram does look a bit lopsided, but it may not be obvious that this is necessarily an unfair coin. Maybe we're being fooled by a statistical fluctuation. The green bars show the frequencies of various outcomes when we reanalyze the same data as fewer batches, of $M = 100$ flips each. This is a narrower distribution, and it seems clearer that it's not centered on $\xi = 1/2$, but with a total of just a few trials we can hardly trust the counts in each bin. We need a better method than this.

To compute the maximally likely value of ξ, we use the total number of heads, which was $\ell = 347$ out of 800 flips, obtaining $\xi_* = 347/800 \approx 0.43$. Stars on the graph show predicted frequencies corresponding to the Binomial distribution with this ξ and $M = 10$. Indeed, there's a resemblance, but again: How sure are we that the data *aren't* compatible with a fair-coin distribution?

We can answer that question by computing the full posterior distribution for ξ, not just finding its peak ξ_*, and then finding the range of ξ values around ξ_* that accounts for, say, 90% of the area under the curve. Figure 6.2b shows the result of this calculation. For each value of the interval width 2Δ, the graph shows the area under the posterior distribution $\wp(\text{model}_\xi \mid \ell = 347, M = 800)$ between $\xi_* - \Delta$ and $\xi_* + \Delta$. As Δ gets large, this area approaches 1 (because $\wp$ is normalized). Reading the graph shows that 90% of the area lies in the region between $\xi_* \pm 0.027$, that is, between 0.407 and 0.461. This range is also called the 90% **credible interval** on the inferred value of the parameter ξ. In this case, it does not include the value $\xi = 1/2$, so it's unlikely, given these data, that the coin is actually fair.

For comparison, Problem 6.5 discusses some data taken with an ordinary US penny.

$\boxed{T_2}$ *Section 6.3.3′ (page 144) discusses another credible interval, that for the expectation of a*

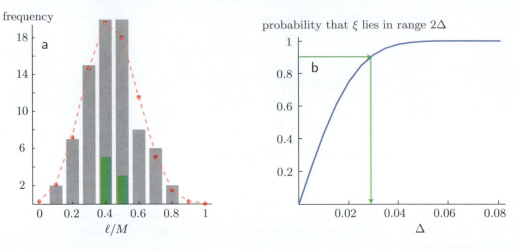

Figure 6.2 Determination of a credible interval. (a) [Experimental data with fit.] *Gray bars:* Observed frequencies of various values of ℓ/M, over 80 trials in each of which a bent coin was flipped 10 times. The frequencies peak at a value of ℓ smaller than 5, suggesting that ξ may be less than 1/2. *Red symbols:* Binomial distribution with ξ equal to the maximally likely value, multiplied by 80 to give a prediction of frequency. *Green bars:* The same 800 coin flips were reinterpreted as 8 "trials" of 100 flips each. This estimated distribution is much narrower than the one for 10-flip trials, and again suggests that the coin is not fair. However, in this presentation, we have very few "trials." (b) [Mathematical function.] The integral of the posterior distribution over a range of values surrounding $\xi_* = 0.43$, that is, the probability that ξ lies within that range. The *arrows* illustrate that, given the data in (a), the probability that the true value of ξ is within 0.027 of the estimate ξ_* equals 90%. The range from $0.43 - 0.027$ to $0.43 + 0.027$ does not include the fair-coin hypothesis $\xi = 1/2$, so that hypothesis is ruled out at the 90% level.

variable known to be Gaussian distributed. It also discusses related ideas from classical statistics and the case of parameters with asymmetric credible intervals.

6.3.4 Summary

The preceding sections were rather abstract, so we should pause to summarize them before proceeding. To model some partially random experimental data, we

1. Choose one or more models and attribute some prior probability to each. If we don't have grounds for assigning a numerical prior, we take it to be Uniform on a set of physically reasonable models.

2. Compute the likelihood ratio of the models to be compared, multiplying it by the ratio of priors, if known.

If the model(s) to be assessed contain one or more parameters, we augment this procedure:

3. For each model family under consideration, find the posterior probability distribution for the parameters.

4. Find the maximally likely value of the parameter(s).

5. Select a criterion, such as 90%, and find a range about the maximally likely value that encloses that fraction of the total posterior probability.

6. If more than one model is being considered, first marginalize each one's posterior probability over the parameter(s),[12] then compare the resulting total probabilities.

---

[12]See Section 3.4.2.

6.4 Biological Applications

6.4.1 Likelihood analysis of the Luria-Delbrück experiment

This chapter's Signpost pointed out a gap in our understanding of the Luria-Delbrück experiment: In Figure 4.8a, the two models being compared appear about equally successful at explaining the data. But when we re-graphed the data on a log scale, the resulting plot made it clear that one model was untenable (Figure 4.8b). How can the *presentation* of data affect its *meaning*?

Likelihood analysis provides us with a more objective criterion. In Problem 6.9, you'll apply the reasoning in the preceding section to the Luria-Delbrück data, obtaining an estimated value for the likelihood ratio between two models for the appearance of resistance in bacteria. It is true that one graph makes it easier to see why you'll get the result, but the result stands on its own: Even if we initially thought that the "Lamarckian" model was, say, 1000 times more probable than the "Darwinian" one, that prior is overwhelmed by the huge likelihood ratio favoring the latter model.

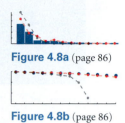

Figure 4.8a (page 86)

Figure 4.8b (page 86)

6.4.2 Superresolution microscopy

6.4.2.1 On seeing

Many advances in science have involved seeing what was previously invisible. For example, the development of lenses led to microscopes, but microscope imaging was long limited by the available staining techniques—without appropriate staining agents, specific to each type of organelle, little could be seen.

Another, less obvious, problem with imaging concerns *not* seeing the riotous confusion of objects that do not interest us at the moment. Recently many forms of fluorescent molecules have been developed to address this problem. A fluorescent molecule ("fluorophore"[13]) absorbs light of one color, then reemits light of a different color. **Fluorescence microscopy** involves attaching a fluorophore specifically to the molecular actor of interest and illuminating the sample with just the specific wavelength of light that excites that fluorophore. The resulting image is passed through a second color filter, which blocks all light except that with the wavelengths given off by the fluorophore. This approach removes many distracting details, allowing us to see only what we wish to see in the sample.

6.4.2.2 Fluorescence imaging at one nanometer accuracy

A third limitation of microscopy involves the fact that most molecular actors of interest are far smaller than the wavelength of visible light. More precisely, a subwavelength object, like a single macromolecule, appears as a blur, indistinguishable from an object a few hundred nanometers in diameter. Two such objects, if they are too close to each other, will appear fused—we say that a light microscope's **resolution** is at best a few hundred nanometers. Many clever approaches have been discovered to image below this **diffraction limit**, but each has particular disadvantages. For example, the electron microscope damages whatever it views, so that we do not see molecular machines actively performing their normal cellular tasks. X-ray crystallography requires that a single molecule of interest be removed from its context, purified, and crystallized. Scanning-probe and near-field optical microscopy generally require a probe in physical contact, or nearly so, with the object being studied; this restriction eliminates the possibility to see cellular devices in situ. How can we image

---

[13] More precisely, a fluorophore may also be a *part* of a molecule, that is, a functional group.

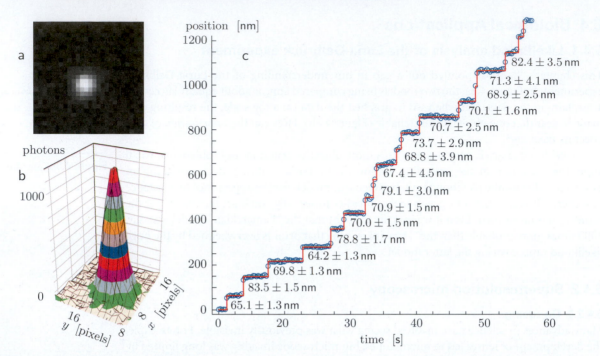

a

photons

b

1000

0

16
y [pixels]   8

16
x [pixels]   8

position [nm]

1200

1000

800

600

400

200

0

c

82.4 ± 3.5 nm
71.3 ± 4.1 nm
68.9 ± 2.5 nm
70.1 ± 1.6 nm
70.7 ± 2.5 nm
73.7 ± 2.9 nm
68.8 ± 3.9 nm
67.4 ± 4.5 nm
79.1 ± 3.0 nm
70.9 ± 1.5 nm
70.0 ± 1.5 nm
78.8 ± 1.7 nm
64.2 ± 1.3 nm
69.8 ± 1.3 nm
83.5 ± 1.5 nm
65.1 ± 1.3 nm

0   10   20   30   40   50   60

time [s]

Figure 6.3 [Experimental data.] **FIONA imaging.** (a) One frame from a video micrograph of the movement of a single fluorescent dye attached to the molecular motor protein myosin-V. Each camera pixel represents 86 nm in the system, so discrete, 74 nm steps are hard to discern in the video (see Media 8). (b) Another representation of a single video frame. Here the number of light blips collected in each pixel is represented as height. The text argues that the center of this distribution can be determined to accuracy much better than the value suggested by its spread. (c) The procedure in the text was applied to each frame of a video. A typical trace reveals a sequence of ≈ 74 nm steps. The horizontal axis is time; the vertical axis is position projected onto the line of average motion. Thus, the steps appear as vertical jumps, and the pauses between steps as horizontal plateaux. [Courtesy Ahmet Yildiz; see also Yildiz et al., 2003.]

cellular components with high spatial resolution, without damaging radiation, with a probe that is far (many thousands of nanometers) from the object under study?

A breakthrough in this impasse began with the realization that, for some problems, we do not need to form a full image. For example, molecular motors are devices that convert "food" (molecules of ATP) into mechanical steps. In order to learn about the stepping mechanism in a particular class of motors, it's enough to label an individual motor with a fluorophore and then watch the motion of that one point of light as the motor steps across a microscope's field of view.[14] For that problem, we don't really need to resolve two nearby points. Instead, we have a single source of light, a fluorophore attached to the motor, and we wish to determine its position accurately enough to detect and measure individual steps.

In concrete detail, there is a camera that contains a two-dimensional grid of light-detecting elements. Each such **pixel** on the grid corresponds to a particular position in the sample. When we illuminate the sample, each pixel in our camera begins to record discrete blips.[15] Our problem is that, even if the source is fixed in space and physically far

---

[14]See Media 7 for a video micrograph of molecular motors at work. It's essential to confirm that the attached fluorophore has not destroyed or modified the function of the motor, prior to drawing any conclusions from such experiments.

[15]See Section 3.2.1 (page 36).

smaller than the region assigned to each pixel, nevertheless many pixels will receive blips: The image is blurred (Figure 6.3a). Increasing the magnification doesn't help: Then each pixel corresponds to a smaller physical region, but the blips are spread over a larger number of pixels, with the same limited spatial resolution as before.

To make progress, notice that although the pixels fire at random, they have a definite probability distribution, called the **point spread function** of the microscope (Figure 6.3b). If we deliberately move the sample by a tiny, known amount, the smeared image changes only by a corresponding shift. So we need to measure the point spread function only *once*; thereafter, we can think of the true location of the fluorophore, (μ_x, μ_y), as parameters describing a family of hypotheses, each with a known likelihood function—the shifted point spread functions. Maximizing the likelihood over these parameters thus tells what we want to know: Where is the source?

Let's simplify by considering only one coordinate, x. Figure 6.3b shows that the point spread function is approximately a Gaussian, with measured variance σ^2 but unknown center μ_x. We would like to find μ_x to an accuracy much better than σ. Suppose that the fluorophore yields M blips before it either moves or stops fluorescing, and that the locations of the camera pixels excited by the blips correspond to a series of apparent positions $x_1, \ldots, x_M$. Then the log-likelihood function is[16] (see Section 6.2.3)

$$\ln \wp(x_1, \ldots, x_M \mid \mu_x) = \sum_{i=1}^{M} \left[-\tfrac{1}{2} \ln(2\pi\sigma^2) - (x_i - \mu_x)^2/(2\sigma^2) \right]. \tag{6.6}$$

We wish to maximize this function over μ_x, holding σ and all the data $\{x_1, \ldots, x_M\}$ fixed.

Your Turn 6A

Show that our best estimate of μ_x is the sample mean of the apparent positions: $\mu_{x*} = \bar{x} = (1/M) \sum x_i$.

That result may not be very surprising, but now we may also ask, "How *good* is this estimate?" Equation 3.23 (page 58) already gave one answer: The sample mean has variance σ^2/M, or standard deviation $\sigma/\sqrt{M}$. Thus, if we collect a few thousand blips, we can get an estimate for the true position of the fluorophore that is significantly better than the width σ of the point spread function. In practice, many fluorophores undergo "photobleaching" (that is, they break) after emitting about one million blips. It is sometimes possible to collect a substantial fraction of these blips, and hence to reduce the corresponding standard deviation to just one nanometer, leading to the name fluorescence imaging at one nanometer accuracy, or FIONA. Yildiz and coauthors collected about 10 000 light blips per video frame, enabling a localization accuracy one hundred times smaller than the width of their point spread function.

We can get a more detailed prediction by reconsidering the problem from the likelihood viewpoint. Rearranging Equation 6.6 gives the log likelihood as

$$\text{const} - \tfrac{1}{2\sigma^2} \sum_i ((x_i)^2 - 2x_i\mu_x + (\mu_x)^2) = \text{const}' - \tfrac{M}{2\sigma^2}(\mu_x)^2 + \tfrac{1}{2\sigma^2}2M\bar{x}\mu_x$$

$$= \text{const}'' - \tfrac{M}{2\sigma^2}(\mu_x - \bar{x})^2. \tag{6.7}$$

---

[16] As Nick pointed out on page 127, the likelihood function has dimensions $\mathbb{L}^M$, so strictly speaking we cannot take its logarithm. However, the offending $\ln \sigma$ terms in this formula all cancel when we compute *ratios* of likelihood functions or, equivalently, the differences of log likelihood.

The constant in the second expression includes the term with the sum of x_i^2. This term does not depend on μ_x, so it is "constant" for the purpose of optimizing over that desired quantity. Exponentiating the third form of Equation 6.7 shows that it is a Gaussian. Its variance equals σ^2/M, agreeing with our earlier result.

Yildiz and coauthors applied this method to the successive positions of the molecular motor myosin-V, obtaining traces like those in Figure 6.3c. Each such "staircase" plot shows the progress of a single motor molecule. The figure shows the motion of a motor that took a long series of rapid steps, of length always near to 74 nm. Between steps, the motor paused for various waiting times. Chapter 7 will study those waiting times in greater detail.

T_2 *Section 6.4.2.2′ (page 146) outlines a more complete version of the analysis in this section, for example, including the effect of background noise and pixelation.*

6.4.2.3 Localization microscopy: PALM/FPALM/STORM

Methods like FIONA are useful for pinpointing the location of a small light source, such as a fluorescent molecule attached to a protein of interest. Thus, Section 6.4.2.2 showed that, if our interest lies in tracking the change in position of one such object from one video frame to the next, FIONA can be a powerful tool. But the method involved the assumption that, in each frame, the light arriving at the microscope came from a *single* point source.

Generally, however, we'd like to get an *image*, that is, a representation of the positions of *many* objects. For example, we may wish to see the various architectural elements in a cell and their spatial relationships. If those objects are widely separated, then the FIONA method may be applied to each one individually: We mark the objects of interest with an appropriate fluorescent tag and model the light distribution as the sum of several point spread functions, each centered on a different unknown location. Unfortunately, the accuracy of this procedure degrades rapidly if the objects we are viewing are spaced more closely than the width of the point spread function. But a modified technique can overcome this limitation.

When we think of a visual image, we normally imagine many objects, all emitting or reflecting light *simultaneously.* Similarly, when a microscopic object of interest is tagged with ordinary fluorophores, they all fluoresce together. But suppose that we have a static scene, tagged by molecules whose fluorescence can be *switched* on and off. We could then

1. Switch "on" only a few widely separated fluorophores, so that their point spread functions are well separated;
2. Localize each such "on" tag by maximizing likelihood (Section 6.4.2.2);
3. Begin building an image by placing dots at each of the inferred tag locations;
4. Extinguish the "on" tags;
5. Switch "on" a different set of tags; and
6. Repeat the process until we have built up a detailed image.

Figure 6.4 represents these steps symbolically.

Implementing the procedure just sketched requires some kind of switchable fluorophore. The key idea is that some molecules don't fluoresce at all, others fluoresce reliably, but yet another class of molecules have different internal states, some fluorescent and others not. Molecules in the third class may spontaneously transition between "on" and "off." Interestingly, however, a few can be triggered to make those transitions by exposure to light—they are "photoactivatable." Both fluorescent proteins and other kinds of organic dyes have been found with this property. When illuminated with light of the appropriate wavelength, each individual fluorescent tag has a fixed probability per unit time to pop into its "on" state, so soon a randomly selected subset are "on." The experimenter then

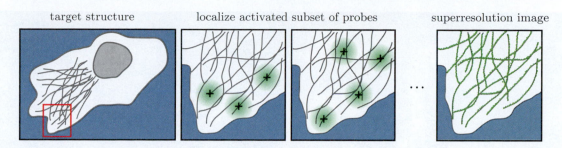

Figure 6.4 [Sketches.] **The idea behind localization microscopy.** A structure of interest is labeled by fluorescent tags, but at any moment, most of the tags are "off" (dark). A few tags are randomly selected and turned "on," and FIONA-type analysis is used to find each one's location to high accuracy, generating scattered dots whose locations are saved (*crosses*). Those tags are turned "off" (or allowed to photobleach), and the procedure is repeated. Finally, the complete image is reconstructed from the dots found in each iteration. [Courtesy Mark Bates, Dept. of NanoBiophotonics, Max Planck Institute for Biophysical Chemistry. See also Bates et al., 2008.]

stops the photoactivation, begins illuminating the sample with the wavelength that induces fluorescence, and performs steps **2–3** given above. Eventually the "on" fluorophores either bleach (lose their fluorescence permanently) or else switch back "off," either spontaneously or by command. Then it's time for another round of photoactivation.

The first variants of this method were called photoactivated localization microscopy (PALM), stochastic optical reconstruction microscopy (STORM), and fluorescence photoactivated localization microscopy (FPALM). All of these methods were similar, so we'll refer to any of them as **localization microscopy**.[17] Later work extended the idea in many ways, for example, allowing imaging in three dimensions, fast imaging of moving objects, imaging in live cells, and independent tagging of different molecular species.[18]

$\boxed{T_2}$ *Section 6.4.2.3' (page 147) describes mechanisms for photoactivation and mentions other superresolution methods.*

6.5 An Extension of Maximum Likelihood Lets Us Infer Functional Relationships from Data

Often we are interested in discovering a relation between some aspect of an experiment x that we control (a "knob on the apparatus," or **independent variable**) and some other aspect y that we measure (a "response" to "turning the knob," or **dependent variable**). A complex experiment may even have multiple variables of each type.

One straightforward approach to this problem is to conduct a series of independent trials with various settings of x, measure the corresponding values of y, and compute their correlation coefficient.[19] This approach, however, is limited to situations where we believe that the quantities have a simple, straight-line relation. Even if we do wish to study a linear model, the correlation coefficient by itself does not answer questions such as the credible interval of values for the slope of the relation.

We could instead imagine holding x fixed to one value x_1, making repeated measurements of y, and finding the sample mean $\bar{y}(x_1)$. Then we fix a different x_2, again measure the distribution of y values, and finally draw a straight line between $(x_1, \bar{y}(x_1))$ and $(x_2, \bar{y}(x_2))$

---

[17] Localization microscopy, in turn, belongs to a larger realm of "superresolution microscopy."
[18] See the front cover of this book.
[19] See Section 3.5.2' (page 60).

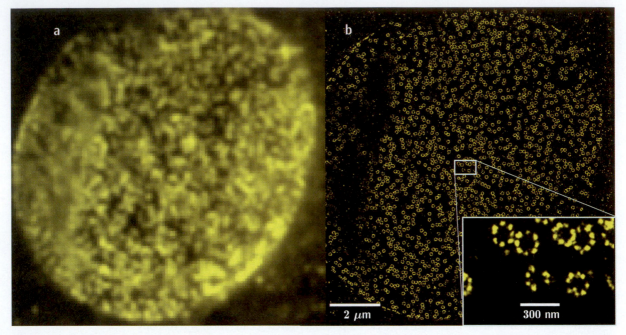

Figure 6.5 [Micrographs.] **STORM imaging.** The images show the nucleus of a kidney cell from the frog *Xenopus laevis*. The nucleus has many openings to the cytoplasm. These "nuclear pore complexes" have an intricate structure; to visualize them, one specific protein (GP210) has been tagged with a fluorophore via immunostaining. (a) Conventional fluorescence microscopy can only resolve spatial features down to about 200 nm, and hence the individual pore complexes are not visible. The coarse pixel grid corresponds to the actual pixels in the camera used to make the image. Magnifying the image further would not improve it; it is unavoidably blurred by diffraction. (b) Superresolution optical microscopy techniques such as STORM allow the same sample, fluorescing with the same wavelength light, to be imaged using the same camera but with higher spatial resolution. For example, in the magnified *inset*, the 8-fold symmetric ring-like structure of the nuclear pore complexes can be clearly seen. [Courtesy Mark Bates, Dept. of NanoBiophotonics, Max Planck Institute for Biophysical Chemistry.]

(Figure 6.6a). We could then estimate our uncertainty for the slope and intercept of that line by using what we found about the uncertainties of the two points anchoring it.

The procedure just described is more informative than computing the correlation coefficient, but we can do still better. For example, making measurements at just two x values won't help us to *evaluate* the hypothesis that x and y actually do have a linear relation. Instead, we should spread our observations across the entire range of x values that interest us. Suppose that when we do this, we observe a generally linear trend in the data with a uniform spread about a straight line (Figure 6.6b). We would like some objective procedure to estimate that line's slope and intercept, and their credible intervals. We'd also like to compare our fit with alternative models.

Suppose that we take some data at each of several typical x values and find that, for each fixed x, the observed y values have a Gaussian distribution about some expectation $\mu_y(x)$, with variance σ^2 that does not depend on x. If we have reason to believe that $\mu_y(x)$ depends on x via a linear function $\mu_y(x) = Ax + B$, then we'd like to know the best values of A and B. Let $\text{model}_{A,B}$ denote the hypothesis that these parameters have particular values, or in other words,

$$\wp(y \mid x, \text{model}_{A,B}) = \wp_{\text{gauss}}(y; Ax + B, \sigma) \quad \text{for unknown parameters } A, B.$$

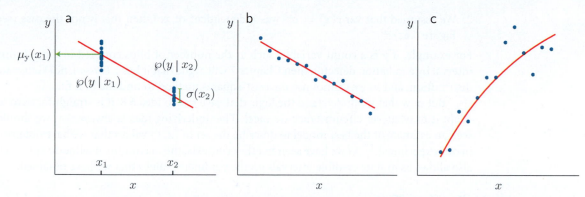

Figure 6.6 [Simulated data, with fits.] **Illustrations of some ideas in data fitting.** (a) Many measurements of a dependent variable y are taken at each of two values of an independent variable x. (b) Many measurements of y are taken, each at a different value of x. (c) As (b), but illustrating the possibilities that the locus of expectations $\bar{y}(x)$ may be a nonlinear function (*curve*) and that the spread of data may depend on the value of x.

The likelihood for M independent measurements is then

$$\wp(y_1, \ldots, y_M \mid x_1, \ldots, x_M, \text{model}_{A,B})$$

$$= \wp_{\text{gauss}}(y_1; Ax_1 + B, \sigma) \times \cdots \times \wp_{\text{gauss}}(y_M; Ax_M + B, \sigma),$$

or

$$\ln \wp(y_1, \ldots, y_M \mid x_1, \ldots, x_M, \text{model}_{A,B})$$

$$= \sum_{i=1}^{M} \left[-\tfrac{1}{2} \ln(2\pi\sigma^2) - (y_i - Ax_i - B)^2/(2\sigma^2) \right].$$

To find the optimal fit, we maximize this quantity over A and B holding the x_i and y_i fixed. We can neglect the first term in square brackets, because it doesn't depend on A or B. In the second term, we can also factor out the overall constant $(2\sigma^2)^{-1}$. Finally, we can drop the overall minus sign and seek the *minimum* of the remaining expression:

> *Under the assumptions stated, the best-fitting line is the one that minimizes the*
> **chi-square statistic** $\sum_{i=1}^{M}(y_i - Ax_i - B)^2/\sigma^2$. $\qquad$ (6.8)

Because we have assumed that every value of x gives y values with the same variance, we can even drop the denominator when minimizing this expression.

Idea 6.8 suggests a more general procedure called **least-squares fitting**. Even if we have a physical model for the experiment that predicts a *non*linear relation $\langle y(x) \rangle = F(x)$ (for example, an exponential, as in Figure 6.6c), we can still use it, simply by substituting $F(x)$ in place of $Ax + B$ in the formula.

Idea 6.8 is easy to implement; in fact, many software packages have this functionality built in. But we must remember that such turnkey solutions make a number of restrictive assumptions that must be checked before the results can be used:

1. We assumed that, for fixed x, the variable y was Gaussian distributed. Many experimental quantities are not distributed in this way.

2. We assumed that var $y(x) = \sigma^2$ was independent of x. Often this is not the case (see Figure 6.6c).[20]

For example, if y is a count variable, such as the number of blips counted in a fixed time interval by a radiation detector, then Chapter 7 will argue that it has a *Poisson*, not Gaussian, distribution, and so we cannot use the least-squares method to find the best fit.

But now that we understand the logic that gave rise to Idea 6.8, it is straightforward to adapt it to whatever circumstance we meet. The underlying idea is always that we should seek an estimate of the best model to describe the set of (x, y) values that we have obtained in our experiment.[21] As we have seen in other contexts, the maximum-likelihood approach also allows us to state credible intervals on the parameter values that we have obtained.

Your Turn 6B

Suppose that each $y(x_i)$ is indeed Gaussian distributed, but with nonuniform variance $\sigma^2(x_i)$, which you have measured. Adapt Idea 6.8 to handle this case.

We have finally arrived at an objective notion of curve fitting. The "looks good" approach used earlier in this book has the disadvantage that the apparent goodness of a fit can depend on stylistic matters, such as whether we used a log or linear scale when plotting. Maximizing likelihood avoids this pitfall.

$\boxed{T_2}$ *Section 6.5′ (page 147) discusses what to do if successive measurements are not independent.*

THE BIG PICTURE

Returning to this chapter's Signpost, we have developed a quantitative method to choose between models based on data—though not to "confirm" a model in an absolute sense. Section 6.3.4 summarizes this approach. When a family of theories contains one or more continuous parameters, then we can also assign a credible interval to their value(s). Here, again, the posterior distribution (perhaps approximated by the likelihood) was helpful; it gives us a more objective guide than just changing parameter values until the fit no longer "looks good."

The method involves computing the likelihood function (or the posterior, if a prior is known); it does not depend on any particular way of presenting the data. Some data points will be more important than others. For example, in the Luria-Delbrück experiment, the outlier points are enormously unlikely in one of the two models we considered. Maximizing likelihood automatically accounts for this effect: The fact that outliers were indeed observed generates a huge likelihood penalty for the "Lamarckian" model. The method also automatically downweights less reliable parts of our data, that is, those with larger $\sigma(x)$.

When data are abundant, then generally the likelihood ratio overwhelms any prior probability estimates we may have had;[22] when data are scarce or difficult to obtain, then the prior becomes more important and must be chosen more carefully.

---

[20]The magnificent word "heteroscedasticity" was invented to describe this situation.
[21]See Problems 6.9 and 7.13.
[22]This is the content of Idea 6.3 (page 128).

KEY FORMULAS

- *Ratio of posterior probabilities:* The posterior ratio describes the relative successes of two models at explaining a single dataset. It equals the likelihood ratio times the prior ratio (Equation 6.2, page 127).
- *Fitting data:* Sometimes we measure response of an experimental system to imposed changes of some independent variable x. The measured response y (the dependent variable) has a probability distribution $\wp(y \mid x)$ that depends on x. The expectation of this distribution is some unknown function $F(x)$. To find which among some family of such functions fits the data best, we maximize the likelihood over the various proposed F's (including a prior if we have one).
  Sometimes it's reasonable to suppose that for each x, the distribution of y is Gaussian about $F(x)$, with variance that is some known function $\sigma(x)^2$. Then the result of maximizing likelihood is the same as that of minimizing the "chi-square statistic"

$$\sum_{\text{observations } i} \frac{(y_i - F(x_i))^2}{\sigma(x_i)^2}$$

over various trial functions F, holding the experimental data (x_i, y_i) fixed. If all of the σ's are equal, then this method reduces to least-squares fitting.

FURTHER READING

Semipopular:
Silver, 2012; Wheelan, 2013.

Intermediate:
Section 6.2 draws on the presentations of many working scientists, including Berendsen, 2011; Bloomfield, 2009; Bolker, 2008; Cowan, 1998; Jaynes & Bretthorst, 2003; Klipp et al., 2009, chapt. 4; Sivia & Skilling, 2006; Woodworth, 2004.
For a sobering look at current practices in the evaluation of statistical data, see Ioannidis, 2005.
Superresolution microscopy: Mertz, 2010.

Technical:
Linden et al., 2014.
FIONA imaging: Selvin et al., 2008; Toprak et al., 2010. Precursors to this method include Bobroff, 1986; Cheezum et al., 2001; Gelles et al., 1988; Lacoste et al., 2000; Ober et al., 2004; Thompson et al., 2002.
FIONA applied to molecular stepping: Yildiz et al., 2003.
Localization microscopy: Betzig et al., 2006; Hess et al., 2006; Lidke et al., 2005; Rust et al., 2006; Sharonov & Hochstrasser, 2006. Reviews, including other superresolution methods: Bates et al., 2008; Bates et al., 2013; Hell, 2007; Hell, 2009; Hinterdorfer & van Oijen, 2009, chapt. 4; Huang et al., 2009; Mortensen et al., 2010; Small & Parthasarathy, 2014. Important precursors include Betzig, 1995 and Dickson et al., 1997.
On fitting a power law from data: Clauset et al., 2009; Hoogenboom et al., 2006; White et al., 2008.
For a fitting algorithm that is as flexible and efficient as least squares, but suitable for Poisson-distributed data, see Laurence & Chromy, 2010.

$\boxed{T_2}$ **Track 2**

6.2.1′ Cross-validation

Nick mentioned a potential pitfall when attempting to infer a model from data (Section 6.2): We may contrive a model tailored to the data we happen to have taken; such a model can then represent the observed data very well, yet still fail to reflect reality. This chapter takes the viewpoint that we only consider models with a reasonable physical foundation based on general knowledge (of physics, chemistry, genetics, and so on), a limited set of actors (such as molecule types), and a limited number of parameters (such as rate constants). We can think of this dictum as a prior on the class of acceptable models. It does not completely eliminate the need to worry about overfitting, however, and so more advanced methods are often important in practice.

For example, the method of **cross-validation** involves segregating part of the data for exploratory analysis, choosing a model based on that subset, then applying it to the remaining data (assumed to be independent of the selected part). If the chosen model is equally successful there, then it is probably not overfit. For more details, see Press and coauthors (2007) and Efron and Gong (1983).

$\boxed{T_2}$ **Track 2**

6.2.4′a Binning data reduces its information content

Experimental data are often "binned"; that is, the observed range of a continuous quantity is divided into finite intervals. Each observed value is classified by noting the bin into which it falls, and counts are collected of the number of instances for each. Binning is useful for generating graphical representations of data, and we have used it often for that purpose. Still, it may seem a mysterious art to know what's the "right" bin size to use. If our bins are too narrow, then each contains so few instances that there is a lot of Poisson noise in the counts; see, for example, Figure 4.9. If they are too wide, then we lose precision and may even miss features present in the dataset. In fact, even the "best" binning scheme always destroys some of the information present in a collection of observations.

Figure 4.9 (page 91)

So it is important to note that *binning is often unnecessary* when carrying out a likelihood test. Each independently measured value has a probability of occurring under the family of models in question and the likelihood is just the product of those values. For example, in Equation 6.6 (page 135) we used the actual observed data $x_1, \ldots, x_N$ and not some bin populations.

Certainly it may happen that we are presented with someone else's data that have already been binned and the original values discarded. In such cases, the best we can do is to compute an approximate likelihood by manufacturing data points from what was given to us. For example, if we know that the bin centered at x_0 contains N_{x_0} observations, we can calculate the likelihood by pretending that N_{x_0} observations were made that all yielded exactly x_0, and so on for the other values. If the bin is wide, it may be slightly better to imagine that each of the N_{x_0} observations fell at a different value of x, uniformly spaced throughout the width of the bin.

You'll need to apply an approach of this sort to Luria and Delbrück's binned data in Problem 6.9.

6.2.4′b Odds

Equation 6.2 expresses the ratio of posterior probabilities in terms of a ratio of priors. Probability ratios are also called **odds**. Odds are most often quoted for a proposition versus its opposite. For example, the statement that, "The odds of a recurrence after one incident of an illness are 3:2" means that

$$\mathcal{P}(\text{recurrence} \mid \text{one incident})/\mathcal{P}(\text{no recurrence} \mid \text{one incident}) = 3/2.$$

To find the corresponding *probability* of recurrence after one incident, solve $\mathcal{P}/(1-\mathcal{P}) = 3/2$ to find $\mathcal{P} = 3/5$.

T_2 Track 2

6.3.2′a The role of idealized distribution functions

The discussion in Section 6.2 may raise a question about the status of the special distributions we have been using, for example, $\mathcal{P}_{\text{pois}}(\ell; \mu)$. On one hand, we have introduced such symbols as exact, explicit mathematical functions, in this case a function of ℓ and μ. But on the other hand, Section 6.2 introduced a considerably looser seeming, more subjective sense to the symbol $\mathcal{P}$ ("Nick's approach to probability"). How can these two very different viewpoints coexist? And how should we think about the parameter appearing after the semicolon in this and the other idealized distributions? The answer appeared implicitly in Section 6.3, but we can make the discussion more general.

In the approach of Section 6.2, events are understood as logical propositions. Still using the Poisson distribution as an example, we are interested in the propositions

$$E = (\text{system is well described by some Poisson distribution})$$

and

$$E_\mu = (\text{system is well described by the Poisson distribution with particular value } \mu).$$

The meaning of proposition E_μ is precisely that the probability distribution of the random variable ℓ, given E_μ, is given by the explicit expression $e^{-\mu}\mu^\ell/(\ell!)$.

Suppose that we observe a single value of ℓ. We want the posterior estimate of the probabilities of various parameter values, given the observed data and the assumption E. The generalized Bayes formula (Equation 3.27, page 60) lets us express this as

$$\wp(E_\mu \mid \ell \text{ and } E) = \mathcal{P}(\ell \mid E_\mu)\wp(E_\mu \mid E)/\mathcal{P}(\ell \mid E). \tag{6.9}$$

The first factor on the right side is the formula $e^{-\mu}\mu^\ell/(\ell!)$. The next factor is a prior on the value of μ. The denominator does not depend on the value of μ.

So the role of the idealized distribution is that it enters Equation 6.9, which tells us how to nail down the parameter values(s) that are best supported by the data, and their credible interval(s). The main text implicitly used this interpretation.

6.3.2'b Improved estimator

If we take seriously the interpretation of the posterior as a probability density, then we may prefer to estimate ξ based on its *expectation,* not its maximally likely value. In the coin-flip example, we can compute this quantity from Equation 6.5 (page 130):

$$\langle \xi \rangle = (\ell + 1)/(M + 2).$$

Although this expression is not quite the same as $\xi_* = \ell/M$, the two do agree for large enough sample size.

$\boxed{T_2}$ **Track 2**

6.3.3'a Credible interval for the expectation of Gaussian-distributed data

Here is another situation that arises in practice. Suppose that a large control group of animals grows to an average size at maturity of 20 cm. A smaller experimental group of 30 animals is similar to the control group in every respect known to be relevant, except that they are fed a dietary supplement. This group had measured sizes at maturity of $L_1, \dots, L_{30}$. Is this group drawn from a distribution with expectation significantly different from that of the control group?

As with other parameter estimation problems, the essence of this one is that, even if the intervention had *no* effect, nevertheless any sample of 30 animals from the original distribution would be unlikely to have sample mean exactly equal to the expectation. We would therefore like to know the *range* of credible values for the experimental group's (unknown) expectation, μ_{ex}, and specifically whether that range includes the control group's value. This, in turn, requires that we find the posterior distribution $\wp(\mu_{ex} \mid L_1, \dots, L_{30})$. To find it, we use the Bayes formula:

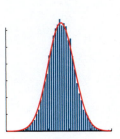

Figure 5.4c (page 108)

$$\wp(\mu_{ex} \mid L_1, \dots, L_{30}) = \int_0^\infty d\sigma \, A\wp_{gauss}(L_1 \mid \mu_{ex}, \sigma) \cdots \wp_{gauss}(L_{30} \mid \mu_{ex}, \sigma)\wp(\mu_{ex}, \sigma).$$

$$(6.10)$$

To obtain Equation 6.10, we used some general knowledge (that body length is a trait that is generally Gaussian distributed; see Figure 5.4c) and the mathematical result that a Gaussian distribution is fully specified by its expectation (μ_{ex}) and variance (σ). If we have no prior knowledge about the values of these parameters, then we can take their prior distribution to be constant. We were not given the value of σ, nor were we asked to determine it, so we have marginalized it to obtain a distribution for μ_{ex} alone. As usual, A is a constant that will be chosen at the end of the calculation to ensure normalization.

The integral appearing in Equation 6.10 looks complicated, until we make the abbreviation $B = \sum_{i=1}^{30}(L_i - \mu_{ex})^2$. Then we have

$$\wp(\mu_{ex} \mid L_1, \dots, L_{30}) = A' \int_0^\infty d\sigma \, (\sigma)^{-30} e^{-B/(2\sigma^2)},$$

where A' is another constant. Changing variables to $s = \sigma/\sqrt{B}$ shows that this expression equals yet another constant times $B^{(1-30)/2}$, or

$$\wp(\mu_{ex} \mid L_1, \ldots, L_{30}) = A'' \Big(\sum_{i=1}^{30} (L_i - \mu_{ex})^2 \Big)^{-29/2}. \qquad (6.11)$$

This distribution has power-law falloff.

We obtained Equation 6.11 by using a very weak ("uninformative") prior. Another, almost equally agnostic, choice we could have made is the "Jeffreys prior," $\wp(\mu_{ex}, \sigma) = 1/\sigma$, which yields a similar result called **Student's t distribution**[23] (Sivia & Skilling, 2006). But in any concrete situation, we can always do better than either of these choices, by using relevant prior information such as the results of related experiments. For example, in the problem as stated it is reasonable to suppose that, although the expectation of L may have been shifted slightly by the intervention, nevertheless the variance, far from being totally unknown, is similar to that of the control group. Then there is no need to marginalize over σ, and the posterior becomes a simple Gaussian, without the long tails present in Equation 6.11.

However we choose the prior, we can then proceed to find a credible interval for μ_{ex} and see whether it contains the control group's value.

6.3.3′b Confidence intervals in classical statistics

The main text advocated using the posterior probability density function as our expression of what we know from some data, in the context of a particular family of models. If we want an abbreviated form of this information, we can summarize the posterior by its most-probable parameter value (or alternatively the expectation of the parameter), along with a credible interval containing some specified fraction of the total probability.

Some statisticians object that evaluating the posterior requires more information than just the class of models and the observed data: It also depends on a choice of a prior, for which we have no general recipe. Naively attempting to use a Uniform prior can bring us problems, including the fact that re-expressing the parameter in a different way may change whether its distribution is Uniform (Nick's point in Section 6.2.3).[24]

An alternative construction is often used, called the "confidence interval," which does not make use of any prior. Instead, one version of this construction requires that we choose an "estimator," or recipe for obtaining an estimate of the parameter in question from experimental data. For example, in our discussion of localization microscopy we believed our data to be independent draws from a Gaussian distribution with known variance but unknown center μ; then a reasonable estimator might be the sample mean of the observed values. The procedure is then[25]

a. Entertain various possible "true" values for the sought parameter μ. For each such value, use the likelihood function to find a range of potential observed data that are "likely," and the corresponding range of values obtained when the estimator is applied to each one. In other words, for each possible value of μ, find the smallest region on the space of possible estimator values that corresponds to 95% of the total probability in $\mathcal{P}(\text{data} \mid \mu)$.

b. The confidence interval is then the set of all values of the parameter μ for which the data that we actually observed are "likely," that is, for which the estimator applied to that dataset give a value within the region found in step **a**.

[23] Not "*the student's*" t distribution, because "Student" was the pen name of W. S. Gosset.

[24] For our coin flip/carcinogen example, the Uniform prior on the range from 0 to 1 that we used is certainly reasonable. But we may have some information about the person who offered us the coin, or prior experience with other chemicals similar to the suspected carcinogen, that may modify the prior.

[25] See Cowan, 1998; Roe, 1992.

Thus, the confidence interval includes each value of μ for which, had it been the true value, then the actual data would not be too atypical.

The construction just outlined is a more precise form of Nora's ideas in Section 6.2.3, and, as Nick pointed out, it requires some prescription to choose the "likely" region of potentially observed data. It also requires that we choose the "right" estimator, without telling us how to do so. And the choice matters, particularly when available data are limited.

Moreover, suppose that our observation consists of just two measurements, x_1 and x_2, and that instead of a Gaussian, we believe that the data have a Cauchy distribution with unknown center. An obvious estimator to use is again the sample mean, $(x_1 + x_2)/2$. But basing our confidence interval on just that statistic discards some information. Surely if x_1 is close to x_2, we should feel more confident in our prediction than if not. The confidence interval method can be refined to get an answer embodying this intuition, but the analysis is rather subtle. The posterior distribution automatically gives the correct result.

The confidence interval method also has difficulties when there are multiple parameters, but we are only interested in the value of one of them. In the method of the posterior distribution, it is straightforward to handle the remaining "nuisance parameters": We just marginalize $\mathcal{P}(\alpha_1, \alpha_2, \ldots \mid \mathsf{data})$ over the uninteresting parameters by integrating them.

6.3.3′c Asymmetric and multivariate credible intervals

The posterior distribution for a model parameter may not be symmetric about its maximum. In such a case, we may get a more meaningful credible interval by finding the *narrowest* range of parameter values that encloses a specified fraction of its probability weight.

Another generalization of the concept of credible interval concerns the case of multiple parameters, that is, a vector $\boldsymbol{\alpha}$. In this case, it is useful to define an ellipsoid around $\boldsymbol{\alpha}_*$ that encloses most of the probability (see Press et al., 2007, chapt. 15).

$\boxed{T_2}$ **Track 2**

6.4.2.2′ More about FIONA

The main text considered randomness in blip arrival locations, but there are other sources of error in FIONA and localization microscopy. For example, the camera used to take images of the scene does not record the exact location at which each blip deposited its energy. Rather, the camera is divided into pixels and only records that a particular pixel received a blip somewhere in its sensitive region.

Suppose that the exact location is x and the pixels are a square grid of side a. Then we can divide x into a discrete part ja, where j is an integer, plus a fractional part Δ lying between $\pm a/2$. If a is larger than the size of the point spread function σ, then Δ is approximately Uniformly distributed throughout its range, and independent of j. The apparent location ja of the blip then has variance equal to the sum $\mathrm{var}(x) + \mathrm{var}(\Delta) = \sigma^2 + a^2/12$. Hence the inferred location when N blips have been recorded, $x_{\mathrm{est}} = \bar{j}a$, has variance $(\sigma^2 + a^2/12)/N$. This formula generalizes the result in Section 6.4.2.2 by accounting for pixelation.

For a more careful and accurate derivation, which also includes other sources of noise such as a uniform background of stray light, see Mortensen and coauthors (2010) and the review by Small and Parthasarathy (2014). The simple formula given in Your Turn 6A (page 135) is not a very good estimator when such realistic details are included, but the general method of likelihood maximization (with a better probability model) continues to work.

T_2' **Track 2**

6.4.2.3′ More about superresolution

An example of a photoactivatable molecule, called Dronpa, undergoes a *cis-trans* photoiso-merization, which takes it between "on" and "off" states (Andresen et al., 2007). Still other fluorophores are known that fluoresce in both states, but with different peak wavelengths that can be selectively observed.

Other superresolution methods involving switching of fluorophores have also emerged, including stimulated emission depletion (STED), as well as purely optical methods like structured illumination (Hell, 2007).

T_2' **Track 2**

6.5′ What to do when data points are correlated

Throughout this book, we mainly study repeated measurements that we assume to be independent. For such situations, the appropriate likelihood function is just a product of simpler ones. But many biophysical measurements involve partially correlated quantities. For example, in a time series, successive measurements may reflect some kind of system "memory." One such situation was the successive locations of a random walker.

Another such situation arises in electrical measurements in the brain. When we record electrical potential from an extracellular electrode, the signal of interest (activity of the nearest neuron) is overlaid with a hash of activity from more distant neurons, electrical noise in the recording apparatus, and so on. Even if the signal of interest were perfectly repeated each time the neuron fired, our ability to recognize it would still be hampered by this noise. Typically, we record values for the potential at a series of M closely spaced times, perhaps every 0.1 ms, and ask whether that signal "matches" one of several candidates, and if so, which one. To assess the quality of a match, we need to account for the noise.

One way to proceed is to subtract each candidate signal in turn from the actual observed time series, and evaluate the likelihood that the residual is drawn from the same distribution as the noise. If each time slot were statistically independent of the others, this would be a simple matter: We would just form the product of M separate factors, each the pdf of the noise evaluated on the residual signal at that time. But the noise is correlated; assuming otherwise misrepresents its true high-dimensional pdf. Before we can solve our inference problem, we need a statistical model characterizing the noise.

To find such a model, as usual we'll propose a family and choose the best one by examining N samples of pure noise (no signals from the neuron of interest) and maximizing likelihood. Each sample consists of M successive measurements. One family we might consider consists of Gaussian distributions: If $\{x_1, \ldots, x_M\}$ is an observed residual signal, then the independent-noise hypothesis amounts to a likelihood function

$$\wp_{\text{noise}}(x_1, \ldots, x_M \mid \sigma) = A \exp\left[-\tfrac{1}{2}(x_1{}^2 + \cdots + x_M{}^2)/\sigma^2\right], \quad \text{uncorrelated noise model}$$

where $A = (2\pi\sigma^2)^{-N/2}$. To go beyond this, note that the exponential contains a quadratic function of the variables x_i, and we may replace it by a more general class of such functions—those that are not diagonal. That is, let

$$\wp_{\text{noise}}(x_1, \ldots, x_M \mid \mathbf{S}) = A \exp\left[-\tfrac{1}{2}\boldsymbol{x}^{\text{t}}\mathbf{S}\boldsymbol{x}\right]. \quad \text{correlated noise model} \qquad (6.12)$$

In this formula, the matrix **S** plays the role of σ^{-2} in the ordinary Gaussian. It has units inverse to those of x^2, so we might guess that the best choice to represent the noise, given N samples, would be the inverse of the **covariance matrix**:[26]

$$\mathbf{S} = \left(\overline{\mathbf{x}\mathbf{x}^{\mathrm{t}}} \right)^{-1}.$$

Your Turn 6C

a. Obtain a formula for the normalization constant A in Equation 6.12.
b. Then maximize the likelihood function, and confirm the guess just made.

S is an $M \times M$ matrix that gives the best choice of generalized Gaussian distribution to represent the noise; it can then be used to formulate the needed likelihood function. For more details, see Pouzat, Mazor, and Laurent (2002).

Recently, neural recording methods have been advanced by the construction of multi-electrode arrays, which listen to many nearby locations simultaneously. Such recordings also have *spatial* correlations between the potentials measured on nearby electrodes. Methods similar to the one just given can also be used to "spatially decorrelate" those signals, before attempting to identify their content.

---

[26]Compare the covariance as defined in Section 3.5.2′b (page 61).

PROBLEMS

6.1 Many published results are wrong

The widespread practice of ignoring negative results is a problem of growing concern in biomedical research. Such results are often never even submitted for publication, under the assumption that they would be rejected. To see why this is a problem, consider a study that tests 100 hypotheses, of which just 10 are true. Results are reported only for those hypotheses supported by the data at the "95% confidence level." This statement is an estimate of the false-positive rate of the study; it says that, even if a particular hypothesis were false, inevitable randomness in the data would nevertheless create an erroneous impression that it is true in 5% among many (imagined) repetitions of the experiment.

Suppose furthermore that the experiment was designed to have a moderately low false-negative rate of 20%: Out of every 10 true hypotheses tested, at most 2 will be incorrectly ruled out because their effects are not picked up in the data. Thus, the study finds and reports eight of the true hypotheses, missing two because of false negatives.

a. Of the remaining 90 hypotheses that are false, about how many will spuriously appear to be confirmed? Add this to the 8 true positive results.

b. What fraction of the total reported positive results are then false?

c. Imagine another study with the same "gold standard" confidence level, but with a higher false-negative rate of 60%, which is not unrealistic. Repeat (a,b).

6.2 Effect of a prior

Suppose that we believe a measurable quantity x is drawn from a Gaussian distribution with known variance σ_x^2 but unknown expectation μ_x. We have some prior belief about the value of μ_x, which we express by saying that its most probable value is zero, but with a variance S^2; more precisely, we suppose the prior distribution to be a Gaussian with that expectation and variance. We make a single experimental measurement of x, which yields the value 0.5. Now we want the new (posterior) estimate of the distribution of μ_x.

To make a specific question, suppose that σ_x and S both equal 0.1. Make the abbreviations $A(\mu_x) = \exp(-(\mu_x)^2/(2S^2))$ and $B(\mu_x) = \exp(-(x - \mu_x)^2/(2\sigma_x^2))$. Thus, A is the prior, times a constant that does not depend on μ_x; B is the likelihood function, again multiplied by something independent of μ_x.

a. Show that the product AB is also a Gaussian function of μ_x, and find its parameter values. Normalize this function, and call the result $C_1(\mu_x)$.

b. Repeat, but this time suppose that $\sigma_x = 0.6$; this time, call the normalized product $C_2(\mu_x)$.

c. Use your results to draw some qualitative conclusions about the respective effects of the prior and the likelihood on the posterior distribution (see Section 6.2.4). Are those conclusions more generally applicable?

d. Make a graph showing $A(\mu_x)$ as a solid black line, the two different functions $B(\mu_x)$ as solid colored lines, and $C_{1,2}(\mu_x)$ as dashed lines with corresponding colors. Show how your answer to (c) appears graphically.

6.3 Horse kicks

Extensive data are available for the number of deaths between 1875 and 1895 of cavalry soldiers kicked by their own horses. The simplest hypothesis is that each soldier has a fixed probability per unit time of being killed in this way.

One dataset provides the number of casualties in each of 20 years for 14 different army units, each with the same large number of soldiers. That is, this dataset consists of 280 numbers. We can summarize it as a set of frequencies ("instances") whose sum equals 280:

casualties	instances
0	144
1	91
2	32
3	11
4	2
5 or more	0

a. Write a general expression for the probability that in one year any given unit will suffer ℓ such casualties.

b. Your formula in (a) contains a parameter. Obtain the best estimate for this parameter, based on the data, by maximizing likelihood. That is, suppose that we have no prior belief about the value of the parameter. [*Hint:* Generalize from Section 6.3.1, which found the maximum-likelihood estimate of the expectation in the case where we believe that the data follow a Binomial distribution, and Section 6.4.2.2, which discussed the case where we believe that the data follow a Gaussian distribution.] Plot the data along with your theoretical distribution, using the best-estimate parameter value.

c. Section 6.3.3 (page 130) gave a technique to estimate the credible interval of parameter values. The idea is to examine the probability of the parameter given the data, and find a symmetric range about the maximum that encloses, say, 90% of the total probability. Graph the posterior probability of the parameter's value, and *estimate* a range that includes nearly all the probability.

6.4 Diffusion constant from Brownian motion

Problem 4.5 introduced a model for diffusion in which the final position of a random walker is the sum of many two-dimensional displacements, each with $\Delta x = (\pm d, \pm d)$. Section 5.3.1 argued that, for a large number of steps, the pdf for the x displacement will approach a Gaussian, and similarly for the y displacement. Because x and y are independent, the joint distribution of $x = (x, y)$ will approach a **2D Gaussian**:

$$\wp(x) = \frac{1}{2\pi\sigma^2} e^{-(x^2+y^2)/(2\sigma^2)}.$$

The parameter σ depends on the size of the particle, the nature of the surrounding fluid, and the elapsed time. In Problem 3.4 you explored whether experimental data really have this general form.

a. Under the hypothesis just stated about the pdf of these data, find the likelihood function for the parameter σ in terms of a set of x vectors.

b. Obtain Dataset 4, which contains Jean Perrin's data points shown in Figure 3.3c. Find the best estimate for σ.

c. The quantity $\sigma^2/(2T)$, where T is the elapsed time, is called the particle's **diffusion coefficient**. Evaluate this quantity, given that T was 30 s in Perrin's experiment.

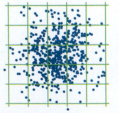

Figure 3.3c (page 39)

6.5 Credible interval

a. Six hundred flips of a coin yielded 301 <u>heads</u>. Given that information, compute the 90% credible interval for the coin fairness parameter ξ by using the method outlined in the text. Assume a uniform prior for ξ. [*Hint:* Your computer math package may balk at integrating the likelihood function. To help it out, first find the location ξ_* of the peak analytically. Next consider the function $f(\xi) = (\xi/\xi_*)^\ell((1-\xi)/(1-\xi_*))^{M-\ell}$. This is the likelihood, divided by its peak value. It still is not normalized, but at least it never exceeds 1.]

b. Six out of 25 animals fed a suspected carcinogen developed a particular cancer. Find the 90% credible interval for the probability ξ to develop the disease, and comment on whether this group is significantly different from a much larger control group, in which $\xi_{control}$ was 17%. Again assume a Uniform prior.

c. This time, suppose that both the control and the experimental groups were of finite size, say, 25 individuals each. Describe a procedure for assessing your confidence that their distributions are different (or not).

6.6 Count fluctuations

Suppose that you look through your microscope at a sample containing some fluorescent molecules. They are individually visible, and they drift in and out of your field of view independently of one another. They are few enough in number that you can count how many are in your field of view at any instant. You make 15 measurements, obtaining the counts $\ell =$ 19, 19, 19, 19, 26, 22, 17, 23, 14, 25, 28, 27, 23, 18, and 26. You'd like to find the best possible estimate of the expectation of ℓ, that is, the average value you'd find if you made many more trials.

a. First, you expect that the numbers above were drawn from a Poisson distribution. Check if that's reasonable by testing to see if the above numbers roughly obey a simple property that any Poisson distribution must have.

b. Use a computer to plot the likelihood function for μ, assuming that the above numbers really did come from a Poisson distribution with expectation μ. What is the maximally likely value μ_*? [*Hint:* The likelihood function may take huge numerical values that your computer finds hard to handle. If this happens, try the following: Compute analytically the maximum value of the log likelihood. Subtract that constant value from the log-likelihood function, obtaining a new function that has maximum exactly equal to zero, at the same place where the likelihood is maximum. Exponentiate and graph that function.]

c. Estimate the 90% credible interval for your estimate from your graph. That is, find a range $(\mu_* - \Delta, \mu_* + \Delta)$ about your answer to (b) such that about 90% of the area under your likelihood function falls within this range.

6.7 Gaussian credible interval

Suppose that you have observed M values x_i that you believe are drawn from a Gaussian distribution with known variance but unknown expectation. Starting from Equation 6.7 (page 135), find the 95% credible interval for the expectation.

6.8 $\boxed{T_2}$ Consistency of the Bayes formula

Section 6.2.3 made a claim about the self-consistency of Nick's scheme for revising probabilities. To prove that claim, note that the intermediate estimate (posterior after accounting for data) is[27]

$$\mathcal{P}(\text{model} \mid \text{data}) = \mathcal{P}(\text{data} \mid \text{model})\mathcal{P}(\text{model})/\mathcal{P}(\text{data}).$$

When additional information become available, we construct a more refined posterior by writing a similar formula with data′ in place of data, and with everything conditional on the already-known data:

$$\frac{\mathcal{P}(\text{data}' \mid \text{model and data})\mathcal{P}(\text{model} \mid \text{data})}{\mathcal{P}(\text{data}' \mid \text{data})}.$$

Rearrange this expression to prove that it is symmetric if we exchange data and data′: It doesn't matter in what order we account for multiple pieces of new information.

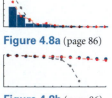

Figure 4.8a (page 86)

Figure 4.8b (page 86)

6.9 T_2 Luria-Delbrück data, again

Figure 4.8 shows experimental data from a total of 87 trials of the Luria-Delbrück experiment. The figure also shows an exact evaluation of the expected distribution supposing the "Lamarckian" hypothesis (gray dots), evaluated for a parameter value that gives a good-looking fit, as well as an approximate (simulated) evaluation supposing the "Darwinian" hypothesis (red dots). Estimate the logarithm of the ratio of likelihoods for the two models from the information in the graph. Suppose that you initially thought the Lamarckian hypothesis was, say, five times more probable than the Darwinian. What would you then conclude after the experiment?

6.10 T_2 Fitting non-Gaussian data

This problem uses the same series $\{y_n\}$ that you used in Problem 5.15, and studies the same family of possible distributions as in part (d) of that problem. You may be dissatisfied just guessing values for the parameters that make the graph "look good."

We are exploring a family of Cauchy-like distributions: $\wp_{\text{CL}}(y; \mu_y, \eta) = A/[1 + ((y - \mu_y)/\eta)^\alpha]$. To keep the code simple, in this problem fix $\alpha = 4$, and adjust μ_y and η to find an optimal fit. (The normalization factor A is not free; you'll need to find it for each set of parameter values that you check.)

Choose an initial guess, say, $\mu_y = 0$ and $\eta = 0.02$. If we knew that the distribution was described by $\wp_{\text{CL}}$ and that these values were the right ones, then the probability that the observed time series would have arisen would be the product $\wp_{\text{CL}}(y_1; \mu_y, \eta)\wp_{\text{CL}}(y_2; \mu_y, \eta) \cdots$. To make this easier to work with, instead compute its logarithm L. Write a code that computes L for many values of η, starting with 0.015 and ending with 0.035, and also for several values of μ_y. That is, your code will have two nested loops over the desired values of η and μ_y; inside them will be a third loop over each of the data points, which accumulates the required sum. Plot your L as a function of η and μ_y, and find the location of its maximum. Then repeat for the hypothesis of a Gaussian distribution of weekly log-changes, and compare the best-fit Cauchy-like distribution with the best-fit Gaussian model.

---

[27] Here we use $\mathcal{P}$ as a generic symbol for either the continuous or discrete case.

Poisson Processes

7

The objective of physics is to establish new relationships between seemingly unrelated, remote phenomena.
—Lev D. Landau

7.1 Signpost

Many key functions in living cells are performed by devices that are themselves individual molecules. These "molecular machines" generally undergo discrete steps, for example, synthesizing or breaking down some other molecules one by one. Because they are so small, they must do their jobs in spite of (or even with the help of) significant randomness from thermal motion. If we wish to understand how they work, then, we must characterize their behavior in probabilistic terms. Even with this insight, however, the challenges are daunting. Imagine an automobile engine far smaller than the wavelength of light: How can we get "under the hood" and learn about the mechanisms of such an engine?

More specifically, we will look at one aspect of this Focus Question:

Biological question: How do you detect an invisible step in a molecular motor cycle?

Physical idea: The waiting-time distributions of individual molecular motors can provide evidence for a physical model of stepping.

7.2 The Kinetics of a Single-Molecule Machine

Some molecular motors have two "feet," which "walk" along a molecular "track." The track is a long chain of protein molecules (such as **actin** or **tubulin**). The feet[1] are subunits of the motor with binding sites that recognize specific, regularly spaced sites on the track. When the energy molecule ATP is present, another binding site on the foot can bind an ATP

---

[1] For historical reasons, the feet are often called "heads"!

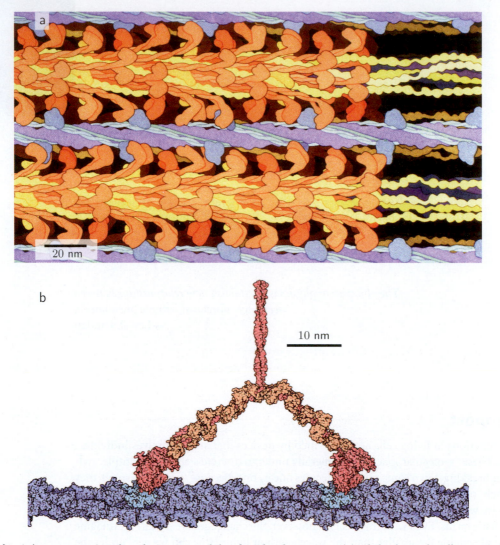

Figure 7.1 [Artist's reconstructions based on structural data.] **Molecular motors.** (a) Skeletal muscle cells contain bundles of the motor protein myosin-II (*orange*). These are interspersed with long filaments composed of the protein actin (*blue*). When activated, the myosin motors in the bundle consume ATP and step along the actin filaments, dragging the red bundle rightward relative to the blue tracks and hence causing the muscle cell to contract. (The thin snaky molecule shown in *yellow* is titin, a structural protein that keeps the actin and myosin filaments in proper alignment.) (b) The myosin-V molecule has two "legs," which join its "feet" to their common "hip," allowing it to span the 36 nm separation between two binding sites (*light blue*) on an actin filament (*blue*). [(a,b) Courtesy David S Goodsell.]

molecule. Clipping off one of the phosphate groups on the ATP yields some chemical bond energy, which is harnessed to unbind the foot from its "track" and move it in the desired direction of motion, where it can, in turn, find another binding site. In this way the motor takes a step, typically of a few nanometers but for certain motors much longer.

Figure 7.1a shows a schematic of the arrangement of many molecular motors, ganged together to exert a significant total force in our skeletal muscles. Other motors operate singly, for example, to transport small cargo from one part of a cell to another. In order to be useful, such a motor must be able to take many steps without falling off its track; that is, it must be highly **processive**. Myosin-V, a motor in this class, is known to have a structure with two identical feet (Figure 7.1b). It is tempting to guess that myosin-V achieves its processivity (up to 50 consecutive steps in a run) by always remaining bound by one foot while the other one takes a step, just as we walk with one foot always in contact with the ground.[2]

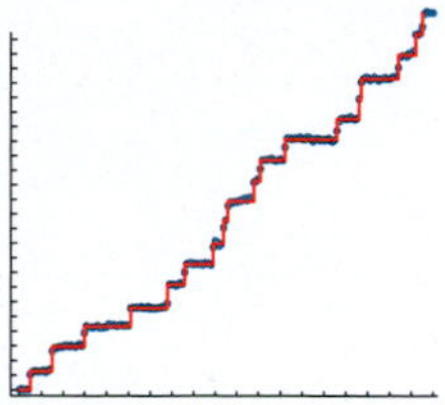

Chapter 6 introduced myosin-V and described how Yildiz and coauthors were able to visualize its individual steps via optical imaging. As shown in Figure 6.3c, the motor's position as a function of time looks like a staircase. The figure shows an example with rapid rises of nearly uniform height, corresponding to 74 nm steps. But the *widths* of the stairs in that figure, corresponding to the waiting times (pauses) between steps, are quite nonuniform. Every individual molecule studied showed such variation.

We may wish to measure the speed of a molecular motor, for example, to characterize how it changes if the molecule is modified. Perhaps we wish to study a motor associated with some genetic defect, or an intentionally altered form that we have engineered to test a hypothesis about the function of a particular element. But what does "speed" mean? The motor's progress consists of sudden steps, spread out between widely variable pauses. And yet, the overall trend of the trace in Figure 6.3c does seem to be a straight line of definite slope. We need to make this intuition more precise.

To make progress, imagine the situation from the motor's perspective. Each step requires that the motor bind an ATP molecule. ATPs are available, but they are greatly outnumbered by other molecules, such as water. So the motor's ATP-binding domain is bombarded by molecular collisions at a very high rate, but almost all collisions are not "productive"; that is, they don't lead to a step. Even when an ATP does arrive, it may fail to bind, and instead simply wander away.

The discussion in the previous paragraph suggests a simple physical model: We imagine that collisions occur every Δt, that each one has a tiny probability ξ to be productive, and that every collision is independent of the others. After an unproductive collision, the motor is in the same internal state as before. We also assume that after a productive collision, the internal state resets; the motor has no memory of having just taken a step. Viewed from the outside, however, its position on the track has changed. We'll call this position the system's **state variable**, because it gives all the information relevant for predicting future steps.

$\boxed{T_2}$ *Section 7.2′ (page 171) gives more details about molecular motors.*

Figure 6.3c (page 134)

7.3 Random Processes

Before we work out the predictions of the physical model, let's think a bit more about the nature of the problem. We can replicate our experiment, with many identical myosin-V molecules, each in a solution with the same uniform ATP concentration, temperature, and so on. The output of each trial is not a single number, however; instead, it is an entire *time series of steps* (the staircase plot). Each step advances the molecule by about the same distance; thus, to describe any particular trial, we need only state the list of times $\{t_1, t_2, \ldots, t_N\}$ when steps occurred on that trial. That is, each trial is a draw from a probability distribution whose sample space consists of increasing sequences of time values. A random system with this sort of sample space is called a **random process**.

---

[2]Borrowing a playground metaphor, many authors instead refer to this mechanism as "hand-over-hand stepping."

The pdf on the full sample space is a function of all the many variables t_α. In general, quite a lot of data is needed to estimate such a multidimensional distribution. But the physical model for myosin-V proposed in the preceding section gives rise to a special kind of random process that allows a greatly reduced description: Because the motor is assumed to have no memory of its past, we fully specify the process when we state the collision interval Δt and productive-step probability ξ. The rest of this chapter will investigate random processes with this Markov property.[3]

7.3.1 Geometric distribution revisited

We are considering a physical model of molecular stepping that idealizes each collision as independent of the others, and also supposes them to be simple Bernoulli trials. We (temporarily) imagine time to be a discrete variable that can be described by an integer i (labeling which "time slot"). We can think of our process as reporting a string of step/no-step results for each time slot.[4]

Let E_* denote the event that a step happened at time slot i. Then to characterize the discrete-time stepping process, we can find the probability that, given E_*, the *next* step takes place at a particular time slot $i + j$, for various positive integers j. Call this proposition "event E_j." We seek the conditional probability $\mathcal{P}(\mathsf{E}_j \mid \mathsf{E}_*)$.

More explicitly, E_* is the probability that a step occurred at slot i, *regardless of what happened on other slots*. Thus, many elementary outcomes all contribute to $\mathcal{P}(\mathsf{E}_*)$. To find the conditional probability $\mathcal{P}(\mathsf{E}_j \mid \mathsf{E}_*)$, then, we must evaluate $\mathcal{P}(\mathsf{E}_j \text{ and } \mathsf{E}_*)/\mathcal{P}(\mathsf{E}_*)$.[5]

- The denominator of this fraction is just ξ. Even if this seems clear, it is worthwhile to work through the logic, in order to demonstrate how our ideas fit together.

  In an interval of duration T, there are $N = T/\Delta t$ time slots. Each outcome of the random process is a string of N Bernoulli trials (step/no-step in time slot $1, \dots, N$). E_* is the subset of all possible outcomes for which there was a step at time slot i (see Figures 7.2a–e). Its probability, $\mathcal{P}(\mathsf{E}_*)$, is the sum of the probabilities corresponding to each elementary outcome in E_*.

  Because each time slot is independent of the others, we can factor $\mathcal{P}(\mathsf{E}_*)$ into a product and use the rearrangement trick in Equation 3.14 (page 49). For each time slot prior to i, we don't care what happens, so we sum over both possible outcomes, yielding a factor of $\big(\xi + (1 - \xi)\big) = 1$. Time slot i gives a factor of ξ, the probability to take a step. Each time slot following i again contributes a factor of 1. All told, the denominator we seek is

$$\mathcal{P}(\mathsf{E}_*) = \xi. \tag{7.1}$$

- Similarly in the numerator, $\mathcal{P}(\mathsf{E}_j \text{ and } \mathsf{E}_*)$ contains a factor of 1 for each time slot prior to i, and a factor of ξ representing the step at i. It also has $j - 1$ factors of $(1 - \xi)$ representing *no* step for time slots $i + 1$ through $i + j - 1$, another ξ for the step at time slot $i + j$, and then factors of 1 for later times:

$$\mathcal{P}(\mathsf{E}_j \text{ and } \mathsf{E}_*) = \xi(1 - \xi)^{j-1}\xi. \tag{7.2}$$

---

[3]See Section 3.2.1 (page 36).
[4]This situation was introduced in Section 3.4.1.2 (page 47).
[5]See Equation 3.10 (page 45).

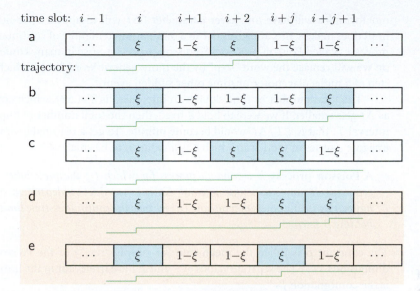

time slot: $i-1$ i $i+1$ $i+2$ $i+j$ $i+j+1$

Figure 7.2 [Diagrams.] **Graphical depiction of the origin of the Geometric distribution.** (a–e) Examples of time series, and their contributions to $\mathcal{P}(E_*)$, the probability that a step occurs at time slot i (Equation 7.1). *Colored boxes* represent time slots in which an event ("blip") occurred. *Green staircases* represent the corresponding motions if the blips are steps of a molecular motor. That is, they are graphs of the state variable (motor position) versus time, analogous to the real data in Figure 6.3c (page 134). (d–e) Examples of contributions to $\mathcal{P}(E_j \text{ and } E_*)$, the probability that, in addition, the *next* blip occurs at time slot $i+j$, for the case $j=3$ (Equation 7.2). The terms shown in (d–e) differ only at position $i+j+1$, which is one of the "don't care" positions. Thus, their sum is $\cdots \xi(1-\xi)(1-\xi)\xi(\xi + 1 - \xi) \cdots$.

The conditional probability is the quotient of these two quantities. Note that it does not depend on i, because shifting everything in time does not affect how long we must wait for the next step. In fact, $\mathcal{P}(E_j \mid E_*)$ is precisely the Geometric distribution (Equation 3.13):

$$\mathcal{P}(E_j \mid E_*) = \xi(1-\xi)^{j-1} = \mathcal{P}_{\text{geom}}(j;\xi), \text{ for } j = 1,2,\ldots. \qquad (3.13)$$

Like the Binomial, Poisson, and Gaussian distributions, this one, too, has its roots in the Bernoulli trial.

7.3.2 A Poisson process can be defined as a continuous-time limit of repeated Bernoulli trials

The Geometric distribution is useful in its own right, because many processes consist of discrete attempts that either "succeed" or "fail." For example, an animal may engage in isolated contests to establish dominance or catch prey; its survival may involve the number of attempts it must make before the next success.

But often it's not appropriate to treat time as discrete. For example, as far as motor stepping is concerned, nothing interesting is happening on the time scale Δt. Indeed, the motor molecule represented by the trace in Figure 6.3c generally took a step every few seconds. This time scale is enormously slower than the molecular collision time Δt, because the vast majority of collisions are unproductive. This observation suggests that we may gain a simplification if we consider a *limit*, $\Delta t \to 0$. If such a limit makes sense, then

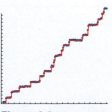

Figure 6.3c (page 134)

our formulas will have one fewer parameter (Δt will disappear). We now show that the limit does make sense, and gives rise to a one-parameter family of continuous-time random processes called Poisson processes.[6] Poisson processes arise in many contexts, so from now on we will replace the word "step" by the more generic word "blip," which could refer to a step of a molecular motor or some other sudden event.

The total number of time slots in an interval T is $T/(\Delta t)$, which approaches infinity as Δt gets smaller. If we were to hold ξ fixed, then the total number of blips expected in the interval T, that is, $\xi T/\Delta t$, would become infinite. To get a reasonable limit, then, we must imagine a series of models in which ξ is *also* taken to be small:

> A **Poisson process** *is a random process for which (i) the probability of a blip occurring in any small time interval Δt is $\xi = \beta\Delta t$, independent of what is happening in any other interval, and (ii) we take the continuous-time limit $\Delta t \to 0$ holding β fixed.* $\qquad$ (7.3)

The constant β is called the **mean rate** (or simply "rate") of the Poisson process; it has dimensions $1/\mathbb{T}$. The separate values of ξ and Δt are irrelevant in the limit; all that matters is the combination β.

Your Turn 7A

Suppose that you examine a random process. Taking $\Delta t = 1\,\mu\text{s}$, you conclude that the condition in Idea 7.3 is satisfied with $\beta = 5/\text{s}$. But your friend takes $\Delta t = 2\,\mu\text{s}$. Will your friend agree that the process is Poisson? Will she agree about the value of β?

It's important to distinguish the Poisson *process* from the Poisson *distribution* discussed in Chapter 4. Each draw from the Poisson distribution is a single integer; each draw from the Poisson process is a sequence of real numbers $\{t_\alpha\}$. However, there is a connection. Sometimes we don't need all the details of arrival times given by a random process; we instead want a more manageable, reduced description.[7] Two very often used reductions of a random process involve its waiting time distribution (Section 7.3.2.1) and its count distribution (Section 7.3.2.2). For a Poisson *process*, we'll find that the second of these reduced descriptions follows a Poisson *distribution*.

7.3.2.1 Continuous waiting times are Exponentially distributed

The interval between successive blips is called the **waiting time** (or "dwell time"), t_w. We can find its distribution by taking the limit of the corresponding discrete-time result (Section 7.3.1 and Figure 7.3).

The pdf of the waiting time is the discrete distribution divided by Δt:[8]

$$\wp(t_\text{w}) = \lim_{\Delta t \to 0} \frac{1}{\Delta t}\mathcal{P}_\text{geom}(j;\xi). \qquad (7.4)$$

In this formula, $t_\text{w} = (\Delta t)j$ and $\xi = (\Delta t)\beta$, with t_w and β held fixed as $\Delta t \to 0$. To simplify Equation 7.4, note that $1/\xi \gg 1$ because Δt approaches zero. We can exploit that

---

[6] In some contexts, a signal that follows a Poisson process is also called "shot noise."

[7] For example, often our experimental dataset isn't extensive enough to deduce the full description of a random process, but it does suffice to characterize one or more of its reduced descriptions.

[8] See Equation 5.1 (page 98).

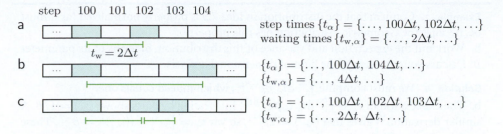

Figure 7.3 [Diagrams.] **Waiting times.** Three of the same time series as in Figure 7.2. This time we imagine the starting time slot to be number 100, and illustrate the absolute blip times t_α as well as the relative (waiting) times $t_{w,\alpha}$.

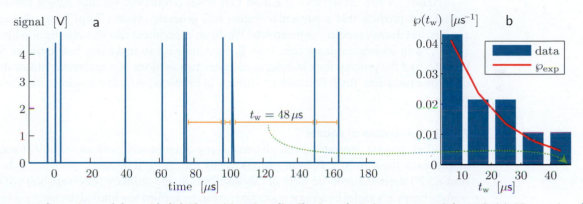

Figure 7.4 [Experimental data with fit.] **The waiting time distribution of a Poisson process (Idea 7.5).** (a) Time series of 11 blips. The *orange arrows* indicate 4 of the 10 waiting times between successive blips. The *green arrow* connects one of these to the corresponding point on the horizontal axis of a graph of $\wp(t_w)$. (b) On this graph, *bars* indicate estimates of the pdf of t_w inferred from the 10 waiting times in (a). The *curve* shows the Exponential distribution with expectation equal to the sample mean of the experimental t_w values. [Data courtesy John F Beausang (Dataset 8).]

fact by rearranging slightly:

$$\wp(t_w) = \lim_{\Delta t \to 0} \frac{1}{\Delta t} \xi (1 - \xi)^{(t_w/\Delta t)-1} = \lim_{\Delta t \to 0} \frac{\xi}{\Delta t} \left((1 - \xi)^{(1/\xi)} \right)^{(t_w \xi/\Delta t)} (1 - \xi)^{-1}.$$

Taking each factor in turn:

- $\xi/\Delta t = \beta$.
- The middle factor involves $(1 - \xi)^{(1/\xi)}$. The compound interest formula[9] says that this expression approaches e^{-1}. It is raised to the power $t_w \beta$.
- The last factor approaches 1 for small ξ.

With these simplifications, we find a family of continuous pdfs for the interstep waiting time:

> The waiting times in a Poisson process are distributed according to the **Exponential distribution** $\wp_{\exp}(t_w; \beta) = \beta e^{-\beta t_w}$.
(7.5)

Figure 7.4 illustrates this result with a very small dataset.

---

[9]See Equation 4.5 (page 76).

Example a. Confirm that the distribution in Idea 7.5 is properly normalized (as it must be, because the Geometric distribution has this property).

b. Work out the expectation and variance of this distribution, in terms of its parameter β. Discuss your answers in the light of dimensional analysis.

Solution a. We must compute $\int_0^\infty \mathrm{d}t_\mathrm{w}\, \beta e^{-\beta t_\mathrm{w}}$, which indeed equals one.

b. The expectation of t_w is $\int_0^\infty \mathrm{d}t_\mathrm{w}\, t_\mathrm{w} \beta e^{-\beta t_\mathrm{w}}$. Integrating by parts shows $\langle t_\mathrm{w} \rangle = 1/\beta$. A similar derivation[10] gives that $\langle t_\mathrm{w}^2 \rangle = 2\beta^{-2}$, so $\mathrm{var}\, t_\mathrm{w} = \langle t_\mathrm{w}^2 \rangle - (\langle t_\mathrm{w} \rangle)^2 = \beta^{-2}$. These results make sense dimensionally, because $[\beta] \sim \mathbb{T}^{-1}$.

Figure 7.4 also illustrates a situation that arises frequently: We may have a physical model that predicts that a particular system will generate events ("blips") in a Poisson process, but doesn't predict the mean rate. We do an experiment and observe the blip times $t_1, \ldots, t_N$ in an interval from time 0 to T. Next we wish to make our best estimate for the rate β of the process, for example, to compare two versions of a molecular motor that differ by a mutation. You'll find such an estimate in Problem 7.6 by maximizing a likelihood function.

7.3.2.2 Distribution of counts

A random process generates a complicated, many-dimensional random variable; for example, each draw from a Poisson process yields the entire time series $\{t_1, t_2, \ldots\}$. Section 7.3.2.1 derived a reduced form of this distribution, the ordinary (one-variable) pdf of waiting times. It's useful because it is simple and can be applied to a limited dataset to obtain the best-fit value of the one parameter β characterizing the full process.

We can get another useful reduction of the full distribution by asking, "How many blips will we observe in a fixed, finite time interval T_1?" To approach the question, we again begin with the discrete-time process, regarding the interval T_1 as a succession of $M_1 = T_1/\Delta t$ time slots. The total number of blips, ℓ, equals the sum of M_1 Bernoulli trials, each with probability $\xi = \beta \Delta t$ of success. In the continuous-time limit ($\Delta t \to 0$), the distribution of ℓ values approaches a Poisson distribution,[11] so

> *For a Poisson process with mean rate β, the probability of getting ℓ blips in any time interval T_1 is $\mathcal{P}_\mathrm{pois}(\ell; \beta T_1)$.* $\qquad$ (7.6)

Δt does not appear in Idea 7.6 because it cancels from the expression $\mu = M_1 \xi = \beta T_1$. Figure 7.5 illustrates this result with some experimental data.

The quantity ℓ/T_1 is different in each trial, but Idea 7.6 states that its expectation (its value averaged over many observations) is $\langle \ell/T_1 \rangle = \beta$; this fact justifies calling β the mean rate of the Poisson process.

We can also use Idea 7.6 to estimate the mean rate of a Poisson process from experimental data. Thus, if blip data have been given to us in an aggregated form, as counts in each of a series of time bins of duration T_1, then we can maximize likelihood to determine a best-fit value of $T_1\beta$, and from this deduce β.[12]

---

[10] See the Example on page 55.

[11] See Section 4.3.2 (page 75).

[12] See Problem 6.3.

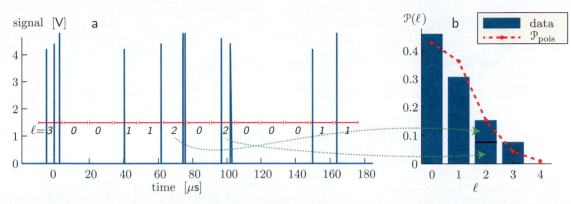

Figure 7.5 [Experimental data with fit.] **The count distribution of a Poisson process over fixed intervals (Idea 7.6).** (a) The same 11 blips shown in Figure 7.4a. The time interval has been divided into equal bins, each of duration $T_1 = 13$ s (*red*); the number of blips in each bin, ℓ, is given beneath its bin indicator. (b) On this graph, *bars* indicate estimates of the probability distribution of ℓ from the data in (a). *Green arrows* connect the instances of $\ell = 2$ with their contributions to the bar representing this outcome. The *red dots* show the Poisson distribution with expectation equal to the sample mean of the observed ℓ values. [Data courtesy John F Beausang (Dataset 8).]

Your Turn 7B

Alternatively, we may consider a single trial, but observe it for a long time T. Show that, in this limit, ℓ/T has expectation β and its relative standard deviation is small.

In the specific context of molecular motors, the fact you just proved explains the observation that staircase plots, like the one in Figure 6.3c, appear to have definite slope in the long run, despite the randomness in waiting times.

Figure 7.6a represents symbolically the two reduced descriptions of the Poisson process derived in this section.

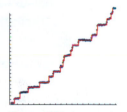

Figure 6.3c (page 134)

7.3.3 Useful Properties of Poisson processes

Two facts about Poisson processes will be useful to us later.

7.3.3.1 Thinning property

Suppose that we have a Poisson process with mean rate β. We now create another random process: For each time series drawn from the first one, we accept or reject each blip based on independent Bernoulli trials with probability ξ_{thin}, reporting only the times of the accepted blips. The **thinning property** states that the new process is also Poisson, but with mean rate reduced from β to $\xi_{\text{thin}}\beta$.

To prove this result, divide time into slots Δt so small that there is negligible probability to get two or more blips in a slot. The first process has probability $\beta\Delta t$ to generate a blip in any slot. The product rule says that in the thinned process, every time slot is again a Bernoulli trial, but with probability of a blip reduced to $(\beta\Delta t)\xi_{\text{thin}}$. Thus, the new process fulfills the condition to be a Poisson process, with mean rate $\xi_{\text{thin}}\beta$ (see Figures 7.6b1,b2).

7.3.3.2 Merging property

Suppose that we have two independent Poisson processes, generating distinct types of blips. For example, Nora may randomly throw blue balls at a wall at mean rate β_1, while Nick

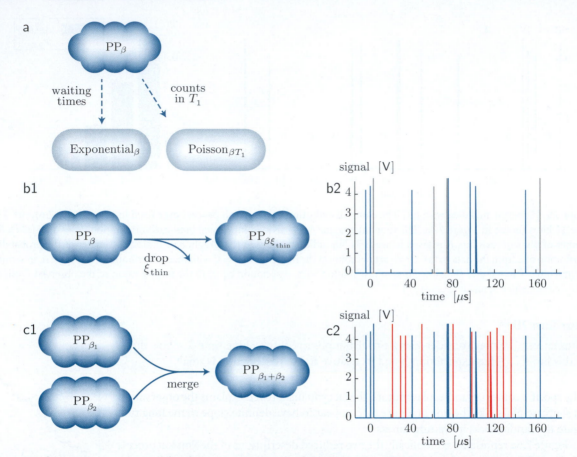

Figure 7.6 [Diagrams; experimental data.] **Some operations involving Poisson processes.** (a) A Poisson process (*upper bubble*) gives rise to two simpler reduced descriptions: Its distribution of waiting times is Exponential, whereas the distribution of blip counts in any fixed interval is Poisson (Figures 7.4, 7.5). (b1,c1) Graphical depictions of the thinning and merging properties of Poisson processes. (b2) The same data as in the two preceding figures have been thinned by randomly rejecting some blips (*gray*), with probability $\xi_{thin} = 1/2$. The remaining blips again form a Poisson process, with mean rate reduced by half. (c2) The same data have been merged with a second Poisson process with the same mean rate (*red*). The complete set of blips again forms a Poisson process, with mean rate given by the sum of the mean rates of the two contributing processes.

randomly throws red balls at the same wall at mean rate β_2. We can define a "merged process" that reports the arrival times of *either* kind of ball. The **merging property** states that the merged process is itself Poisson, with mean rate $\beta_{tot} = \beta_1 + \beta_2$. To prove it, again divide time into small slots Δt and imagine an observer who merely *hears* the balls hitting the target. Because Δt is small, the probability of getting two balls in the same time slot is negligible. Hence, the addition rule says that, in any short interval Δt, the probability of hearing a thump is $(\beta_1 \Delta t) + (\beta_2 \Delta t)$, or $\beta_{tot} \Delta t$ (see Figures 7.6c1,c2).

Example Consider three independent Poisson processes, with mean rates β_1, β_2, and β_3. Let $\beta_{tot} = \beta_1 + \beta_2 + \beta_3$.
a. After a blip of *any* type, what's the distribution of waiting times till the next event of *any* type?

b. What's the probability that any particular blip will be of type 1?

Solution   a. This can be found by using the merging property and the waiting-time distribution (Idea 7.5, page 159). Alternatively, divide time into slots of duration Δt. Let's find the probability of *no* blip during a period of duration t_w (that is, $M = t_w/\Delta t$ consecutive slots). For very small Δt, the three blip outcomes become mutually exclusive, so the negation, addition, and product rules yield

$$\mathcal{P}(\text{none during } t_w) = (1 - \beta_{\text{tot}}\Delta t)^M = \big(1 - (\beta_{\text{tot}} t_w/M)\big)^M = \exp(-\beta_{\text{tot}} t_w).$$

The probability of such a period with no blip, followed by a blip of any type, is

$$\mathcal{P}(\text{none during } t_w) \times \big(\mathcal{P}(\text{type 1 during } \Delta t) + \cdots + \mathcal{P}(\text{type 3 during } \Delta t)\big) = \exp(-\beta_{\text{tot}} t_w)\beta_{\text{tot}}\Delta t.$$

Thus, the pdf of the waiting time for any type blip is $\beta_{\text{tot}} \exp(-\beta_{\text{tot}} t_w)$, as predicted by the merging property.

b. We want $\mathcal{P}(\text{blip of type 1 in } \Delta t \mid \text{blip of any type in } \Delta t) = (\beta_1\Delta t)/(\beta_{\text{tot}}\Delta t) = \beta_1/\beta_{\text{tot}}$.

Your Turn 7C

Connect the merging property to the count distribution (Idea 7.6) and the Example on page 80.

7.3.3.3 Significance of thinning and merging properties

The two properties just proved underlie the usefulness of the Poisson process, because they ensure that, in some ways, it behaves similarly to a purely regular sequence of blips:

- Imagine a long line of professors passing a turnstile exactly once per second. You divert every third professor through a door on the left. Then the diverted stream is also regular, with a professor passing the left door once every three seconds. The thinning property states that a particular kind of *random* arrival (a Poisson process), subject to a *random* elimination (a Bernoulli trial), behaves similarly (the new process has mean rate reduced by the thinning factor).

- Imagine two long lines of professors, say of literature and chemistry, respectively, converging on a single doorway. Individuals in the first group arrive exactly once per second; those in the second group arrive once every two seconds. The stream that emerges through the door has mean rate $(1\,\text{s})^{-1} + (2\,\text{s})^{-1}$. The merging property states that a particular class of *random* processes has an even nicer property: They merge to form a new random process *of the same kind*, with mean rate again given by the sum of the two component rates.

In a more biological context,

- Section 7.2 imagined the stepping of myosin-V as a result of two sequential events: First an ATP molecule must encounter the motor's ATP-binding site, but then it must also bind and initiate stepping. It's reasonable to model the first event as a Poisson process, because most of the molecules surrounding the motor are not ATP and so cannot generate a step. It's reasonable to model the second event as a Bernoulli trial, because even when an ATP does encounter the motor, it must overcome an activation barrier to bind; thus, some fraction of the encounters will be nonproductive. The thinning property leads us to expect that the complete stepping process will itself be Poisson, but with a mean rate lower than the ATP collision rate. We'll see in a following section that this expectation is correct.

- Suppose that two or more identical enzyme molecules exist in a cell, each continually colliding with other molecules, a few of which are substrates for a reaction that the enzymes catalyze. Each enzyme then emits product molecules in a Poisson process, just as in the motor example. The merging property leads us to expect that the *combined* production will also be a Poisson process.

7.4 More Examples

7.4.1 Enzyme turnover at low concentration

Molecular motors are examples of **mechanochemical** enzymes: They hydrolyze ATP and generate mechanical force. Most enzymes instead have purely chemical effects, for example processing substrate molecules into products.[13] At low substrate concentration, the same reasoning as in Section 7.2 implies that the successive appearances of product molecules will follow a Poisson process, with mean rate reflecting the substrate concentration, enzyme population, and binding affinity of substrate to enzyme. Chapter 8 will build on this observation.

T_2 *Section 7.5.1′ a (page 171) describes some finer points about molecular turnovers.*

7.4.2 Neurotransmitter release

Nerve cells (neurons) mainly interact with each other by chemical means: One neuron releases **neurotransmitter** molecules from its "output terminal" (**axon**), which adjoins another neurons's "input terminal" (**dendrite**). Electrical activity in the first neuron triggers this release, which in turn triggers electrical activity in the second. A similar mechanism allows neurons to stimulate muscle cell contraction. When it became possible to monitor the electric potential across a muscle cell membrane,[14] researchers were surprised to find that it was "quantized": Repeated, identical stimulation of a motor neuron led to muscle cell responses with a range of peak amplitudes, and the pdf of those amplitudes consisted of a series of discrete bumps (Figure 7.7). Closer examination showed that the bumps were at integer multiples of a basic response strength. Even in the absence of any stimulus, there were occasional blips resembling those in the first bump. These observations led to the discoveries that

- Neurotransmitter molecules are packaged into bags (**vesicles**) within the nerve axon, and these vesicles all contain a roughly similar amount of transmitter. A vesicle is released either completely or not at all.
- Thus, the amount of transmitter released in response to any stimulus is roughly an integer multiple of the amount in one vesicle. Even in the absence of stimulus, an occasional vesicle can also be released "accidentally," leading to the observed spontaneous events.
- The electrical response in the muscle cell (or in another neuron's dendrite) is roughly linearly proportional to the total amount of transmitter released, and hence to the number ℓ of vesicles released.

---

[13]See Section 3.2.3 (page 40).

[14]The work of Katz and Miledi discussed earlier examined a much more subtle feature, the effect of discrete openings of ion channels in response to bathing a dendrite, or a muscle cell, in a fixed concentration of neurotransmitter (Section 4.3.4, page 78).

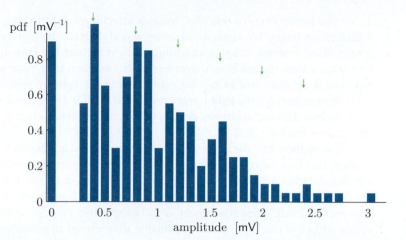

Figure 7.7 [Experimental data.] **Electrical response at a neuromuscular junction.** The bars give the estimated pdf of response amplitude, from a total of 198 stimuli. The horizontal axis gives the amplitudes (peak voltage change from rest), measured in a muscle cell in response to a set of identical stimuli applied to its motor neuron. Bumps in the distribution of amplitudes occur at 0, 1, 2, . . . , 6 times the mean amplitude of the spontaneous electrical events (*arrows*). There is some spread in each bump, mostly indicating a distribution in the number of neurotransmitter molecules packaged into each vesicle. The narrow peak near zero indicates failures to respond at all. [Data from Boyd & Martin, 1956.]

Separating the histogram in Figure 7.7 into its constituent peaks, and computing the area under each one, gave the estimated probability distribution of ℓ in response to a stimulus. This analysis showed that, at least in a certain range of stimuli, ℓ is Poisson distributed.[15] More generally,

> *If the exciting neuron is held at a constant membrane potential, then neurotransmitter vesicles are released in a Poisson process.*

7.5 Convolution and Multistage Processes

7.5.1 Myosin-V is a processive molecular motor whose stepping times display a dual character

Figure 6.3c shows many steps of the single-molecule motor myosin-V. This motor is highly processive: Its two "feet" rarely detach simultaneously, allowing it to take many consecutive steps without ever fully unbinding from its track. The graph shows each step advancing the motor by sudden jumps of roughly 74 nm. Interestingly, however, only about one quarter of the individual myosin-V molecules studied had this character. The others *alternated* between short and long steps; the *sum* of the long and short step lengths was about 74 nm. This division at first seemed mysterious—were there two distinct kinds of myosin-V molecules? Was the foot-over-foot mechanism wrong?

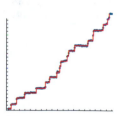

Figure 6.3c (page 134)

Yildiz and coauthors proposed a simpler hypothesis to interpret their data:

> *All the myosin-V molecules are in fact stepping in the same way along their actin tracks. They merely differ in where the fluorescent marker, used to image the stepping, is attached to the myosin-V molecule.* (7.7)

---

[15]You'll explore this claim in Problem 7.14.

To see the implications of this idea, imagine attaching a light to your hip and walking in a dark room, taking 1 m steps. An observer would then see the flashlight advancing in 1 m jumps. Now, however, imagine attaching the light to your left *knee*. Each time your right foot takes a step, the left knee moves less than 1 m. Each time your left foot takes a step, however, it detaches and swings forward, moving the light by *more* than 1 m. After any two consecutive steps, the light has always moved the full 2 m, regardless of where the light was attached. This metaphor can explain the alternating stride observed in some myosin-V molecules—but is it right?

Now suppose that the light is attached to your left *ankle*. This time, the shorter steps are so short that they cannot be observed at all. All the observer sees are 2 m jumps when your left foot detaches and moves forward. The biochemical details of the fluorescent labeling used by Yildiz and coauthors allowed the fluorophore to bind in any of several locations, so they reasoned that "ankle attachment" could happen in a subpopulation of the labeled molecules. Although this logic seemed reasonable, they wanted an additional, more quantitative prediction to test it.

To find such a prediction, first recall that molecular motor stepping follows a Poisson process, with mean rate β depending on the concentration of ATP.[16] Hence, the pdf of interstep waiting times should be an Exponential distribution.[17] In fact, the subpopulation of myosin-V motors with alternating step lengths really does obey this prediction (see Figure 7.8a), as do the kinetics of many other chemical reactions. But for the other subpopulation (the motors that took 74 nm steps), the prediction fails badly (Figure 7.8b).

To understand what's going on, recall the hypothesis of Yildiz and coauthors for the nature of stepping in the 74 nm population, which is that the first, third, fifth, ... steps are *not visible*. Therefore, what appears to be the αth interstep waiting time, $t'_{w,\alpha}$, is actually the *sum* of two consecutive waiting times:

$$t'_{w,\alpha} = t_{w,2\alpha} + t_{w,2\alpha-1}.$$

Even if the true waiting times are Exponentially distributed, we will still find that the apparent waiting times t'_w have a different distribution, namely, the convolution.[18] Thus,

$$\wp_{t'_w}(t'_w) = \int_0^{t'_w} dx \, \wp_{exp}(x; \beta) \times \wp_{exp}(t'_w - x; \beta), \tag{7.8}$$

where x is the waiting time for the first, invisible, substep.

Example a. Explain the limits on the integral in Equation 7.8.
b. Do the integral.
c. Compare your result qualitatively with the histograms in Figures 7.8a,b.
d. Discuss how your conclusion in (c) supports Idea 7.7.

Solution a. x is the waiting time for the invisible first substep. It can't be smaller than zero, nor can it exceed the specified total waiting time t'_w for the first and second substeps.

---

[16] See Section 7.2.
[17] See Idea 7.5 (page 159).
[18] See Section 4.3.5 (page 79).

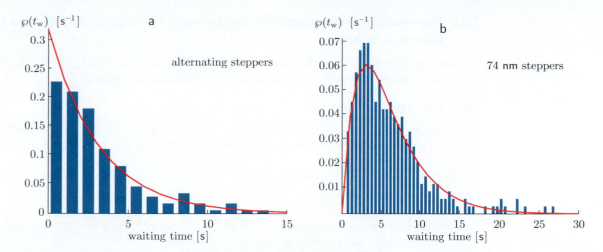

Figure 7.8 [Experimental data with fits.] **The stepping of molecular motors.** (a) Estimated pdf of the waiting times for the subpopulation of myosin-V molecules that displayed alternating step lengths, superimposed on the expected Exponential distribution (see Problem 7.8). (b) Similar graph for the other subpopulation of molecules that displayed only long steps, superimposed on the distribution derived in the Example on page 166. The shape of the curve in (b) is the signature of a random process with two alternating types of substep. Each type of substep has Exponentially distributed waiting times with the same mean rate as in (a), but only one of them is visible. [Data from Yildiz et al., 2003.]

b.

$$\beta^2 \int_0^{t'_\mathrm{w}} \mathrm{d}x \, \exp\!\left(-\beta x - \beta(t'_\mathrm{w} - x)\right) = \beta^2 e^{-\beta t'_\mathrm{w}} \int_0^{t'_\mathrm{w}} \mathrm{d}x = \beta^2 t'_\mathrm{w} e^{-\beta t'_\mathrm{w}}.$$

c. The function just found falls to zero at $t'_\mathrm{w} \to 0$ and $t'_\mathrm{w} \to \infty$. In between these extremes, it has a bump. The experimental data in Figure 7.8b have the same qualitative behavior, in contrast to those in panel (a).

d. The hypothesis under study predicted that behavior, because Figure 7.8b shows that the molecules with unimodal step length distributions are also the ones for which the hypothesis says that half the steps are invisible.

In fact, fitting the histogram in Figure 7.8a leads to a value for the mean rate β, and hence to a completely unambiguous prediction (no further free parameters) for the histogram in Figure 7.8b. That prediction was confirmed.[19] Yildiz and coauthors concluded that the correlation between which sort of step lengths a particular molecule displayed (bimodal versus single-peak histogram of step lengths) and which sort of stepping kinetics it obeyed (Exponential versus other) gave strong support for the model of myosin-V as stepping foot-over-foot.[20]

T_2 *Section 7.5.1′ (page 171) discusses more detailed descriptions of some of the processes introduced in this chapter.*

---

[19]See Problem 7.8.
[20]Later experiments gave more direct evidence in favor of this conclusion; see Media 11.

7.5.2 The randomness parameter can be used to reveal substeps in a kinetic scheme

The previous section discussed the probability density function $\wp(t_w) = \beta^2 t_w e^{-\beta t_w}$, which arose from a sequential process with two alternating substeps, each Poisson.

Your Turn 7D

a. Find the expectation and variance of t_w in this distribution. [*Hint:* If you recall where this distribution came from, then you can get the answers with an *extremely* short derivation.]

b. The **randomness parameter** is defined as $\langle t_w \rangle / \sqrt{\mathrm{var}\, t_w}$; compute it. Compare your result with the corresponding quantity for the Exponential distribution (Idea 7.5, page 159).

c. Suggest how your answers to (b) could be used to invent a practical method for discriminating one- and two-step processes experimentally.

7.6 Computer Simulation

7.6.1 Simple Poisson process

We have seen that, for a simple process, the distribution of waiting times is Exponential.[21] This result is useful if we wish to ask a computer to simulate a Poisson process, because with it, we can avoid stepping through the vast majority of time slots in which nothing happens. We just generate a series of Exponentially distributed intervals $t_{w,1}, \ldots$, then define the time of blip α to be $t_\alpha = t_{w,1} + \cdots + t_{w,\alpha}$, the accumulated waiting time.

A computer's basic random-number function has a Uniform, not an Exponential, distribution. However, we can convert its output to get what we need, by adapting the Example on page 106. This time the transformation function is $G(t_w) = e^{-\beta t_w}$, whose inverse gives $t_w = -\beta^{-1} \ln y$.

Your Turn 7E

a. Think about how the units work in the last formula given.

b. Try this formula on a computer for various values of β, making histograms of the results.

7.6.2 Poisson processes with multiple event types

We'll need a slight extension of these ideas, called the **compound Poisson process**, when we discuss chemical reactions in Chapter 8. Suppose that we wish to simulate a process consisting of two types of blip. Each type arrives independently of the other, in Poisson processes with mean rates β_a and β_b, respectively. We could simulate each series separately and merge the lists, sorting them into a single ascending sequence of blip times accompanied by their types (a or b).

There is another approach, however, that runs faster and admits a crucial generalization that we will need later. We wish to generate a single list $\{(t_\alpha, s_\alpha)\}$, where t_α are the event times (continuous), and s_α are the corresponding event types (discrete).

---

[21]See Idea 7.5 (page 159).

Example a. The successive differences of t_α values reflect the waiting times for *either* type of blip to happen. Find their distribution.
b. Once something happens, we must ask *what* happened on that step. Find the discrete distribution for each s_α.

Solution a. By the merging property, the distribution is Exponential, with $\beta_{tot} = (\beta_a + \beta_b)$.[22]
b. It's a Bernoulli trial, with probability $\xi = \beta_a/(\beta_a + \beta_b)$ to yield an event of type a.

We already know how to get a computer to draw from each of the required distributions. Doing so gives the solution to the problem of simulating the compound Poisson process:

Your Turn 7F

Write a short computer code that uses the result just found to simulate a compound Poisson process. That is, your code should generate the list $\{(t_\alpha, s_\alpha)\}$ and represent it graphically.

THE BIG PICTURE

This chapter concludes our formal study of randomness in biology. We have moved conceptually from random systems that yield one discrete value at a time, to continuous single values, and now on to random processes, which yield a whole time series. At each stage, we found biological applications that involve both characterizing random systems and deciding among competing hypotheses. We have also seen examples, like the Luria-Delbrück experiment, where it was important to be able to simulate the various hypotheses in order to find their predictions in a precise, and hence falsifiable, form.

Chapter 8 will apply these ideas to processes involving chemical reactions.

KEY FORMULAS

- *Exponential distribution:* In the limit where $\Delta t \to 0$, holding fixed β and t_w, the Geometric distribution with probability $\xi = \beta \Delta t$ approaches the continuous form $\wp_{exp}(t_w; \beta) = \beta \exp(-\beta t_w)$. The expectation of the waiting time is $1/\beta$; its variance is $1/\beta^2$. The parameter β has dimensions $\mathbb{T}^{-1}$, as does $\wp_{exp}$.

- *Poisson process:* A random process is a random system, each of whose draws is an increasing sequence of numbers ("blip times"). The Poisson process with mean rate β is a special case, with the properties that (*i*) any infinitesimal time slot from t to $t + \Delta t$ has probability $\beta \Delta t$ of containing a blip, and (*ii*) the number in one such slot is statistically independent of the number in any other (nonoverlapping) slot.
The waiting times in a Poisson process are Exponentially distributed, with expectation β^{-1}.
For a Poisson process with mean rate β, the probability of getting ℓ blips in any time interval of duration T_1 is Poisson distributed, with expectation $\mu = \beta T_1$.

- *Thinning property:* When we randomly eliminate some of the blips in a Poisson process with mean rate β, by subjecting each to an independent Bernoulli trial, the remaining blips form another Poisson process with $\beta' = \xi_{thin}\beta$.

---

[22] See Section 7.3.3.2 (page 161).

- *Merging property:* When we combine the blips from two Poisson processes with mean rates β_1 and β_2, the resulting time series is another Poisson process with $\beta_{\text{tot}} = \beta_1 + \beta_2$.
- *Alternating-step process:* The convolution of two Exponential distributions, each with mean rate β, is not itself an Exponential; its pdf is $t_{\text{w}} \beta^2 e^{-\beta t_{\text{w}}}$.
- *Randomness parameter:* The quantity $\langle t_{\text{w}} \rangle / \sqrt{\text{var}\, t_{\text{w}}}$ can be estimated from experimental data. If the data form a simple Poisson process, then this quantity will be equal to one; if on the contrary the blips in the data reflect two or more obligatory substeps, each of which has Exponentially-distributed waiting times, then this quantity will be larger than one.

FURTHER READING

Semipopular:
Molecular machines: Hoffmann, 2012.

Intermediate:
Allen, 2011; Jones et al., 2009; Wilkinson, 2006.
Molecular motors: Dill & Bromberg, 2010, chapt. 29; Nelson, 2014, chapt. 10; Phillips et al., 2012, chapt. 16; Yanagida & Ishii, 2009.

Technical:
Jacobs, 2010.
Yildiz et al., 2003.
Kinetics of other enzymes and motors: Hinterdorfer & van Oijen, 2009, chapts. 6–7.

T_2 **Track 2**

7.2′ More about motor stepping

Section 7.2 made some idealizations in order to arrive at a simple model. Much current research involves finding more realistic models that are complex enough to explain data, simple enough to be tractable, and physically realistic enough to be more than just data summaries.

For example, after a motor steps, the world contains one fewer ATP (and one more each of ADP and phosphate, the products of ATP hydrolysis). Our discussion implicitly assumed that so much ATP is available, and the solution is so well mixed, that depletion during the course of an experiment is negligible. In some experiments, this is ensured by constantly flowing fresh ATP-bearing solution into the chamber; in cells, homeostatic mechanisms adjust ATP production to meet demand.[23]

In other words, we assumed that both the motor *and its environment* have no memory of prior steps. Chapter 8 will develop ideas relevant for situations where this Markov property may not be assumed.

We also neglected the possibility of backward steps. In principle, a motor could bind an ADP and a phosphate from solution, step backward, and emit an ATP. Inside living cells, the concentration of ATP is high enough, and those of ADP and phosphate low enough, that such steps are rare.

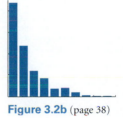

Figure 3.2b (page 38)

T_2 **Track 2**

The main text stated that enzyme turnovers follow a Poisson process. Also, Figure 3.2b suggests that the arrivals of the energy packets we call "light" follow such a random process. Although these statements are good qualitative guides, each needs some elaboration.

7.5.1′a More detailed models of enzyme turnovers

Remarkably, enzymes do display long-term "memory" effects, as seen in these examples:

- The model of myosin-V stepping discussed in the main text implicitly assumed that the motor itself has no "stopwatch" that affects its binding probability based on recent history. However, immediately after a binding event, there is a "dead time" while the step is actually carried out. During this short time, the motor cannot initiate another step. The time bins of Figure 7.8 are too long to disclose this phenomenon, but it has been seen.

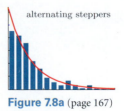

Figure 7.8a (page 167)

- An enzyme can get into substates that can persist over many processing cycles, and that have different mean rates from other substates. The enzyme cycles through these substates, giving its apparent mean rate a long-term drift. More complex Markov models than the one in the main text are needed to account for this behavior (English et al., 2006).

7.5.1′b More detailed models of photon arrivals

Actually, only laser light precisely follows a Poisson process. "Incoherent" light, for example, from the Sun, has more complicated photon statistics, with some autocorrelation.

---

[23] See Chapter 9.

PROBLEMS

7.1 Ventricular fibrillation

A patient with heart disease will sometimes enter "ventricular fibrillation," leading to cardiac arrest. The following table shows data on the fraction of patients failing to regain normal heart rhythm after attempts at defibrillation by electric shock, in a particular clinical trial:

number of attempts	fraction persisting in fibrillation
1	0.37
2	0.15
3	0.07
4	0.02

Assume that with 0 attempts there are no spontaneous recoveries. Also assume that the probability of recovery on each attempt is independent of any prior attempts. Suggest a formula that roughly matches these data. If your formula contains one or more parameters, estimate their values. Make a graph that compares your formula's prediction with the data above. What additional information would you need in order to assert a credible interval on your parameter value?

7.2 Basic properties of $\mathcal{P}_{geom}$

a. Continue along the lines of Your Turn 3D to find the expectation and variance of the Geometric distribution. [*Hint:* You can imitate the Example on page 77. Consider the quantity

$$\frac{d}{d\xi} \sum_{j=0}^{\infty} (1 - \xi)^j.$$

Evaluate this quantity in two different ways, and set your expressions equal to each other. The resulting identity will be useful.]

b. Discuss how your answers to (a) behave as ξ approaches 0 and 1, and how these behaviors qualitatively conform to your expectations.

c. Review Your Turn 3D (page 48), then modify it as follows. Take the Taylor series expansion for $1/(1-z)$, multiply by $(1 - z^K)$, and simplify the result (see page 19). Use your answer to find the total probability that, in the Geometric distribution, the first "success" occurs at *or before* the Kth attempt.

d. Now take the continuous-time limit of your results in (a) and compare them with the corresponding facts about the Exponential distribution (see the Example on page 160).

7.3 Radiation-induced mutation

Suppose that we maintain some single-cell organisms under conditions where they don't divide. Periodically we subject them to a dose of radiation, which sometimes induces a mutation in a particular gene. Suppose that the probability for a given individual to form a mutation after a dose is $\xi = 10^{-3}$, regardless how many doses have previously been given. Let j be the number of doses after which a particular individual develops its first mutation.

a. State the probability distribution of the random variable j.

b. What is the expectation, $\langle j \rangle$?

c. Now find the variance of j.

7.4 Winning streaks via simulation

Section 7.3.1 found a formula for the number of attempts we must make before "success" in a sequence of independent Bernoulli trials. In this problem, you'll check that result by a computer simulation. Simulation can be helpful when studying more complex random processes, for which analytic results are not available.

Computers are very fast at finding patterns in strings of symbols. You can make a long string of N random digits by successively appending the string `"1"` or `"0"` to a growing string called `flipstr`. Then you can ask the computer to search `flipstr` for occurrences of the substring `"1"`, and report a list of all the positions in the long string that match it. The differences between successive entries in this list are related to the length of runs of consecutive `"0"` entries. Then you can tabulate how often various waiting times were observed, and make a histogram.

Before carrying out this simulation, you should try to guess what your graph will look like. Nick reasoned, "Because <u>heads</u> is a rare outcome, once we get a <u>tails</u> we're likely to get a lot of them in a row, so short strings of zeros will be less probable than medium-long strings. But eventually we're bound to get a <u>heads</u>, so *very* long strings of zeros are also less common than medium-long strings. So the distribution should have a bump." Think about it—is that the right reasoning?

Now get your answer, as follows. Write a simple simulation of the sort described above, with $N = 1000$ "attempts" and $\xi = 0.08$. Plot the frequencies of appearance of strings of various lengths, both on regular and on semilog axes. Is this a familiar-looking probability distribution? Repeat with $N = 50\,000$.

7.5 Transformation of exponential distribution

Suppose that a pdf is known to be of Exponential form, $\wp_t(t) = \beta \exp(-\beta t)$. Let $y = \ln(t/(1\,\mathrm{s}))$ and find the corresponding function $\wp_y(y)$. Unlike the Exponential, the transformed distribution has a bump, whose location y_* tells something about the rate parameter β. Find this relation.

7.6 Likelihood analysis of a poisson process

Suppose that you measure a lot of waiting times from some random process, such as the stepping of a molecular motor. You believe that these times are draws from an Exponential distribution: $\wp(t) = Ae^{-\beta t}$, where A and β are constants. But you don't know the values of these constants. Moreover, you only had time to measure six steps, or five waiting times $t_1, \ldots, t_5$, before the experiment ended.[24]

a. A and β are not independent quantities: Express A in terms of β. State some appropriate units for A and for β.

b. Write a symbolic expression for the likelihood of any particular value of β, in terms of the measured data $t_1, \ldots, t_5$.

c. Find the maximum-likelihood estimate of the parameter β; give a short derivation of your formula.

7.7 Illustrate thinning property

a. Obtain Dataset 3, which gives blip arrival times from a sensitive light detector in dim light. Have a computer find the waiting times between events, and histogram them.

---

[24] Perhaps the motor detached from its track in the middle of interval #6.

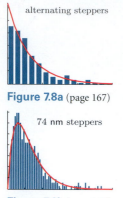

Figure 7.8a (page 167)

Figure 7.8b (page 167)

b. Apply an independent Bernoulli trial to each event in (a), which accepts 60% of them and rejects the rest. Again histogram the waiting times, and comment.

7.8 Hidden steps in myosin-V

If you haven't done Problem 7.6, do it before this problem. Figure 7.8 shows histograms of waiting times for the stepping of two classes of fluorescently labeled myosin-V molecules. The experimenters classified each motor molecule that they observed, according to whether it took steps of two alternating lengths or just a single length. For each class, they reported the frequencies for taking a step after various waiting times. For example, 39 motor steps were observed with t_w between 0 and 1 s.

a. Obtain Dataset 9, and use it to generate the two histograms in Figure 7.8.

b. Section 7.5.1 (page 165) proposed a physical model for this class of motors, in which the waiting times were distributed according to an Exponential distribution. Use the method in Problem 7.6 to infer from the data the value of β for the molecules that took steps of alternating lengths. [*Hint:* The model assumes that all steps are independent, so the *order* in which various waiting times were observed is immaterial. What matters is just the number of times that each t_w was observed. The data have been binned; Dataset 9 contains a list whose first entry $(0.5, 39)$ means that the bin centered on 0.5 s contained 39 observed steps. Make the approximation that all 39 of these steps had t_w exactly equal to 0.5 s (the middle of the first bin), and so on.]

c. Graph the corresponding probability density function superimposed on the data. To make a proper comparison, rescale the pdf so that it becomes a prediction of the frequencies.

d. Section 7.5.1 also proposed that in the other class of molecules, half the steps were unobserved. Repeat (b–c) with the necessary changes.

e. Compare the values of β that you obtained in (b,d). If they are similar (or dissimilar), how do you interpret that?

f. Now consider a different hypothesis that says that each observed event is the last of a series of m sequential events, each of which is an independent, identical Poisson process. (Thus, (d) considered the special case $m = 2$.) Without doing the math, qualitatively what sort of distribution of wait times $\wp(t_w)$ would you expect for $m = 10$?

7.9 Asymmetric foot-over-foot cycle

Suppose that some enzyme reaction consists of two steps whose waiting times are independent, except that they must take place in strict alternation: $A_1 B_1 A_2 B_2 A_3 \cdots$. For example, the enzyme hexokinase alternates between cleaving a phosphate from ATP and transferring it to glucose. Or we could study a motor that walks foot-over-foot, but unlike the main text we won't assume equal rate constants for each foot.

Successive pauses are statistically independent. The pause between an A step and the next B step is distributed according to $\wp_{AB}(t) = \beta e^{-\beta t_w}$, where β is a constant with dimensions $\mathbb{T}^{-1}$. The pause between a B step and the next A step is similarly distributed, but with a different mean rate β'. Find the probability density function for the time between two successive A steps.

7.10 Staircase plot

a. Use a computer to simulate 30 draws from the Exponential distribution with mean rate $0.3\,\mathrm{s}^{-1}$. Call the results $w(1), \ldots, w(30)$. Create a list with the cumulative sums, then duplicate them and append a 0, to get $0,\ w(1),\ w(1),\ w(1)+w(2),$

w(1) + w(2), ... Create another list x with entries 0, 0, step, step, 2*step, 2*step, ..., where step=37, and graph the w's versus x. Interpret your graph by identifying the entries of w with the interstep waiting times of length step.

b. Actually, your graph is not quite a realistic simulation of the observed steps of myosin-V. Adapt your code to account for the alternating step lengths observed by Yildiz and coauthors in one class of fluorescently labeled motor molecules.

c. This time adapt your code to account for the non-Exponential distribution of waiting times observed in the other class of motors. Does your graph resemble some data discussed in the main text?

7.11 Thinning via simulation

Take the list of waiting times from your computer simulation in Your Turn 7E, and modify it by deleting some blips, as follows. Walk through the list, and for each entry $t_{w,i}$ make a Bernoulli trial with some probability ξ_*. If the outcome is <u>heads</u>, move to the next list entry; otherwise, delete the entry $t_{w,i}$ and add its value to that of the next entry in the list. Run this modified simulation, histogram the outcomes, and so check the thinning property (Section 7.3.3.1).

7.12 Convolution via simulation

a. Use the method in Section 7.6.1 (page 168) to simulate draws from the Exponential distribution with expectation 1 s.

b. Simulate the random variable $z = x + y$, where x and y are independent random variables with the distribution used in (a). Generate a lot of draws from this distribution, and histogram them.

c. Compare your result in (b) with the distribution found in the Example on page 166.

d. Next simulate a random variable defined as the sum of *50* independent, Exponentially distributed variables. Comment on your result in the light of Problem 5.7 (page 119).

7.13 $\boxed{T_2}$ Fit count data

Radioactive tagging is important in many biological assays. A sample of radioactive substance furnishes another physical system found to produce blips in a Poisson process.

Suppose that we have a radioactive source of fixed intensity, and a detector that registers individual radiation particles emitted from the source. The average rate at which the detector emits blips depends on its distance L to the source. We measure the rate by holding the detector at a series of fixed distances $L_1, \ldots, L_N$. At each distance, we count the blips on the detector over a fixed time $\Delta T = 15$ s and record the results.

Our physical model for these data is the inverse-square law: We expect the observed number of detector blips at each fixed L to be drawn from a Poisson distribution with expectation equal to A/L^2 for some constant A.[25] We wish to test that model. Also we would like to know the constant of proportionality A, so that we can use it to deduce the rate for any value L (not just the ones that we measured). In other words, we'd like to summarize our data with an **interpolation formula**.

a. One way to proceed might be to plot the observed number of blips y versus the variable $x = (L)^{-2}$, then lay a ruler along the plot in a way that passes through $(0, 0)$ and roughly

---

[25]This constant reflects the intensity of the source, the duration of each measurement, and the size and efficiency of the detector.

tracks the data points. Obtain Dataset 10 and follow this procedure to estimate A as the slope of this line.

b. A better approach would make an objective fit to the data. Idea 6.8 (page 139) is not applicable to this situation—why not?

c. But the logic leading to Idea 6.8 is applicable, with a simple modification. Carry this out, plot the log-likelihood as a function of A, and choose the optimal value. Your answer from (a) gives a good starting guess for the value of A; try various values near that. Add the best-fit line according to maximum likelihood to the plot you made in (a).

d. You can estimate the integral of the likelihood function by finding its sum over the range of A values you graphed in (c) and normalizing. Use those values to estimate a 95% credible interval for the value of A.

Comment: You may still be asking, "But is the best fit *good*? Is the likelihood *big enough* to call it good?" One way to address this is to take your best-fit model, use it to generate lots of *simulated datasets* by drawing from appropriate Poisson distributions at each x_i, calculate the likelihood function for each one, and see if the typical values thus obtained are comparable to the best-fit likelihood you found by using the real data.

Figure 7.7 (page 165)

7.14 $\boxed{T_2}$ **Quantized neurotransmitter release**

The goal of this problem is to predict the data in Figure 7.7 with no fitting parameters. First obtain Dataset 11, which contains binned data on the frequencies with which various peak voltage changes were observed in a muscle cell stimulated by a motor neuron. In a separate measurement, the authors also studied spontaneous events (no stimulus), and found the sample mean of the peak voltage to be $\mu_V = 0.40\,\text{mV}$ and its estimated variance to be $\sigma^2 = 0.00825\,\text{mV}^2$.

The physical model discussed in the text states that each response is the sum of ℓ independent random variables, which are the responses caused by the release of ℓ vesicles. Each of these constituents is itself assumed to follow a Gaussian distribution, with expectation and variance given by those found for spontaneous events.

a. Find the predicted distribution of the responses for the class of events with some definite value of ℓ.

b. The model also assumes that ℓ is itself a Poisson random variable. Find its expectation μ_ℓ by computing the sample mean of all the responses in the dataset, and dividing by the mean response from a single vesicle, μ_V.

c. Take the distributions you found in (a) for $\ell > 0$, scale each by $\mathcal{P}_{\text{pois}}(\ell; \mu_\ell)$, and add them to find an overall pdf. Plot this pdf.

d. Superimpose the estimated pdf obtained from the data on your graph from (c).

e. The experimenters also found in this experiment that, in 18 out of 198 trials, there was no response at all. Compute $\mathcal{P}_{\text{pois}}(0; \mu_\ell)$ and comment.

PART III

Control in Cells

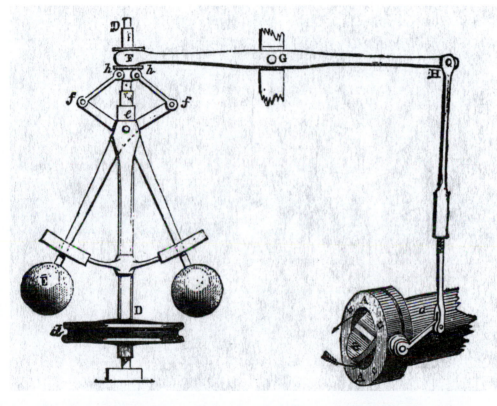

The centrifugal governor, a mechanical feedback mechanism. [From *Discoveries and inventions of the nineteenth century*, by R Routledge, 13th edition, published 1900.]

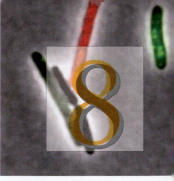

Randomness in Cellular Processes

I think there is a world market for maybe five computers.
—Thomas Watson, Chairman of IBM, 1943

8.1 Signpost

Earlier chapters have emphasized that randomness pervades biology and physics, from subcellular actors (such as motor proteins), all the way up to populations (such as colonies of bacteria). Ultimately, this randomness has its origin in physical processes, for example, thermal motion of molecules. Although it may sound paradoxical, we have found that it is possible to characterize randomness precisely and reproducibly, sometimes with the help of physical models.

This chapter will focus on the particular arena of cellular physiology. A living cell is made of molecules, so those molecules implement all its activities. It is therefore important to understand in what ways, and to what extent, those activities are random.

The Focus Question is

Biological question: How and when will a collection of random processes yield overall dynamics that are nearly predictable?

Physical idea: Deterministic collective behavior can emerge when the copy number of each actor is large.

8.2 Random Walks and Beyond

8.2.1 Situations studied so far

8.2.1.1 Periodic stepping in random directions

One of the fundamental examples of randomness given in Chapter 3 was Brownian motion.[1] Earlier sections discussed an idealization of this kind of motion as a **random walk**: We imagine that an object periodically takes a step to the left or right, with length always equal to some constant d. The only state variable is the object's current position. This simple random process reproduces the main observed fact about Brownian motion, which is that the mean-square deviation of the displacement after many steps is proportional to the square root of the elapsed time.[2] Still, we may worry that much of the chaos of real diffusion is missing from the model. Creating a more realistic picture of random walks will also show us how to model the kinetics of chemical reactions.

8.2.1.2 Irregularly timed, unidirectional steps

We studied another kind of random process in Chapter 7: the stepping of a processive molecular motor, such as myosin-V. We allowed for randomness in the step times, instead of waiting for some fictitious clock to tick. But the step displacements themselves were predictable: The experimental data showed that they are always in the same direction, and of roughly the same length.

8.2.2 A more realistic model of Brownian motion includes both random step times and random step directions

One way to improve the Brownian motion model is to combine the two preceding ideas (random step times and random directions). As usual, we begin by simplifying the analysis to a single spatial dimension. Then one reasonable model would be to say that there is a certain fixed probability per unit time, β, of a small suspended particle being kicked to the left. Independently, there is also a fixed probability per unit time β of the particle being kicked to the right. Chapter 7 discussed one way to simulate such a process:[3] We first consider the merged Poisson process with mean rate 2β, and draw a sequence of waiting times from it. Then for each of these times, we draw from a Bernoulli trial to determine whether the step at that time was rightward or leftward. Finally, we make cumulative sums of the steps, to find the complete simulated trajectory as a function of time.

Figure 8.1 shows two examples of the outcome of such a simulation. Like the simpler random walks studied earlier, this one has the property that after enough steps are taken, a trajectory can end up arbitrarily far from its starting point. Even if several different walkers all start at the same position x_{ini} (variance is zero for the starting position), the variance of their positions after a long time, $\text{var}(x(t))$, grows without bound as t increases. There is no limiting distribution of positions as time goes to infinity.

---

[1] See point **5** on page 36, and follow-up points **5a** and **5b**.
[2] See Problem 4.5.
[3] See Section 7.6.2 (page 168).

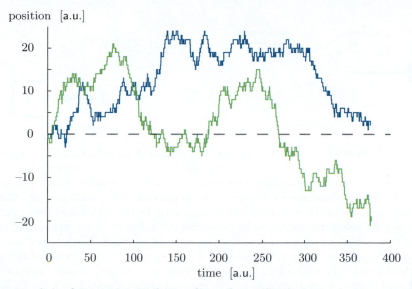

Figure 8.1 [Computer simulation.] **Computer simulation of a random walk.** The curves show two runs of the simulation. In each case, 400 steps were taken, with Exponentially distributed waiting times and mean rate of one step per unit time. Each step was of unit length, with direction specified by the outcome of a Bernoulli trial with equal probability to step up or down ($\xi = 1/2$). See Problem 8.1.

8.3 Molecular Population Dynamics as a Markov Process

Brownian motion is an example of a Markov process:[4] In order to know the probability distribution of the position at time t, all we need to know is the actual position at any *one* time t' prior to t. Any additional knowledge of the actual position at a time t'' earlier than t' gives us no additional information relevant to $\wp_{x(t)}$. If a random process has a limited amount of "state" information at any time, and the property that knowing this state at one t' completely determines the pdf of possible states at any later time t, then the process is called "Markov."

The examples discussed in Section 8.2 (stepping with irregular directions, times, or both) all have the Markov property. Chemical reactions in a well-mixed system, although also Markovian, have an additional complication:[5] They, too, occur after waiting times that reflect a Poisson process, but with a rate that depends on the concentrations of the reacting molecules.

The following discussion will introduce a number of variables, so we summarize them here:

Δt	time step, eventually taken to zero
ℓ_i	number of molecules at time $t_i = (\Delta t)i$, a random variable
ℓ_{ini}	initial value, a constant
β_s	mean rate of mRNA synthesis, a constant
$\beta_\varnothing$	mean rate of mRNA clearance (varies over time)
$k_\varnothing$	clearance rate constant
ℓ_*	steady final value

---

[4]See Section 3.2.1 (page 36).

[5]See point **5c** on page 40.

Figure 8.2 Example of a birth-death process. (a) [Schematic.] A gene (*wide arrow*) directs the synthesis of messenger RNA, which is eventually degraded (cleared) by enzymatic machinery in a cell. (b) [Network diagram.] Abstract representation as a network diagram. The *box* represents a state variable of the system, the inventory (number of copies) of some molecular species X, labeled by its name. Incoming and outgoing *black arrows* represent processes (biochemical reactions) that increase or decrease the inventory. Substrate molecules needed to synthesize X are assumed to be maintained at a fixed concentration by processes not of interest to us; they are collectively represented by the symbol $\emptyset$. Product molecules arising from the clearance of X are assumed not to affect the reaction rates; they, too, are collectively represented by a $\emptyset$. The enzymatic machinery that performs both reactions, and even the gene that directs the synthesis, are not shown at all. The two arrows are assumed to be irreversible reactions, a reasonable assumption for many cellular processes. In situations when this may not be assumed, the reverse reactions will be explicitly indicated in the network diagram by separate arrows. The *dashed arrow* is an influence line indicating that the rate of clearance depends on the level at which X is present. This particular dependence is usually tacitly assumed, however; henceforth we will not explicitly indicate it.

8.3.1 The birth-death process describes population fluctuations of a chemical species in a cell

To make these ideas concrete, imagine a very simple system called the **birth-death process** (Figure 8.2). The system involves just two chemical reactions, represented by arrows in panel (b), and one state variable, represented by the box. The state variable is the number ℓ of molecules of some species X; the processes modify this number.

The "synthesis" reaction is assumed to have fixed probability per unit time β_s to create (synthesize) new molecules of X. Such a reaction is called **zeroth order** to indicate that its mean rate is assumed to be independent of the numbers of other molecules present (it's proportional to those numbers raised to the power zero). Strictly speaking, no reaction can be independent of every molecular population. However, some cellular processes, in some situations, are effectively zeroth order, because the cell maintains roughly constant populations of the needed ingredients (substrate molecules and enzymes to process them), and the distributions of those molecules throughout the cell are unchanging in time.[6]

The other reaction, "clearance," has probability per unit time $\beta_\emptyset$ to eliminate an X molecule, for example, by converting it to something else. Unlike β_s, however, $\beta_\emptyset$ is assumed to depend on ℓ, via[7]

$$\beta_\emptyset = k_\emptyset \ell. \tag{8.1}$$

---

[6] $\boxed{T_2}$ We are also assuming that the population of product molecules is too small to inhibit additional production. In vitro experiments with molecular motors are another case where the zeroth-order assumption is reasonable, because substrate molecules (ATP) in the chamber are constantly replenished, and product molecules (ADP) are constantly being removed. Thus, Chapter 7 implicitly assumed that no appreciable change in the concentrations occurs over the course of the experiment.

[7] Reaction rates in a cell may also change with time due to changes in cell volume; see Section 9.4.5. Here we assume that the volume is constant.

We can think of this formula in terms of the merging property: Each of ℓ molecules has its own independent probability per unit time to be cleared, leading to a merged process for the overall population to decrease by one unit.

The constant of proportionality $k_\emptyset$ is called the **clearance rate constant**. This kind of reaction is called **first order**, because Equation 8.1 assumes that its rate is proportional to the first power of ℓ; for example, the reaction stops when the supply of X is exhausted ($\ell = 0$).

The birth-death process is reminiscent of a fundamental theme in cell biology: A gene, together with the cell's transcription machinery, implements the first arrow of Figures 8.2a,b by synthesizing messenger RNA (mRNA) molecules, a process called **transcription**. If the number of copies of the gene and the population of RNA polymerase machines are both fixed, then it seems reasonable to assume that this reaction is effectively zeroth order. The box in the figure represents the number of RNA molecules present in a cell, and the arrow on the right represents their eventual destruction. Certainly this picture is highly simplified: Cells also duplicate their genes, regulate their transcription, divide, and so on. We will add those features to the physical model step by step. For now, however, we consider only the two processes and one inventory shown in Figure 8.2. We would like to answer questions concerning both the overall development of the system, and its variability from one trial to the next.

We can make progress understanding the birth-death process by making an analogy: It is *just another kind of random walk.* Instead of wandering in ordinary space, the system wanders in its *state space,* in this case the number line of nonnegative integers ℓ. The only new feature is that, unlike in Section 8.2.2, one of the reaction rates is not a constant (see Equation 8.1). In fact, the value of that mean rate at any moment is itself a random variable, because it depends on ℓ. Despite this added level of complexity, however, the birth-death process still has the Markov property. To show this, we now find how the pdf for $\ell(t)$ depends on the system's prior history.

As usual, we begin by slicing time into slots of very short duration Δt, so that slot i begins at time $t_i = (\Delta t)i$, and by writing the population as ℓ_i instead of $\ell(t)$. During any slot i, the most probable outcome is that nothing new happens, so ℓ is unchanged: $\ell_{i+1} = \ell_i$. The next most probable outcomes are that synthesis, or clearance, takes place in the time slot. The probability of two or more reactions in Δt is negligible, for small enough Δt, and so we only need to consider the cases where ℓ changes by ± 1, or not at all. Expressing this reasoning in a formula,

$$\mathcal{P}(\ell_{i+1} \mid \ell_1, \dots, \ell_i) = \begin{cases} (\Delta t)\beta_s & \text{if } \ell_{i+1} = \ell_i + 1; \quad \text{(synthesis)} \\ (\Delta t)k_\emptyset\ell_i & \text{if } \ell_{i+1} = \ell_i - 1; \quad \text{(clearance)} \\ 1 - (\Delta t)(\beta_s + k_\emptyset\ell_i) & \text{if } \ell_{i+1} = \ell_i; \quad \text{(no reaction)} \\ 0 & \text{otherwise.} \end{cases} \quad (8.2)$$

The right-hand side depends on ℓ_i, but not on $\ell_1, \dots, \ell_{i-1}$, so Equation 8.2 defines a Markov process. We can summarize this formula in words:

The birth-death process resembles a compound Poisson process during the waiting time between any two consecutive reaction steps. After each step, however, the mean rate of the clearance reaction can change. (8.3)

Given some starting state ℓ_* at time zero, the above characterization of the birth-death process determines the probability distribution for any question we may wish to ask about the state at a later time.

8.3.2 In the continuous, deterministic approximation, a birth-death process approaches a steady population level

Equation 8.2 looks complicated. Before we attempt to analyze the behavior arising from such a model, we should pause to get some intuition from an approximate treatment.

Some chemical reactions involve huge numbers of molecules. In this case, ℓ is a very large integer, and changing it by one unit makes a negligible relative change. In such situations, it makes sense to pretend that ℓ is actually continuous. Moreover, we have seen in several examples how large numbers imply small relative fluctuations in a discrete random quantity; so it seems likely that in these situations we may also pretend that ℓ varies deterministically. Restating Equation 8.2 with these simplifications yields the **continuous, deterministic approximation**, in which ℓ changes with time according to[8]

$$\frac{d\ell}{dt} = \beta_s - k_\emptyset \ell. \tag{8.4}$$

Example Explain why Equation 8.4 emerges from Equation 8.2 in this limit.

Solution First compute the expectation of ℓ_{i+1} from Equation 8.2:

$$\langle \ell_{i+1} \rangle = \sum_{\ell_i} \mathcal{P}(\ell_i) \left[(\ell_i + 1)(\Delta t)\beta_s + (\ell_i - 1)(\Delta t)k_\emptyset \ell_i + \ell_i \big(1 - \Delta t(\beta_s + k_\emptyset \ell_i)\big) \right].$$

Now subtract $\langle \ell_i \rangle$ from both sides and divide by Δt, to find

$$\frac{\langle \ell_{i+1} \rangle - \langle \ell_i \rangle}{\Delta t} = \beta_s - k_\emptyset \langle \ell_i \rangle.$$

Suppose that the original distribution has small relative standard deviation. Because ℓ is large, and the spread in its distribution only increases by less than 1 unit in a time step (Equation 8.2), the new distribution will also be sharply peaked. So we may drop the expectation symbols, recovering Equation 8.4.

To solve Equation 8.4, first notice that it has a steady state when the population ℓ equals $\ell_* = \beta_s/k_\emptyset$. This makes us suspect that the equation might look simpler if we change variables from ℓ to $x = \ell - \ell_*$, and indeed it becomes $dx/dt = -k_\emptyset x$, whose solution is $x(t) = Be^{-k_\emptyset t}$ for any constant B. Choosing B to ensure $\ell_{\text{ini}} = 0$ (initially there are no X molecules) yields the particular solution

$$\ell(t) = (\beta_s/k_\emptyset)\big(1 - e^{-k_\emptyset t}\big). \tag{8.5}$$

---

[8]We previously met Equation 8.4 in the context of virus dynamics (Chapter 1).

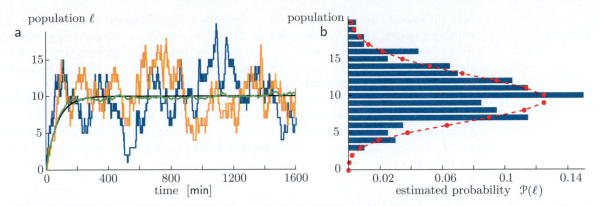

Figure 8.3 [Computer simulations.] **Behavior of a birth-death process.** (a) The *orange and blue traces* show two simulated time series (see Your Turns 8A–8B(a)). The *green trace* shows, at each time, the sample mean of the population ℓ over 200 such instances (see Problem 8.2). The *black curve* shows the corresponding solution in the continuous, deterministic approximation (Equation 8.4). (b) After the system comes to steady state, there is a broad distribution of ℓ values across instances (*bars*). The *red dots* show the Poisson distribution with $\mu = \beta_s/k_\emptyset$ for comparison (see Your Turn 8C).

Thus, initially the number of X molecules rises linearly with time, but then it levels off (**saturates**) as the clearance reaction speeds up, until a steady state is reached at $\ell(t \to \infty) = \ell_*$. The black curve in Figure 8.3a shows this solution.

8.3.3 The Gillespie algorithm

To get beyond the continuous, deterministic approximation, recall one of the lessons of the Luria-Delbrück experiment (Section 4.4, page 81): It is sometimes easier to *simulate* a random system than to derive analytic results. We can estimate whatever probabilities we wish to predict by running the simulation many times and making histograms of the quantities of interest.

Idea 8.3 suggests an approach to simulating the birth-death process, by modifying our simulation of the compound Poisson process in Section 7.6.2 (page 168). Suppose that ℓ has a known value at time zero. Then,

1. Draw the first waiting time $t_{w,1}$ from the Exponential distribution with rate $\beta_{\text{tot}} = \beta_s + k_\emptyset\ell$.

2. Next, determine which reaction happened at that time by drawing from a Bernoulli trial distribution with probability ξ, where[9]

$$\xi = \frac{(\Delta t)\beta_s}{(\Delta t)\beta_s + (\Delta t)k_\emptyset\ell} = \frac{\beta_s}{\beta_s + k_\emptyset\ell}. \tag{8.6}$$

The probability to increase ℓ is ξ; that to decrease ℓ is $1 - \xi$. The quantities ξ and $1 - \xi$ are sometimes called the "relative propensities" of the two reactions.

3. Update ℓ by adding or subtracting 1, depending on the outcome of the Bernoulli trial.

4. Repeat.

Steps **1–4** are a simplified version of an algorithm proposed by D. Gillespie. They amount to simulating a slightly different compound Poisson process at every time step, because

---

[9] One way to derive Equation 8.6 is to find the conditional probability $\mathcal{P}(\ell \text{ increases} \,|\, \ell \text{ changes})$ from Equation 8.2.

both the overall rate and ξ depend on ℓ, which itself depends on the prior history of the simulated system. This dependence is quite limited, however: Knowing the state at one time determines all the probabilities for the next step (and hence all subsequent steps). That is, *the Gillespie algorithm is a method for simulating general Markov processes,* including the birth-death process and other chemical reaction networks.

The algorithm just outlined yields a set of waiting times $\{t_{w,\alpha}\}$, which we can convert to absolute times by forming cumulative sums: $t_\alpha = t_{w,1} + \cdots + t_{w,\alpha}$. It also yields a set of increments $\{\Delta\ell_\alpha\}$, each equal to ± 1, which we convert to absolute numbers in the same way: $\ell_\alpha = \ell_{ini} + \Delta\ell_1 + \cdots + \Delta\ell_\alpha$. Figure 8.3a shows a typical result, and compares it with the behavior of the continuous, deterministic approximation.

Your Turn 8A

Implement the algorithm just outlined on a computer: Write a function that accepts two input arguments `lini` and `T`, and generates two output vectors `ts` and `ls`. The argument `lini` is the initial number of molecules of X. `T` is the total time to simulate, in minutes. `ts` is the list of t_α's, and `ls` is the list of the corresponding ℓ_α's just after each of the transition times listed in `ts`. Assume that $\beta_s = 0.15/\text{min}$ and $k_\emptyset = 0.014/\text{min}$.

Your Turn 8B

a. Write a "wrapper" program that calls the function you wrote in Your Turn 8A with `lini` $= 0$ and `T` $= 1600$, then plots the resulting `ls` versus the `ts`. Run it a few times, plot the results, and comment on what you see.
b. Repeat with faster synthesis, $\beta_s = 1.5/\text{min}$, but the same clearance rate constant $k_\emptyset = 0.014/\text{min}$. Compare and contrast your result with (a).

Your answer to Your Turn 8B will include a graph similar to Figure 8.3a. It shows molecule number ℓ saturating, as expected, but still it is very different from the corresponding solution in the continuous, deterministic approximation.[10]

The Gillespie algorithm can be extended to handle cases with more than two reactions. At any time, we find the rates for all available reactions, sum them, and draw a waiting time from an appropriate Exponential distribution (step **1** on page 185). Then we find the list of all relative propensities, analogous to Equation 8.6. By definition, these numbers sum to 1, so they define a discrete probability distribution. We select which reaction occurred by drawing from this distribution;[11] then we accumulate all the changes at each time step to find the time course of ℓ.

8.3.4 The birth-death process undergoes fluctuations in its steady state

Figure 8.3 shows that the "steady" (late-time) state of the birth-death process can actually be pretty lively. No matter how long we wait, there is always a finite spread of ℓ values. In fact,

The steady-state population in the birth-death process is Poisson distributed, with expectation $\beta_s/k_\emptyset$. (8.7)

---

[10]You'll find a connection between these two approaches in Problem 8.2.
[11]See the method in Section 4.2.5 (page 73).

Your Turn 8C

Continue Your Turns 8A–8B: Equation 8.5 suggests that the birth-death process will have come to its steady state at the end of $T = 300$ min. Histogram the distribution of final values ℓ_T across 150 trials. What further steps could you take to confirm Idea 8.7?

Despite the fluctuation, the birth-death process exhibits a bit more self-discipline than the original random walk, which never settles down to any steady state (the spread of x values grows without limit, Problem 8.1). To understand the distinction, remember that in the birth-death process there is a "hard wall" at $\ell = 0$; if the system approaches that point, it gets "repelled" by the imbalance between synthesis and clearance. Likewise, although there is no upper bound on ℓ, nevertheless if the system wanders to large ℓ values it gets "pulled back," by an imbalance in the opposite sense.

Idea 8.7 has an implication that will be important later: Because the Poisson distribution's relative standard deviation[12] is $\mu^{-1/2}$, we see that the steady-state population of a molecule will be close to its value calculated with the continuous, deterministic approximation, if that value is large. Indeed, you may have noticed a stronger result in your solution to Your Turn 8B:

> *The continuous, deterministic approximation becomes accurate when molecule numbers are high.* $\qquad$ (8.8)

$\boxed{T_2}$ *Section 8.3.4' (page 195) gives an analytic derivation of Idea 8.7.*

8.4 Gene Expression

Cells create themselves by metabolizing food and making proteins, lipids, and other biomolecules. The basic synthetic mechanism is shown in Figure 8.4: DNA is **transcribed** into a messenger RNA (mRNA) molecule by an enzyme called **RNA polymerase**. Next, the resulting **transcript** is **translated** into a chain of amino acids by another enzyme complex, called the **ribosome**. The chain of amino acids then folds itself into a functioning protein (the **gene product**). The entire process is called **gene expression**. If we create a DNA sequence with two protein-coding sequences next to each other, the polymerase will generate a single mRNA containing both; translation will then create a single amino acid chain, which can fold into a combined **fusion protein**, with two domains corresponding to the two protein sequences, covalently linked into a single object.

Enzymes are themselves proteins (or complexes of protein with RNA or other cofactors). And other complex molecules, such as lipids, are in turn synthesized by enzymes. Thus, gene expression lies at the heart of all cellular processes.

$\boxed{T_2}$ *Section 8.4' (page 197) mentions some finer points about gene expression.*

8.4.1 Exact mRNA populations can be monitored in living cells

Each step in gene expression is a biochemical reaction, and hence subject to randomness. For example, Section 8.3 suggested that it would be reasonable to model the inventory of mRNA from any particular gene via the birth-death process represented symbolically in Figure 8.2. I. Golding and coauthors tested this hypothesis in the bacterium *Escherichia coli*,

---

[12] See Equation 4.7 (page 78).

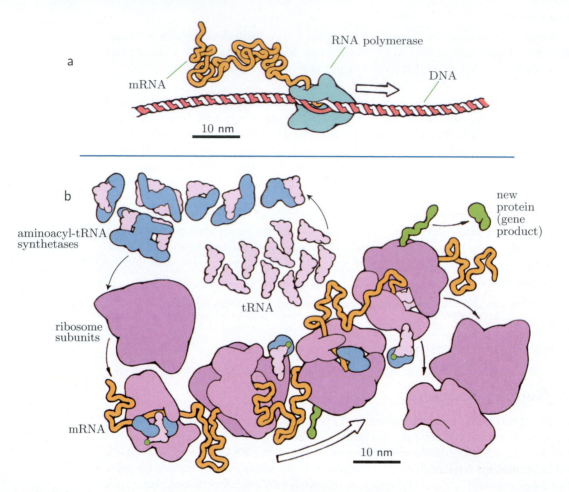

Figure 8.4 [Artist's reconstructions based on structural data.] **Transcription and translation.** (a) Transcription of DNA to messenger RNA by RNA polymerase, a processive enzyme. The polymerase reads the DNA as it walks along it, synthesizing a messenger RNA transcript as it moves. (b) The information in messenger RNA is translated into a sequence of amino acids making up a new protein by the combined action of over 50 molecular machines. In particular, aminoacyl-tRNA synthetases supply transfer RNAs, each loaded with an amino acid, to the ribosomes, which construct the new protein as they read the messenger RNA. [Courtesy David S Goodsell.]

using an approach pioneered by R. Singer. To do so, they needed a way to count the actual number of mRNA molecules in living cells, in real time.

In order to make the mRNA molecules visible, the experimenters created a cell line with an artificially designed gene. The gene coded for a gene product as usual (a red fluorescent protein), but it also had a long, noncoding part, containing 96 copies of a binding sequence. When the gene was transcribed, each copy of the binding sequence folded up to form a binding site for a protein called MS2 (Figure 8.5a). Elsewhere on the genome, the experimenters inserted another gene, for a fusion protein: One domain was a green fluorescent protein (GFP); the other coded for MS2. Thus, shortly after each transcript was produced, it began to glow brightly, having bound dozens of GFP molecules (Figure 8.5b). For each cell studied, the experimenters computed the total fluorescence intensities of all the green

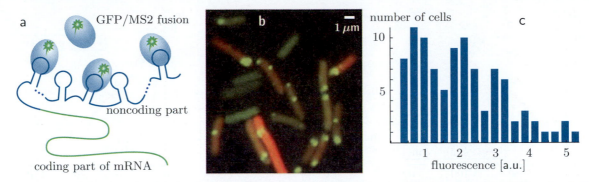

Figure 8.5 Quantification of mRNA levels in individual cells. (a) [Sketch.] Cartoon showing a messenger RNA molecule. The mRNA was designed to fold, creating multiple binding sites for a fusion protein that includes a green <u>f</u>luorescent <u>p</u>rotein (GFP) domain. (b) [Fluorescence micrograph.] Several individual, living bacteria, visualized via their fluorescence. Each *bright green spot* shows the location of one or more mRNA molecules labeled by GFP. The *red* color indicates <u>r</u>ed <u>f</u>luorescent <u>p</u>rotein (RFP), arising from translation of the coding part of the mRNA. (c) [Experimental data.] For each cell, the green fluorescence signal was quantified by finding the total photon arrival rate coming from the green spots only (minus the whole cell's diffuse background). The resulting histogram shows well-separated peaks, corresponding to cells with 1, 2, . . . mRNA molecules (compare Figure 7.7 on page 165). On the horizontal axis, the observed fluorescence intensities have all been rescaled by a common value, chosen to place the first peak near the value 1. Then *all* the peaks were found to occur near integer multiples of that value. This calibration let the experimenters infer the absolute number of mRNA molecules in any cell. [From Golding et al., 2005.]

spots seen in the microscope. A histogram of the observed values of this quantity showed a chain of evenly spaced peaks (Figure 8.5c), consistent with the expectation that each peak represents an integer multiple of the lowest one.[13] Thus, to count the mRNA copies in a cell, it sufficed to measure that cell's fluorescence intensity and identify the corresponding peak.

The experimenters wanted to test the hypothesis that mRNA population dynamics reflects a simple birth-death process. To do so, they noted that such a process is specified by just two parameters, but makes more than two predictions. They determined the parameter values (β_s and $k_\emptyset$) by fitting some of the predictions to experimental data, then checked other predictions.

One such experiment involved suddenly switching on ("inducing") the production of mRNA.[14] In a birth-death process, the number of mRNA molecules, ℓ, averaged over many independent trials, follows the saturating time course given by Equation 8.5.[15] This prediction of the model yielded a reasonable-looking fit to the data. For example, the red trace in Figure 8.6a shows the prediction of the birth-death model with the values $\beta_s \approx 0.15/\text{min}$ and $k_\emptyset \approx 0.014/\text{min}$.

8.4.2 mRNA is produced in bursts of transcription

Based on averages over many cells, then, it may appear that the simple birth-death model is adequate to describe gene expression in *E. coli*. But the ability to count *single molecules in individual cells* gave Golding and coauthors the opportunity to apply a more

---

[13]The intensity of fluorescence per mRNA molecule had some spread, because each mRNA had a variable number of fluorescent proteins bound to it. Nevertheless, Figure 8.5c shows that this variation did not obscure the peaks in the histogram.

[14]Chapter 9 discusses gene switching in greater detail.

[15]See Problem 8.2.

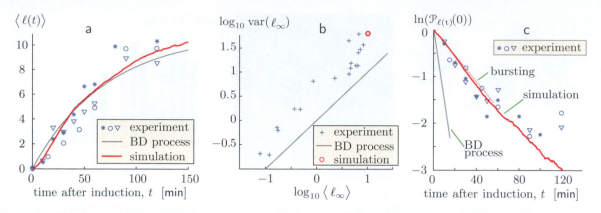

Figure 8.6 [Experimental data with fits.] **Indirect evidence for transcriptional bursting.** (a) *Symbols:* The number of mRNA transcripts in a cell, $\ell(t)$, averaged over 50 or more cells in each of three separate experiments. All of the cells were induced to begin gene expression at a common time, leading to behavior qualitatively like that shown in Figure 8.3a. The *gray curve* shows a fit of the birth-death (BD) process (Equation 8.5, page 184) to data, determining the apparent synthesis rate $\beta_s \approx 0.15$/min and clearance rate constant $k_\circ \approx 0.014$/min. The *red trace* shows the corresponding result from a computer simulation of the bursting model discussed in the text (see also Section 8.4.2'b, page 198). (b) Variance of mRNA population versus sample mean, in steady state. *Crosses:* Many experiments were done, each with the gene turned "on" to different extents. This log-log plot of the data shows that they fall roughly on a line of slope 1, indicating that the Fano factor (var $\ell)/\langle\ell\rangle$ is roughly a constant. The simple birth-death process predicts that this constant is equal to 1 (*gray line*), but the data instead give the value ≈ 5. The *red circle* shows the result of the bursting model, which is consistent with the experimental data. (c) Semilog plot of the fraction of observed cells that have zero copies of mRNA versus elapsed time. *Symbols* show data from the same experiments as in (a). *Gray line:* The birth-death process predicts that initially $\mathcal{P}_{\ell(t)}(0)$ falls with time as $\exp(-\beta_s t)$ (see Problem 8.4). *Dotted line:* The experimental data instead yield initial slope -0.028/min. *Red trace:* Computer simulation of the bursting model. [Data from Golding et al., 2005; see Dataset 12.]

stringent test than the one shown in Figure 8.6a. First, we know that the steady-state mRNA counts are Poisson distributed in the birth-death model,[16] and hence that var $\ell_\infty = \langle\ell_\infty\rangle$. Figure 8.6b shows that the ratio of these two quantities (sample variance/sample mean) really is approximately constant over a wide range of conditions. However, contrary to the prediction of the birth-death model, the value of this ratio (called the **Fano factor**) does *not* equal 1; in this experiment it was approximately 5. The birth-death model also predicts that the fraction of cells with zero copies of the mRNA should initially decrease exponentially with time, as $e^{-\beta_s t}$ (see Problem 8.4). Figure 8.6c shows that this prediction, too, was falsified in the experiment.

These failures of the simplest birth-death model led the experimenters to propose and test a modified hypothesis:

> *Gene transcription in bacteria is a* **bursting process,** *in which the gene makes spontaneous transitions between active and inactive states at mean rates* β_{start} *and* β_{stop}. *Only the active state can be transcribed, leading to bursts of mRNA production interspersed with quiet periods.* (8.9)

More explicitly, β_{start} is the probability per unit time that the gene, initially in the "off" state, will switch to the "on" state. It defines a mean waiting time $\langle t_{\text{w,start}}\rangle = (\beta_{\text{start}})^{-1}$, and similarly for β_{stop} and $t_{\text{w,stop}}$.

---

[16]See Idea 8.7 (page 186).

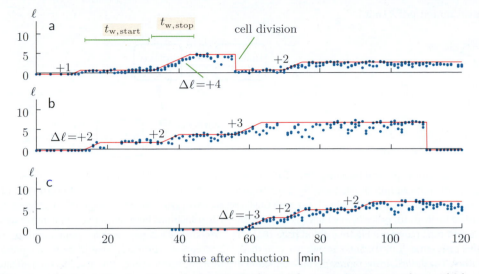

Figure 8.7 [Experimental data.] **Direct evidence for bursting in bacterial gene transcription.** The panels show time courses of ℓ, the population level of a labeled mRNA transcript, in three typical cells. (a) *Dots:* Estimated values of ℓ for one cell. This number occasionally steps downward as the cell divides, because thereafter only one of the two daughter cells' mRNA counts is shown. In this instance, cell division segregated only one of the total five transcripts into the daughter cell selected for further observation (the other four went into the other daughter). The data show episodes when ℓ holds steady (*horizontal segments*), interspersed with episodes of roughly constant production rate (*sloping segments*). The *red line* is an idealization of this behavior. Typical waiting times for transitions to the "on" ($t_{w,start}$) or "off" state ($t_{w,stop}$) are shown, along with the increment $\Delta\ell$ in mRNA population during one episode of transcriptional bursting. (b,c) Observations of two additional individual cells. [Data from Golding et al., 2005.]

The intuition behind the bursting model runs roughly as follows:

1. Each episode of gene activation leads to the synthesis of a variable number of transcripts, with some average value $m = \langle \Delta\ell \rangle$. We can roughly capture this behavior by imagining that each burst contains *exactly* m transcripts. Then the variance of ℓ will be increased by a factor of m^2 relative to an ordinary birth-death process, whereas the expectation will only increase by m. Thus, the Fano factor is larger than 1 in the bursting model, as seen in the data (Figure 8.6b).

2. In the bursting model, the cell leaves the state $\ell = 0$ almost immediately after the gene makes its first transition to the "on" state. Thus, the probability per unit time to exit the $\ell = 0$ state is given by β_{start}. But the initial growth rate of $\langle \ell(t) \rangle$ is given by $\beta_{start} m$, which is a larger number. So the observed initial slopes in panels (a,c) of Figure 8.6 need not be equal, as indeed they are not.

The experimenters tested the bursting hypothesis directly by looking at the time courses of mRNA population in individual cells. Figure 8.7 shows some typical time courses of ℓ. Indeed, in each case the cell showed episodes with no mRNA synthesis, alternating with others when the mRNA population grows at an approximately constant rate.[17] The episodes were of variable duration, so the authors then tabulated the waiting times to transition from

---

[17] $\boxed{T_2}$ The mRNA population in any one cell also dropped suddenly each time that cell divided, because the molecules were partitioned between two new daughter cells, only one of which was followed further. Section 8.4.2'a (page 197) discusses the role of cell division.

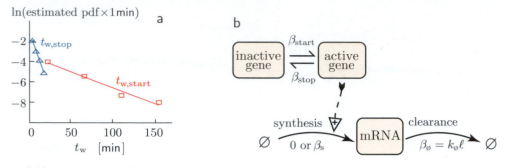

Figure 8.8 **Model for transcriptional bursting.** (a) [Experimental data.] Semilog plot of the estimated probability density for the durations $t_{w,stop}$ of transcription bursts (waiting times to turn off) and of waiting times $t_{w,start}$ to turn on. Fitting the data yielded $\langle t_{w,stop} \rangle \approx 6$ min and $\langle t_{w,start} \rangle \approx 37$ min. [Data from Golding et al., 2005.] (b) [Network diagram.] The bursting hypothesis proposes a modified birth-death process, in which a gene spontaneously transitions between active and inactive states with fixed probabilities per unit time (compare Figure 8.2b on page 182). The *boxes* on the top represent the populations of the gene in its two states (in this case, either 1 or 0). *Solid arrows* between these boxes represent processes that increase one population at the expense of the other. The *dashed arrow* represents an interaction in which one species (here, the gene in its active state) influences the rate of a process (here, the synthesis of mRNA).

"on" to "off" and vice versa, and made separate histograms for each. In the bursting model, when the gene is "on" the probability per unit time to switch off is a constant, β_{stop}. Thus, the model predicts that the waiting times $t_{w,stop}$ will be Exponentially distributed[18] with expectation $(\beta_{stop})^{-1}$, and indeed such behavior was observed (Figure 8.8a). The probability per unit time to switch "on," β_{start}, was similarly found by fitting the distribution of $t_{w,start}$.

The bursting model can be summarized by a network diagram; see Figure 8.8b.

Quantitative checks

The experimental data overconstrain the parameters of the bursting model, so it makes falsifiable predictions.

First, fitting the red data in Figure 8.8a to an Exponential distribution gives $\beta_{start} \approx 1/(37 \text{ min})$. Point **2** above argued that $\ln(\mathcal{P}_{\ell(t)}(0))$ initially falls as $-\beta_{start} t$, and indeed the data in Figure 8.6c do show this behavior, with the same value of β_{start} as was found directly in Figure 8.8a.

Second, Figure 8.6b gives the burst size $m \approx 5$. Point **1** above argued that, if bursts of size m are generated with probability per unit time β_{start}, then we can get the expected number of transcripts by modifying Equation 8.5 to

$$\langle \ell(t) \rangle = \frac{m\beta_{start}}{k_\emptyset} \left(1 - e^{-k_\emptyset t}\right).$$

The only remaining free fitting parameter in this function is $k_\emptyset$. That is, a single choice for this parameter's value predicts the *entire curve* appearing in Figure 8.6a. The figure shows that, indeed, the value $k_\emptyset = 0.014/\text{min}$ gives a function that fits the data.[19] Thus,

---

[18]See Idea 7.5 (page 159).
[19]In fact, Section 8.4.2′a (page 197) will argue that the value of $k_\emptyset$ should be determined by the cells' doubling time, further overconstraining the model's parameters.

the transcriptional bursting hypothesis, unlike the simple birth-death process, can roughly explain all of the data in the experiment.

$\boxed{T_2}$ *This section argued heuristically that it is possible to reconcile all the observations in Figures 8.6a–c and 8.8a in a single model. A more careful analysis, however, requires computer simulation to make testable predictions. Section 8.4.2' b (page 198) describes such a stochastic simulation.*

8.4.3 Perspective

Golding and coauthors followed a systematic strategy for learning more about gene expression:

- Instead of studying the full complex process, they focused on just one step, mRNA transcription.
- They found an experimental technique that let them determine absolute numbers of mRNA, in living cells, in real time.
- They explored the simplest physical model (the birth-death process) based on known actors and the general behavior of molecules in cells.
- They found contact between the experiment and the model by examining reduced statistics, such as the time course of the average of the copy number, its steady-state variance, and the probability that it equals zero. Establishing that contact involved making predictions from the model.
- Comparing these predictions to experimental data was sufficient to rule out the simplest model, so they advanced to the next-simplest one, introducing a new state variable (gene on or off) reminiscent of many other discrete conformational states known elsewhere in molecular biology.
- Although the new model is surely not a complete description, it did make falsifiable predictions that could be more directly tested by experiments designed for that purpose (Figure 8.7), and it survived comparison to the resulting data.

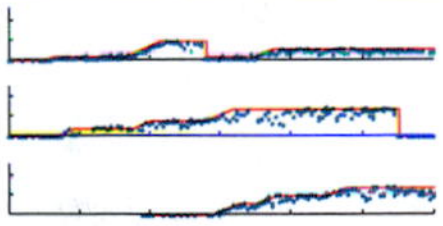

Figure 8.7 (page 191)

Many other groups subsequently documented transcriptional bursting in a wide variety of organisms, including single-cell eukaryotes and even mammals. Even within a single organism, however, some genes are observed to burst while others are not. That is, *transcriptional bursting is a controlled feature* of gene expression, at least in eukaryotes.

Several mechanisms have been proposed that may underlie bursting. Most likely, the complete picture is not simple. But already, this chapter has shown how targeted experiments and modeling succeeded in *characterizing* transcription of a particular gene in a significantly more detailed way than had been previously possible. More recent experiments have also begun to document more subtle aspects of bursting, for example, correlations between transcription bursts of different genes.

8.4.4 Vista: Randomness in protein production

Transcription is just one of many essential cell activities. The general method of fluorescence tagging has also been used to characterize the randomness inherent in protein translation, and in the overall levels of protein populations in cells. In part, protein level fluctuations track mRNA levels, but their randomness can also be increased (for example, by Poisson noise from translation) or suppressed (by averaging over the many mRNA copies created by a single gene).

THE BIG PICTURE

This chapter began by studying random walks in space, such as the trajectories of small diffusing objects in fluid suspension. We then generalized our framework from motion in ordinary space to chemical reactions, which we modeled as random walks on a *state space*. We got some experience handling probability distributions over all possible histories of such systems, and their most commonly used reduced forms. Analogously to our experience deducing a hidden step in myosin-V stepping (Section 7.5.1), we were able to deduce a hidden state transition, leading to the discovery of bursting in bacterial gene expression. Cells must either exploit, learn to live with, or overcome such randomness in their basic processes.

However, we also found situations in which the randomness of gene expression had little effect on the dynamics of mRNA levels, because the overall inventory of mRNA was high.[20] Chapters 9–11 will make this continuous, deterministic approximation as we push forward our study of cellular control networks.

KEY FORMULAS

- *Diffusion:* A small particle suspended in fluid will move in a random walk, due to its thermal motion in the fluid. The mean-square deviation of the particle's displacement, after many steps, is proportional to the elapsed time.
- *Birth-death process:* Let β_s be the synthesis rate, and $k_\emptyset$ the degradation rate constant, for a birth-death process. In the continuous, deterministic approximation the population ℓ of a species X follows $d\ell/dt = \beta_s - \ell k_\emptyset$. One solution to this equation is the one that starts with $\ell(0) = 0$: $\ell(t) = (\beta_s/k_\emptyset)(1 - e^{-k_\emptyset t})$.
- *Stochastic simulation:* The relative propensities for a two-reaction Gillespie algorithm, with reaction rates β_1 and β_2, are $\xi = \beta_1/(\beta_1 + \beta_2)$ and $(1 - \xi)$. (See Equation 8.6.)
- $\boxed{T_2}$ *Master equation:*

$$\frac{\mathcal{P}_{\ell_{i+1}}(\ell) - \mathcal{P}_{\ell_i}(\ell)}{\Delta t} = \beta_s\big(\mathcal{P}_{\ell_i}(\ell - 1) - \mathcal{P}_{\ell_i}(\ell)\big) + k_\emptyset\big((\ell + 1)\mathcal{P}_{\ell_i}(\ell + 1) - \mathcal{P}_{\ell_i}(\ell)\big).$$

FURTHER READING

Semipopular:
Hoagland & Dodson, 1995.

Intermediate:
Klipp et al., 2009, chapt. 7; Otto & Day, 2007; Wilkinson, 2006.
mRNA dynamics: Phillips et al., 2012, chapt. 19.
$\boxed{T_2}$ Master (or Smoluchowski) equations: Nelson, 2014, chapt. 10; Schiessel, 2013, chapt. 5.

Technical:
Gillespie algorithm: Gillespie, 2007; Ingalls, 2013.
Bursting in prokaryotes: Golding et al., 2005; Paulsson, 2005; Taniguchi et al., 2010.
Transcriptional bursting in higher organisms: Raj & van Oudenaarden, 2009; Suter et al., 2011; Zenklusen et al., 2008.

---

[20]See Idea 8.8 (page 187).

T_2

Track 2

8.3.4′ The master equation

We have seen that in the birth-death process, the distribution of system states ℓ is Poisson.[21] We can confirm this observation by inventing and solving the system's "master equation." Similar formulas arise in many contexts, where they are called by other names such as "diffusion," "Fokker-Planck," or "Smoluchowski" equations.

Any random process is defined on a big sample space consisting of all possible *histories* of the state. Treating time as discrete, this means that the sample space consists of sequences $\ell_1, \ldots, \ell_j, \ldots$, where ℓ_j is the population at time t_j. As usual, we'll take $t_i = (\Delta t)i$ and recover the continuous-time version later, by taking the limit of small Δt.

The probability of a particular history, $\mathcal{P}(\ell_1, \ldots)$, is complicated, a joint distribution of many variables. We will be interested in reduced forms of this distribution, for example, $\mathcal{P}_{\ell_i}(\ell)$, the marginal distribution for there to be ℓ molecules of species X at time $(\Delta t)i$, regardless of what happens before or after that time. The Markov property implies that this probability is completely determined if we know that the system is in a definite state at time $i - 1$, so we begin by assuming that.

Imagine making a large number N_{tot} of draws from this random process, always starting the system at time 0 with the same number of molecules, ℓ_{ini} (see Figure 8.9a). That is, we suppose that $\mathcal{P}_{\ell_0}(\ell) = 1$ if $\ell = \ell_{\text{ini}}$, and zero otherwise. We can summarize the notation:

ℓ_i	number of molecules at time $t_i = (\Delta t)i$, a random variable
ℓ_{ini}	initial value, a constant
N_{tot}	number of systems being observed
β_s	mean rate for synthesis
$k_{\o}$	clearance rate constant

Each of these quantities is a constant, except the ℓ_i, each of which is a random variable.

Equivalently, we can express ℓ_{ini} in terms of ℓ in each case, and compare the result with the original distribution:

$$\frac{\mathcal{P}_{\ell_1}(\ell) - \mathcal{P}_{\ell_0}(\ell)}{\Delta t} = \begin{cases} \beta_s & \text{if } \ell = \ell_{\text{ini}} + 1; \\ (\ell + 1)k_{\o} & \text{if } \ell = \ell_{\text{ini}} - 1; \\ -(\beta_s + k_{\o}\ell) & \text{if } \ell = \ell_{\text{ini}}; \\ 0 & \text{otherwise.} \end{cases} \qquad (8.10)$$

Equation 8.10 is applicable to the special case shown in Figure 8.9a.

Next, suppose that initially a fraction q of the N_{tot} systems started out with ℓ_{ini} molecules, but the other $1 - q$ instead start with some other value ℓ'_{ini} (see Figure 8.9b). Thus, the initial distribution is nonzero at just two values of ℓ, so on the next time step the distribution evolves to one that is nonzero on just those two values and their four flanking values, and

---

[21] See Idea 8.7 (page 186).

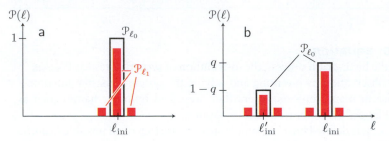

Figure 8.9 [Sketch graphs.] **Time evolution in a birth-death process.** (a) Suppose that a collection of identical systems all have the same starting value of ℓ (*black*). Each of the systems evolves in the next time slot to give a distribution with some spread (*red*). (b) This panel represents an initial distribution of states with *two* values of ℓ. This distribution evolves into one with nonzero probability at *six* values of ℓ.

so on. The six cases that must be considered can all be elegantly summarized as a single formula, called the **master equation**:

$$\frac{\mathcal{P}_{\ell_1}(\ell) - \mathcal{P}_{\ell_0}(\ell)}{\Delta t} = \beta_s\big(\mathcal{P}_{\ell_0}(\ell - 1) - \mathcal{P}_{\ell_0}(\ell)\big) + k_\varnothing\big((\ell + 1)\mathcal{P}_{\ell_0}(\ell + 1) - \ell\mathcal{P}_{\ell_0}(\ell)\big). \quad (8.11)$$

The master equation is actually a chain of many linked equations, one for every allowed value of ℓ. Remarkably, it is no longer necessary to itemize particular cases, as was done in Equation 8.10; this is now accomplished by expressing the right-hand side of Equation 8.11 in terms of the initial distribution $\mathcal{P}_{\ell_0}$.

Example Derive Equation 8.11. Show that it also applies to the case where the initial distribution $\mathcal{P}_{\ell_0}(\ell)$ is arbitrary (not necessarily peaked at just one or two values of ℓ).

Solution As before, it is a bit easier to start by thinking of a finite set of N_{tot} specific trials. Of these, initially about $N_{*,\ell} = N_{tot}\mathcal{P}_{\ell_0}(\ell)$ had ℓ copies of X. (These statements become exact in the limit of large N_{tot}.)

For each value of ℓ, at the next time slot about $N_{*,\ell-1}(\Delta t)\beta_s$ get added to bin ℓ (and removed from bin $(\ell - 1)$).

For each value of ℓ, at the next time slot another $N_{*,\ell+1}(\Delta t)k_\varnothing(\ell + 1)$ get added to bin ℓ (and removed from bin $(\ell + 1)$).

For each value of ℓ, at the next time slot about $N_{*,\ell}(\Delta t)(\beta_s + k_\varnothing\ell)$ get removed from bin ℓ (and added to other bins).

Altogether, then, the number of trials with exactly ℓ copies changes from $N_{*,\ell}$ at time 0 to

$$N_\ell = N_{*,\ell} + \Delta t\big(\beta_s N_{*,\ell-1} + k_\varnothing(\ell + 1)N_{*,\ell+1} - (\beta_s + k_\varnothing\ell)N_{*,\ell}\big).$$

Dividing by $(\Delta t)N_{tot}$ gives the master equation. (Note, however, that for $\ell = 0$ the equation must be modified by omitting its first term.)

The right side of Equation 8.11 consists of a pair of terms for each reaction. In each pair, the positive term represents influx into the state populated by a reaction; the negative term represents the corresponding departures from the state that is depopulated by that reaction.

Our goal was to check Idea 8.7 (page 186), so we now seek a steady-state solution to the master equation. Set the left side of Equation 8.11 equal to zero, and substitute a trial solution of the form $\mathcal{P}_\infty(\ell) = e^{-\mu} \mu^\ell / (\ell!)$.

Your Turn 8D

Confirm that this trial solution works, and find the value of the parameter μ.

The master equation lets us calculate other experimentally observable quantities as well, for example, the *correlation* between fluctuations at different times. To obtain its continuous-time version, we just note that the left side of Equation 8.11 becomes a derivative in the limit $\Delta t \to 0$. In this limit, it becomes a large set of coupled first-order ordinary differential equations, one for each value of ℓ. (If ℓ is a continuous variable, then the master equation becomes a partial differential equation.)

T_2 **Track 2**

8.4′ More about gene expression

1. In eukaryotes, various "editing" modifications also intervene between transcription and translation.
2. Folding may also require the assistance of "chaperones," and may involve the introduction of "cofactors" (extra molecules that are not amino acids). An example is the cofactor retinal, added to an opsin protein to make the light-sensing molecules in our eyes.
3. The gene product may be a complete protein, or just a part of a protein that involves multiple amino acid chains and cofactors.
4. To create a fusion protein, it's not enough to position two genes next to each other: We must also eliminate the first one's "stop codon," so that transcription proceeds to the second one, and ensure that the two genes share the same "reading frame."

T_2 **Track 2**

8.4.2′a The role of cell division

The main text mentioned two processes that could potentially offset the increase of messenger RNA counts in cells (clearance and cell division), and tacitly assumed that both could be summarized via a single rate constant $k_\varnothing$. This is a reasonable assumption if, as discussed in the main text, mRNA molecules make random encounters with an enzyme that degrades them. But in fact, Golding and coauthors found that their fluorescently labeled mRNA constructs were rarely degraded. Instead, in this experiment cell division was the main process reducing concentration.

Upon cell division, the experimenters confirmed that each messenger RNA independently "chooses" which daughter cell it will occupy, similarly to Figure 4.2. Thus, the number passed to a particular daughter is a Binomially distributed random variable. On average,

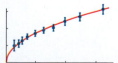

Figure 4.2 (page 73)

this number is one half of the total mRNA population. The bacteria in the experiment were dividing every 50 min. Suppose that we could suddenly shut off synthesis of new mRNA molecules. After the passage of time T, then, the average number will have halved a total of $T/(50\,\text{min})$ times, reducing it by a factor of $2^{-T/(50\,\text{min})}$. Rewriting this result as $\ell(t) = \ell_{\text{ini}} \exp(-k_\emptyset T)$, we find $k_\emptyset = (\ln 2)/(50\,\text{min}) \approx 0.014/\text{min}$.

Making a continuous, deterministic approximation, we just found that about $k_\emptyset \ell\,\mathrm{d}t$ molecules are lost in time $\mathrm{d}t$, so cell division gives rise to a "dilution" effect, similar to clearance but with the value of $k_\emptyset$ given in the previous paragraph. Even if production is nonzero, we still expect that the effect of cell division can be approximated by a continuous loss at rate $k_\emptyset \ell$. The main text shows that the experimental data for $\langle \ell(t) \rangle$ do roughly obey an equation with rate constant $k_\emptyset \approx 0.014/\text{min}$, as was predicted above.[22]

8.4.2′b Stochastic simulation of a transcriptional bursting experiment

Figure 8.8b (page 192)

The main text motivated the transcriptional bursting model (represented symbolically in Figure 8.8b), then gave some predictions of the model, based on rather informal simplifications of the math. For example, cell division was approximated as a continuous, first-order process (see Section (a) above), and the variability of burst sizes was ignored. In addition, there were other real-world complications not even mentioned in the chapter:

- We have implicitly assumed that there is always exactly one copy of the gene in question in the cell. Actually, however, any given gene replicates at a particular time in the middle of a bacterial cell's division cycle. For the experiment we are studying, suppose that gene copy number doubles after about 0.3 of the cell division time, that is, after $(0.3) \times (50\,\text{min})$.

- Moreover, there may be more than one copy of the gene, even immediately after division. For the experiment we are studying, this number is about 2 (So et al., 2011). Suppose that each new copy of the gene is initially "off," and that immediately after division all copies are "off." Because the gene spends most of its time "off," these are reasonable approximations.

- Cell division does not occur precisely every 50 min; there is some randomness.

To do better than the heuristic estimates, we can incorporate every aspect of the model's formulation in a stochastic simulation, then run it many times and extract predictions for the experimentally observed quantities, for any chosen values of the model's parameters (see also So et al., 2011).

The simulation proceeds as follows. At any moment, there are state variables counting the total number of "on" and "off" copies of the gene, the number of messenger RNA molecules present in the cell, and another "clock" variable n describing progress toward division, which occurs when n reaches some preset threshold n_0. A Gillespie algorithm decides among the processes that can occur next:

1. One of the "on" copies of the gene may switch "off." The total probability per unit time for this outcome is β_{stop} times the number of "on" copies at that moment.

2. One of the "off" copies may switch "on." The total probability per unit time for this outcome is β_{start} times the number of "off" copies at that moment.

---

[22]You'll implement this approach to clearance in Problem 8.5. For more details about cell growth, see Section 9.4.5.

3. One of the "on" copies may create a mRNA transcript. The total probability per unit time for this outcome is a rate constant β_s times the number of "on" copies at that moment. (Note that the value of β_s needed to fit the data will not be equal to the value obtained when using the birth-death model.)
4. The "clock" variable n may increment by one unit. The probability per unit time for this outcome is $n_0/(50\ \mathrm{min})$.

The waiting time for the next event is drawn, one of the four reaction types above is chosen according to the recipe in Section 8.3.3, and the system state is updated. Before repeating the cycle, however, the simulation checks for two situations requiring additional actions:

• If the clock variable exceeds $0.3n_0$, then the number of gene copies is doubled before proceeding (gene duplication). The new copies are assumed to be "off." No further doubling will occur prior to cell division.
• If the clock variable exceeds n_0, then the cell divides. The number of gene copies is reset to its initial value, and all are turned "off." To find the number of mRNA molecules passed on to a particular daughter cell, a random number is drawn from the Binomial distribution with $\xi = 1/2$ and M equal to the total number of molecules present.

A simulation following the procedure outlined above yielded the curves shown in Figure 8.6; see Problem 8.7.

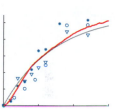

Figure 8.6a (page 190)

8.4.2′c Analytical results on the bursting process
The preceding section outlined a simulation that could be used to make predictions relevant to the experimental data shown in the text. Even more detailed information can be obtained from those data, however: Instead of finding the sample mean and variance in the steady state, one can estimate the entire probability distribution $\mathcal{P}_{\ell(t\to\infty)}$ from data (Golding et al., 2005; So et al., 2011), and compare it to the corresponding distribution found in the simulation.

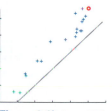

Figure 8.6b (page 190)

If we are willing to make the idealization of treating cell division as a continuous clearance process (see Section (a) above), then there is an alternative to computer simulation: Analytic methods can also be used to predict the distribution starting from the master equation (Raj et al., 2006; Shahrezaei & Swain, 2008; Iyer-Biswas et al., 2009; Stinchcombe et al., 2012). These detailed predictions were borne out in experiments done with bacteria (Golding et al., 2005; So et al., 2011) and eukaryotes (Raj et al., 2006; Zenklusen et al., 2008).

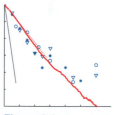

Figure 8.6c (page 190)

PROBLEMS

8.1 Random walk with random waiting times

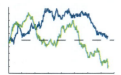

Figure 8.1 (page 181)

a. Implement the strategy outlined in Section 8.2.2 to simulate a random walk in one dimension. Suppose that the steps occur in a compound Poisson process with mean rate $\beta = 1\,\text{s}^{-1}$, and that each step is always of the same length $d = 1\,\mu\text{m}$, but in a randomly chosen direction: $\Delta x = \pm d$ with equal probabilities for each direction. Make a graph of two typical trajectories (x versus t) with total duration $T = 400\,\text{s}$, similar to Figure 8.1.

b. Run your simulation 50 times. Instead of graphing all 50 trajectories, however, just save the ending positions x_T. Then compute the sample mean and variance of these numbers. Repeat for 50 trajectories with durations 200 s, and again with 600 s.

c. Use your result in (b) to guess the complete formulas for $\langle x_T \rangle$ and $\text{var}(x_T)$ as functions of d, β, and T.

d. Upgrade your simulation to two dimensions. Each step is again of length $1\,\mu\text{m}$, but in a direction that is randomly chosen with a Uniform distribution in angle. This time make a graph of x versus y. That is, don't show the time coordinate (but do join successive points by line segments).

e. $\boxed{T_2}$ An animation of the 2D walk is more informative than the picture you created in (c), so try to make one.

8.2 Average over many draws

Continuing Your Turn 8B, write a program that calls your function 150 times, always with $\ell_{\text{ini}} = 0$ and for time from 0 to 300 min. At each value of time, find the average of the population over all 150 trials. Plot the time course of the averages thus found, and comment on the relation between your graph and the result of the continuous, deterministic approximation. [*Hint:* For every trial, and for every value of t from 0 to 300 min, find the step number, α, at which $\texttt{ts(alpha)}$ first exceeds t. Then the value of ℓ after step $\alpha - 1$ is the desired position at time t.]

8.3 Burst number distribution

Consider a random process in which a gene randomly switches between "on" and "off" states, with probability per unit time β_{stop} to switch on→off and β_{start} to switch off→on. In the "off" state, the gene makes no transcripts. In the "on" state, it makes transcripts in a Poisson process with mean rate β_s. Simplify by assuming that both transcription and switching are sudden events (no "dead time").

Obtain analytically the expected probability distribution function for the number $\Delta\ell$ of transcript molecules created in each "on" episode, by taking these steps:

a. After an "on" episode begins, there are two kinds of event that can happen next: Either the gene switches "off," terminating the episode, or else it makes a transcript. Find the probability distribution describing which of these two outcomes happens first.

b. We can think of the events in (a) as "attempts to leave the 'on' state." Some of those attempts "succeed" (the gene switches off); others "fail" (a transcript is made and the gene stays on). The total number of transcripts made in an "on" episode, $\Delta\ell$, is the number of "failures" before the first "success." Use your answer to (a) to find the distribution of this quantity in terms of the given parameters.

c. Find the expectation of the distribution you found in (b) (the average burst size) in terms of the given parameters.

8.4 Probability of zero copies, via simulation

First work Problem 8.2. Now add a few lines to your code to tabulate the number of trials in which the number of copies, ℓ, is still zero after time t, for various values of t. Convert this result into a graph of $\ln \mathcal{P}_{\ell(t)}(0)$ versus time, and compare to the semilog plot of experimental data in Figure 8.6c.

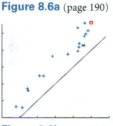

Figure 8.6c (page 190)

8.5 Simulate simplified bursting process

First work Problem 8.2. Now modify your code to implement the transcriptional bursting process (Figure 8.8b). To keep your code fairly simple, assume that (*i*) A cell contains a single gene, which transitions between "on" and "off" states. Initially the gene is "off" and there are zero copies of its mRNA. (*ii*) The cell never grows or divides, but there is a first-order clearance process[23] with rate constant $k_\emptyset = (\ln 2)/(50\,\mathrm{min})$.

Figure 8.8b (page 192)

Take the other rates to be $\beta_{\mathrm{start}} = 1/(37\,\mathrm{min})$, $\beta_{\mathrm{stop}} = 1/(6\,\mathrm{min})$, and $\beta_s = 5\beta_{\mathrm{stop}}$. Run your simulation 300 times, and make graphs of $\langle \ell(t) \rangle$ and $\ln\big(\mathcal{P}_{\ell(t)}(0)\big)$ versus time over the course of 150 minutes. Also compute the Fano factor $\mathrm{var}(\ell_{\mathrm{final}})/\langle \ell_{\mathrm{final}} \rangle$, and comment.

8.6 [T₂] Probability of zero copies, via master equation

Suppose that a molecule is created (for example, a messenger RNA) in a Poisson process with mean rate β_s. There is no clearance process, so the population of the molecule never decreases. Initially there are zero copies, so the probability distribution for the number of molecules ℓ present at time zero is just $\mathcal{P}_{\ell(0)}(0) = 1$; all other $\mathcal{P}_{\ell(0)}(\ell)$ equal zero. Find the value of $\mathcal{P}_{\ell(t)}(0)$ at later times by solving a reduced form of the master equation.

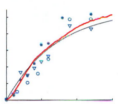

Figure 8.6a (page 190)

8.7 [T₂] Simulate transcriptional bursting

Obtain Dataset 12. Use these experimental data to make graphs resembling those in Figure 8.6. Now write a computer code based on your solution to Problem 8.5, but with the additional realism outlined in Section 8.4.2′b (page 198), and see how well you can reproduce the data with reasonable choices of the model parameters. In particular, try the value $n_0 = 5$, which gives a reasonable amount of randomness in the cell division times.

Figure 8.6b (page 190)

---

[23] [T₂] Section 8.4.2′a (page 197) gives some justification for this approach.

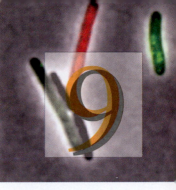

Negative Feedback Control

A good sketch is better than a long speech.
—Napoleon Bonaparte

9.1 Signpost

Living organisms are physical systems that are able to respond appropriately to their unpredictable environment. The mechanisms that they (we) use to acquire information about the environment, process it, and act on it form one of the main threads of this book. There is tremendous selective pressure to do these jobs well. An organism with good control circuitry can find food better than a more dim-witted one (perhaps even finding the latter as its food). It can also evade its own predators, observe and act on early warning signs of environmental change, and so on. Other control circuits allow organisms to shut down some of their capabilities when not needed, in order to direct their resources to other tasks such as reproduction.

Even single-cells can do all of these things. For example, the video clip in Media 4 shows a white blood cell (a neutrophil) as it engages and eventually engulfs a pathogen.[1] This activity can only be called "pursuit," yet the neutrophil has no brain—it's a *single cell*. How can it connect signals arriving at its surface to the motions needed to move toward, and ultimately overtake, the pathogen? *How could anything like that possibly happen at all?*

Clearly we must begin addressing such questions with the simplest systems possible. Thus, Chapters 9–11 will draw inspiration from more intuitive, nonliving examples. It may seem a stretch to claim any link between the sophisticated control mechanisms of cells

---

[1] See also Media 12.

(let alone our own multicellular bodies) and an everyday gadget like a thermostat, but in both realms, the key idea turns out to be *feedback control.*

Each of these chapters will first introduce a situation where control is needed in cells. To gain some intuition, we'll then look at iconic examples of control mechanisms in the physical world. Along the way, we will create some mathematical tools needed to make quantitative predictions based on measured properties of a system. We'll also look at some of the molecular apparatus ("wetware") available in cells that can implement feedback control.

It is now possible to install custom control mechanisms in living cells, leading to a discipline called **synthetic biology**. Three papers, all published in the first year of the 21st century, have become emblematic of this field. Chapters 9–11 will describe their results, along with more recent work. Besides being useful in its own right, synthetic biology tests and deepens our understanding of evolved, natural systems; each of these chapters closes with an example of this sort.

This chapter's Focus Question is

Biological question: How can we maintain a fixed population in a colony of constantly reproducing bacteria?

Physical idea: Negative feedback can stabilize a desired setpoint in a dynamical system.

9.2 Mechanical Feedback and Phase Portraits

9.2.1 The problem of cellular homeostasis

Viewed in terms of its parts list, a cell may seem to be a jumble of molecular machines, all churning out proteins. But any factory needs *management.* How does the cell know how much of each molecule to synthesize? What controls all of those machines? The problem is particularly acute in light of what we found in Chapter 8: The birth-death process is partly random, so there will be deviations from any desired state.

Homeostasis is the generic term for the maintenance of a desired overall state in biology. This section will begin by describing a feedback mechanism that achieves something similar in a mechanical context; we will then use it as the basis for a physical model for cell-biological processes. Later chapters will consider different mechanisms that can lead to switch-like, and even oscillatory, behavior.

9.2.2 Negative feedback can bring a system to a stable setpoint and hold it there

Figure 9.1a shows a **centrifugal governor**. This device became famous when James Watt introduced it into his steam engine in 1788. The principle of the governor is a good starting point for constructing a graphical language that will clarify other feedback systems.

In the figure, the engine spins a shaft (on the left), thereby throwing the two weights outward, by an amount related to the engine's rotation frequency v. Mechanical linkages translate this outward motion into countervailing changes in the engine's fuel supply valve (on the right), resulting in the engine maintaining a particular value of v. Thus,

> *A governor continuously monitors a state variable (the "output"; here, v), compares it with a desired value (the **setpoint**, v_*), and generates a correction to be applied to the input (here, the fuel supply valve setting).*

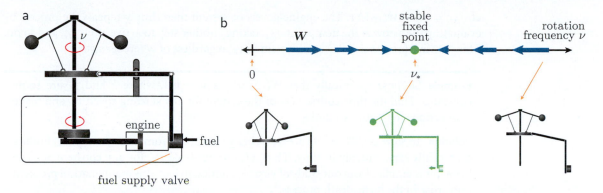

Figure 9.1 **Negative feedback.** (a) [Schematic.] A centrifugal governor controls the speed of an engine by regulating the amount of fuel admitted, so as to maintain a near-constant speed, regardless of changes in load or fuel supply conditions. A shaft connected to the engine spins two masses (*spheres*). As the rotation frequency ν increases, the masses move away from the axle, actuating a set of linkers and reducing the fuel supply. A more realistic drawing appears on page 177. (b) [Phase portrait.] *Top:* Abstract representation of a negative feedback control system. *Blue arrows* represent change per unit time in the rotation frequency ν of an engine. This rate of change depends on the starting value of ν, so the arrows have differing lengths and directions; they amount to a vector field $W(\nu)$. For practical reasons, we cannot draw all infinitely many of these arrows; instead, a sample of several points on the ν axis has been chosen. The engine's setpoint is the value ν_* at which the arrow length is zero (the stable fixed point of W, *green dot*). *Bottom:* The sketches represent the state of the governor at three points on the phase portrait.

Engineers refer to this mechanism as **negative feedback**, because (*i*) the corrective signal reflects the momentary difference between the variable and the desired value but with the opposite sign and (*ii*) the corrective signal is "fed" back from the output to the input of the engine.

Feedback is essential if we want our engine to run at a particular speed, because it will encounter unpredictable load conditions: Under heavy load, we must supply more fuel to maintain a particular setpoint value ν_*. Moreover, the relation between fuel and speed may change over time, for example, as the engine initially warms up or the quality of the fuel changes. Rather than attempting to work out all these effects from first principles, the governor automatically maintains the set speed, at least within a certain operating range.[2]

Figure 9.1b abstracts the essential idea of this device. It shows a line representing various values of the engine's rotation frequency ν. Imagine that, at each point of the line, we superimpose an arrow $W(\nu)$ starting at that point. The length and direction of the arrow indicate how ν would *change* in the next instant of time if it started at the indicated point. We can initially place the system anywhere on that line, then let it respond. For example, we could manually inject some extra fuel to put ν over the setpoint ν_*; the leftward arrows indicate a response that pulls it back down. Or we may suddenly apply a heavy load to reduce ν momentarily; the rightward arrows indicate a response that pulls it back up.

If at some moment the engine's speed is exactly at the setpoint ν_*, then the error signal is zero, the fuel supply valve setting does not change, and ν remains constant. That is, the arrow at the setpoint has length zero: $W(\nu_*) = \mathbf{0}$. More generally, any point where the vector field vanishes is called a **fixed point** of the dynamical system.

If the system starts at some other value ν_0, we can work out the ensuing behavior a short time Δt later: We evaluate the arrow W at ν_0, then move along the line, arriving

---

[2] If we overload the engine, then no amount of fuel will suffice and feedback control will break down.

at $v_{\Delta t} = v_0 + W(v_0)\Delta t$. The engine/governor system then simply repeats the process by evaluating the arrow at the new point $v_{\Delta t}$, taking another step to arrive at $v_{(2\Delta t)}$, and so on. This iterative process will drive the system to v_*, regardless of where it starts.

Example Suppose specifically that $W(v) = -k(v - v_*)$, where k and v_* are some constants. Find the time course $v(t)$ of the system for any starting speed v_0, and make connections to Chapters 1 and 8.

Solution To solve $dv/dt = -k(v - v_*)$, change variables from v to $x = v - v_*$, finding that x falls exponentially in time. Thus, $v(t) = v_* + x_0 e^{-kt}$ for any constant x_0. The behavior is familiar from our study of virus dynamics. In fact, this mathematical problem also arose in the birth-death process.[3]

To find the value of the constant x_0, evaluate the solution at $t = 0$ and set it equal to v_0. This step yields $x_0 = v_0 - v_*$, or

$$v(t) = v_* + (v_0 - v_*)e^{-kt}.$$

The solution asymptotically approaches the fixed-point value v_*, regardless of whether it started above or below that value.

Figure 9.1b is called a one-dimensional **phase portrait** of the control system, because it involves points on a single line.[4] Its main feature is a single **stable fixed point**. The solution to the Example just given explains why it's called "stable"; we'll meet other kinds of fixed points later on.

The same sort of logic used in this section, though with different implementation, underlies how cruise control on a car maintains fixed speed regardless of the terrain, as well as other governor circuits in the devices around us.

9.3 Wetware Available in Cells

We have seen how a simple physical system can self-regulate one of its state variables. Now we must investigate whether there are useful connections between biological and mechanical phenomena, starting with the question of what cellular mechanisms are available to implement feedback and control. This section will draw on background information about cell and molecular biology; for details, see any text on that subject. Throughout, we'll use the continuous, deterministic approximation, introduced in Chapter 8, to simplify the mathematics. This approximation is a good guide when molecule numbers are large.[5]

9.3.1 Many cellular state variables can be regarded as inventories

One lesson from mechanical control is that a cell needs *state variables* that it can use to encode external information, yes/no memories, and other quantities, such as the elapsed time since the last "tick" of an internal clock. The limited, but dynamic, information in these

---

[3]See Equation 1.1 (page 14) and Section 8.3.2 (page 184).
[4]The word "phase" entered the name of this construction for historical reasons; actually, there is no notion of phase involved in the examples we will study.
[5]See Idea 8.8 (page 187).

state variables combines with the vast, but static, information in the cell's genome to create its behavior. The strategy is familiar:

- You bring your limited, but dynamic, personal experience into a vast library filled with static books.
- Your computer records your keystrokes, photos you take, and so on, then processes them with the help of a vast archive of system software that arrived on a read-only medium or download.

In an electronic device, electric potentials are used as state variables. Cells, however, generally use *inventories,* counts of the numbers of various molecular species.[6] Because each cell is surrounded by a membrane that is nearly impermeable to macromolecules, these counts usually remain steady unless actively increased by the import or production of the molecules in question, or actively decreased by their export or breakdown. To understand the dynamics of an inventory, then, we need to model these relatively few processes.

T_2 *Section 9.3.1' (page 234) compares and contrasts this arrangement with electronic circuits, and gives more details about permeability.*

9.3.2 The birth-death process includes a simple form of feedback

We are already familiar with one straightforward kind of negative feedback in cells: The degradation of a molecular species X in Figure 8.2b occurs at a rate depending on the inventory of X. Although Chapter 8 considered the context of mRNA, similar ideas can be applied to the cell's inventory of a protein, because proteins, too, are constantly being degraded (and diluted by cell growth).

Figure 8.2b (page 182)

In fact, the Example on page 206 showed that such a system approaches its fixed point exponentially. Although this feedback does create a stable fixed point, we will now see that cells have found other, more effective control strategies.

9.3.3 Cells can control enzyme activities via allosteric modulation

The production and breakdown of molecules are processes that affect each species' inventory. These jobs are often performed by enzymes, macromolecular machines that catalyze the chemical transformation of other molecules without themselves being used up. Generally, an enzyme binds a specific substrate molecule (or more than one), modifies it, releases the resulting product(s), and repeats as long as substrate is available. For example, the enzyme could join two substrates with a covalent bond, to build something more complex; conversely, some enzymes cut a bond in a substrate, releasing two smaller products. Creating (or destroying) one enzyme molecule has a multiplier effect on the inventories of the substrate and product, because each enzyme can process many substrates before eventually breaking or being cleared by the cell.

In addition to raw molecule counts, cells care about the division of each molecular species into subclasses. A molecule may have different isomers (conformational states), and each isomer may have a different meaning to the cell. The difference may be gross, like popping a bond from the *cis* to the *trans* form, or exceedingly subtle. For example, an enzyme may bind a smaller molecule (a **ligand**) at one binding site; this event in turn flexes the whole macromolecule slightly (Figure 9.2), for example, changing the fit between

---

[6]However, specialized cells such as neurons do make use of electrochemical potentials as state variables (see Sections 4.3.4 and 7.4.2). Electrochemical signaling is much faster than using molecular inventories alone.

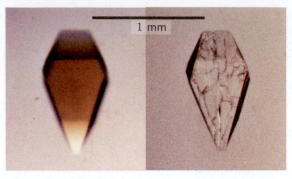

Figure 9.2 [Photographs.] **Allosteric conformational change.** *Left:* A crystal of *lac* repressor molecules bound to their ligand (short strands of DNA containing the repressor's binding sequence). *Right:* When such a crystal is exposed to its effector molecule IPTG, the resulting conformational change is enough to make the crystal structure unstable, and the crystal shatters almost immediately. [Courtesy Helen C Pace; see also Pace et al., 1990.]

a second site and *its* binding partner. Such "action at a distance," or **allosteric interaction**, between parts of a macromolecule can modulate or even destroy its ability to bind the second ligand. In this context, the first ligand is sometimes called an **effector**.

Thus,

> *An enzyme's activity can be controlled by the availability of an effector, via an allosteric interaction. Small changes in the effector's inventory can then translate into changes in the production of those enzymes' products (or the elimination of their substrates) throughout the cell.* (9.1)

Figure 9.3a represents this idea. Later chapters will discuss variations on the theme:

- An effector molecule may itself have multiple conformational states, only one of which stimulates its partner enzyme. Then the enzyme can be controlled by interconversion of the effector between its conformations (Figure 9.3b).
- The effector may not enter the cell at all. Instead, a **receptor** protein is embedded in the cell membrane. An effector molecule can bind to its outer part, allosterically triggering enzymatic activity of its inner part.
- Cells also use other types of modification to control enzyme activity, for example, the attachment and removal of chemical groups such as phosphate (Figure 9.3c).[7] These modifications are in turn carried out by enzymes specialized for that task, which may themselves be controlled, leading to a "cascade" of influences.

The following section will discuss yet another variant, represented by Figures 9.3d,e.

T_2 *Figure 9.3 shows only a few control mechanisms, those that will be discussed later in this book. Section 9.3.3′ (page 234) mentions a few more examples.*

9.3.4 Transcription factors can control a gene's activity

Different cells with the same genome can behave differently; for example, each type of somatic cell in your body carries the complete instructions needed to generate any of the

---

[7]See Chapter 11.

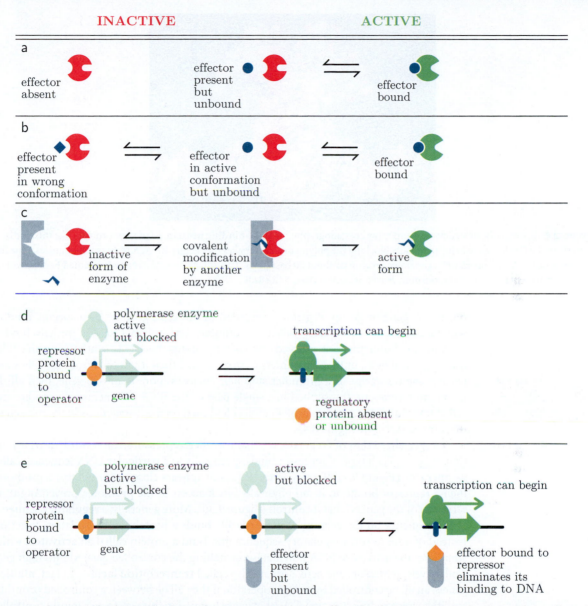

INACTIVE   **ACTIVE**

Figure 9.3 [Cartoons.] **A few mechanisms of enzyme control.** (a) An effector binds to an enzyme and activates it by an allosteric interaction. (b) An effector has two states, only one of which can bind and activate an enzyme. (c) A second enzyme modifies the first one, activating it permanently (or at least until another enzyme removes the modification). (d) A repressor protein controls a polymerase enzyme by binding at or near its promoter. (e) The repressor may itself be controlled by an effector, which modifies its ability to bind to its operator sequence on the cell's DNA.

other types, but they don't all behave the same way, nor even look the same. One source of distinct cell fates is that cells can exert feedback control on gene expression. Such control takes a straightforward form in bacteria.

Section 8.4 (page 187) outlined some standard terminology in gene expression. A bacterial **gene** is a stretch of DNA that is transcribed as a unit and gives rise to a single

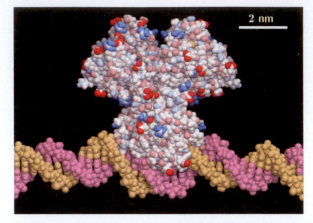

Figure 9.4 [Composite of structures from x-ray crystallography.] **A DNA-binding protein.** Repressor proteins like this one bind directly to the DNA double helix, physically blocking the polymerase that makes messenger RNA. A repressor recognizes a specific sequence of DNA (its "operator"), generally a region of about 20 base pairs. This image depicts a repressor named FadR, involved in the control of fatty acid metabolism in *E. coli*. [Courtesy David S Goodsell.]

Figure 8.4 (page 188)

protein. A gene or group of genes is preceded in the cell's genome by a specific binding sequence for RNA polymerase called a **promoter**. The polymerase will initially bind at a promoter. From there, it proceeds by walking along the DNA, synthesizing the RNA transcript as it goes, until it encounters a "stop" signal; then it unbinds from the DNA and releases the transcript. If a promoter/stop pair encloses more than one gene, then all the intervening genes are transcribed in a single pass of the RNA polymerase. A separate step called translation then creates protein molecules based on the instructions in the messenger RNA transcripts (see Figure 8.4).

A specialized class of "regulatory" proteins, called **repressors**, can control the transcription of genes. Each type of repressor binds specifically to a particular DNA sequence, called its **operator** (Figure 9.4). If an operator sequence appears close to (or within) a promoter, then a repressor bound to it can physically block access to the promoter, preventing transcription of the gene(s) that it controls (Figure 9.3d). More generally, a **regulatory sequence** is any sequence on the genome that specifically binds a protein with regulatory function. Instead of a repressor, a regulatory sequence may bind a protein called an **activator**, which *stimulates* the activity of RNA polymerase, switching the controlled gene(s) into high gear. Collectively, repressors and activators are also called **transcription factors**. In fact, multiple genes can all be controlled by a single operator, if they all lie between a controlled promoter and its corresponding stop signal. A set of genes lumped in this way form a jointly controlled element called an **operon**.

One well-known example of a transcription factor is called the *lac* **repressor**, abbreviated as **LacI**. It binds to a regulatory DNA sequence called the *lac* **operator**, and therefore can control genes downstream of that operator.[8]

In order to control gene expression, it suffices for the cell to control either the inventory of transcription factors or their binding to DNA. The first of these options is rather slow; it requires time to synthesize additional proteins, or to eliminate existing ones. The second option can operate faster. For example, LacI can itself bind a small molecule (effector), inducing an allosteric change that immediately alters the protein's affinity for its operator

---

[8]Chapter 10 will discuss this system in greater detail.

(Figure 9.3e). This short chain of command allows a cell to *sense* conditions (concentration of effector), and to *regulate* production of its metabolic machinery (products of genes controlled by the operator)—precisely the elements needed to implement feedback control.

Figure 9.3e (page 209)

If effector binding inactivates a repressor, then the effector is called an **inducer**, because it permits transcription. Thus, for example, the *lac* repressor will not bind to its operator if it has already bound the inducer molecule allolactose. Two other, similar molecules, called **TMG** and **IPTG**, also have the same effect as allolactose on the *lac* repressor.[9]

In short,

> A cell can modulate the expression of a specific gene, or set of genes, either positively
> or negatively, by sensing the amount and state of a transcription factor specific to
> the regulatory sequence accompanying the gene(s).

T_2 *Section 9.3.4′ (page 235) discusses some details of transcription and activation.*

9.3.5 Artificial control modules can be installed in more complex organisms

A key conclusion from the ideas outlined in the previous section is that the response of a particular gene to each transcription factor is not immutably determined by the gene itself; it depends on whether that gene's promoter contains or is adjacent to a regulatory sequence. Thus, a cell's life processes can be *programmed* (by evolution) or even reprogrammed (by human intervention). Moreover, all living organisms have at least some of their genetic apparatus in common, including the basic motifs of gene→mRNA via transcription, and mRNA→protein via translation. Can we apply insights from bacteria, even specific control mechanisms, all the way up to single-cell eukaryotes, and even to mammals like ourselves?

Figure 9.5 shows vividly that we can. The mouse on the left is an ordinary albino animal. That is, one gene (called *TYR*) necessary for the production of the pigment melanin has a mutation that renders its product protein inoperative. The second mouse shown is a **transgenic** variety: A new gene for the missing enzyme (tyrosinase) has been artificially added to its genome. This "transgene" is expressed, leading to brown hair and skin.

C. Cronin and coauthors created a variant of this transgenic mouse, in which the *TYR* transgene was controlled by the *lac* operator introduced earlier. Because mammals do not produce the *lac* repressor, however, this modification had no effect on gene expression, and the resulting mice were again pigmented. Next, the experimenters created a second transgenic mouse line on the albino background, without the *TYR* transgene but with a transgene for the *lac* repressor. Because there was nothing for LacI to repress, its presence had no effect in these mice.

When the two transgenic mouse strains were bred together, however, some of the offspring were "doubly transgenic"; they contained both modifications. Those individuals still appeared albino: Although they had a functioning *TYR* gene, it was repressed by LacI. But simply feeding them the inducer IPTG in their drinking water removed the repression (switched on the *TYR* gene), leading to brown fur just as in the singly transgenic line!

The fact that a regulatory mechanism found only in *bacteria* can be functional even in *mammals* is a remarkable demonstration of the unity of Life. But we will now return to bacteria, to develop a more quantitative version of the ideas behind gene regulation.

T_2 *Section 9.3.5′ (page 235) discusses natural gene regulation systems in eukaryotes.*

---

[9] The abbreviations stand for thiomethyl-β-D-galactoside and isopropyl β-D-1-thiogalactopyranoside, respectively.

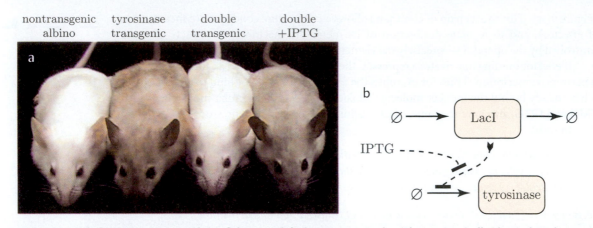

Figure 9.5 **Control of a gene in a mammal.** (a) [Photograph.] The two mice on the right are genetically identical. Both contain a transgene coding for the enzyme tyrosinase, needed to synthesize brown fur pigment. In both, this transgene is controlled by the *lac* operator. Both mice also contain a transgene coding for the *lac* repressor. But they differ visibly because the *TYR* gene has been turned on in the rightmost individual, by introducing the inducer molecule IPTG into its drinking water. [From Cronin et al., 2001.] (b) [Network diagram.] The control strategy used in the experiment (see Section 9.5.1).

9.4 Dynamics of Molecular Inventories

Understanding feedback in a cellular context requires that we represent the words and pictures of the preceding sections by definite formulas.

9.4.1 Transcription factors stick to DNA by the collective effect of many weak interactions

Section 9.3.4 characterized the key step in gene regulation as the specific binding of a transcription factor, which we'll call R, to its regulatory sequence O in the cell's DNA. We wish to represent this binding, ultimately by introducing the concept of a "gene regulation function."

The word "binding" implies that a repressor molecule stops its random thermal motion, becoming immobilized at the regulatory sequence. Unlike synthesis and degradation reactions, however, binding associations are not permanent; they do not involve formation of covalent chemical bonds. Instead, both R and O retain their distinct identities as molecules. Many weak interactions between specific atoms on R and O (such as electrostatic attraction) add up to a significant reduction of their potential energy when they are touching in the proper orientation. But a big enough kick from thermal motion in the environment can still break the association. Because molecular interactions are generally of short range, once R has left O it is likely to wander away completely; later, it or another copy of R may wander back and rebind O. During the time that O is unoccupied, any genes that it controls are available for transcription.

Because thermal motion is random, so is the binding and unbinding of repressors to their regulatory sequences. Thus, cellular control is *probabilistic*. Chapter 8 discussed some of the phenomena that we can expect to find in such systems. Even with just two reactions, matters got rather complex; with the many reactions in a cell, we could easily miss the overall behavior amid all that complexity. But Chapter 8 showed a simplification that emerges when molecule counts are high enough: the continuous, deterministic approximation, where we neglect the stochastic character of number fluctuations. Similarly, we may hope that it will suffice simply to know what fraction of a gene's time is spent in the repressed (or activated) state, or in other words, its activity averaged over time, neglecting the stochastic character of repressor binding. This chapter will work at that coarse level of detail.

9.4.2 The probability of binding is controlled by two rate constants

Imagine a single DNA molecule, containing the regulatory sequence O. The DNA is in solution, in a chamber of volume V. Suppose that the chamber contains just *one* repressor molecule. If we know that initially it is bound to O, then the probability to remain bound at a later time starts equal to 100% and then decreases, because of the possibility of unbinding. That is, after a short time Δt,

$$\mathcal{P}(\text{bound at } \Delta t \mid \text{bound at } t = 0) \approx 1 - (\Delta t)\beta_{\text{off}}. \tag{9.2}$$

The constant β_{off} is called the **dissociation rate**. It represents probability per unit time, so it has dimensions $\mathbb{T}^{-1}$. In chemical notation, this process is written $OR \xrightarrow{\beta_{\text{off}}} O + R$.

    If the repressor is initially *unbound,* then it has no opportunity to stick to its regulatory sequence until random thermal motion brings the two into physical contact. Imagine a small "target" region with volume v surrounding the regulatory sequence, chosen so that outside of it O has no influence on R. The probability to be located in this target region is v/V. If the repressor finds itself in the target region, then there is some probability that it will stick to the regulatory sequence instead of wandering away. Again we suppose that this probability initially changes linearly with time, that is, as $\kappa \Delta t$ for some constant κ. Putting these two ideas together yields[10]

$$
\begin{aligned}
&\mathcal{P}(\text{bound at } \Delta t \mid \text{unbound at } t = 0) \\
&= \mathcal{P}(\text{bound at } \Delta t \mid \text{unbound but in target at } t = 0) \times \mathcal{P}(\text{in the target} \mid \text{unbound}) \\
&= (\kappa \Delta t)(v/V).
\end{aligned} \tag{9.3}
$$

We do not know a priori values for κ and v. But note that they appear only in one combination, their product, which we may abbreviate by the single symbol k_{on} (the **binding rate constant**). Also, because we assumed there was just one repressor molecule in the chamber, the concentration c of repressors is just $1/V$. So all together, the probability to bind in time Δt is

$$\mathcal{P}(\text{bound at } \Delta t \mid \text{unbound at } t = 0) = (k_{\text{on}}c)\Delta t, \tag{9.4}$$

which we indicate by adding another arrow to the reaction: $O + R \underset{\beta_{\text{off}}}{\overset{k_{\text{on}}c}{\rightleftharpoons}} OR$. If there are N repressor molecules, the probability for *any one* of them to bind is proportional to N/V, which is again the concentration.

---

[10] $\boxed{T_2}$ Equation 9.3 is similar to the generalized product rule, Equation 3.25 (page 60).

9.4.3 The repressor binding curve can be summarized by its equilibrium constant and cooperativity parameter

Because concentrations serve as state variables in cells (see Section 9.3.1), we would like to see how they can control other variables. The key observation is that, like any chemical reaction, binding is controlled not only by the **affinity** (stickiness) of the participants for each other, but also by their concentration (availability), as seen in Equation 9.4.

The discussion so far assumed that we knew the binding state at $t = 0$. If that's not the case, we can nevertheless conclude that

$$
\begin{aligned}
&\mathcal{P}(\text{bound at } \Delta t)\\
&= \mathcal{P}(\text{bound at } \Delta t \textbf{ and } \text{unbound at } t = 0) + \mathcal{P}(\text{bound at } \Delta t \textbf{ and } \text{bound at } t = 0)\\
&= \mathcal{P}(\text{bound at } \Delta t \mid \text{unbound at } t = 0)\mathcal{P}(\text{unbound at } t = 0)\\
&\quad + \mathcal{P}(\text{bound at } \Delta t \mid \text{bound at } t = 0)\mathcal{P}(\text{bound at } t = 0).
\end{aligned}
\tag{9.5}
$$

Equations 9.2–9.4 let us simplify this expression to

$$
= (k_{on}c)(\Delta t)\big(1 - \mathcal{P}(\text{bound at } t = 0)\big) + \big(1 - (\Delta t)\beta_{off}\big)\mathcal{P}(\text{bound at } t = 0),
$$

where we used the fact that the probabilities of being bound or unbound must add up to 1.

If we wait for a long time, then the probabilities to be bound or unbound will approach steady-state values. In steady state, $\mathcal{P}(\text{bound})$ becomes time independent, so any terms involving Δt in the preceding formula must cancel:

$$
0 = (k_{on}c)\big(1 - \mathcal{P}(\text{bound})\big) - \beta_{off}\mathcal{P}(\text{bound}).
\tag{9.6}
$$

Solving for $\mathcal{P}(\text{bound})$ now gives

$$
\mathcal{P}(\text{bound}) = (k_{on}c + \beta_{off})^{-1}k_{on}c = \left(1 + \frac{\beta_{off}}{k_{on}c}\right)^{-1}.
\tag{9.7}
$$

It's useful to make the abbreviation

$$
K_d = \beta_{off}/k_{on},
$$

which is called the <u>**dissociation equilibrium constant**</u>. Unlike β_{off} and k_{on}, K_d has no time dimensions;[11] it is a concentration. It describes the intrinsic strength of the binding: The

---

[11] See Your Turn 9A.

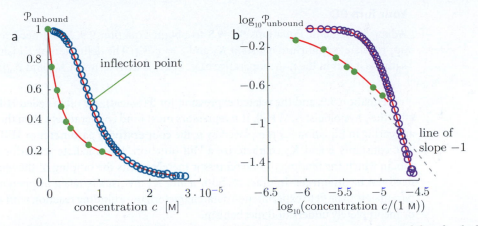

Figure 9.6 [Experimental data with fits.] **Binding curves.** (a) The *dots and circles* are experimental data for the binding curve of a ligand (oxygen) to each of two different proteins. The *curves* are functions of the form given in Your Turn 9D for n chosen to fit the data. One has an inflection point; the other does not. (b) The same functions in (a), but displayed as log-log plots. For comparison, a straight line of slope -1 is shown in *gray*. [Data from Mills et al., 1976 and Rossi-Fanelli & Antonini, 1958; see Dataset 13.]

formula $\mathcal{P}(\text{bound}) = 1/\big(1 + (K_\text{d}/c)\big)$ states that increasing K_d would lower the probability to be bound at any fixed value of c. To get less cluttered formulas, we also define the dimensionless concentration variable $\bar{c} = c/K_\text{d}$. Then the probability to be unbound becomes simply

$$\mathcal{P}(\text{unbound}) = 1 - \frac{1}{1 + \bar{c}^{-1}} = \frac{1}{1 + \bar{c}}. \quad \textbf{noncooperative binding curve} \quad (9.8)$$

The phrase "binding curve" refers to the graph of $\mathcal{P}(\text{unbound})$ as a function of concentration (solid dots in Figure 9.6a). It's one branch of a hyperbola.[12] It's a strictly decreasing function of $\bar{c}$, with no inflection point.

Alternatively, we can imagine a **cooperative binding** model, in which *two* repressor molecules must simultaneously bind, or none. For example, such cooperative behavior could reflect the fact that two regulatory sequences are located next to each other on the DNA, and repressors bound to the two sites can also touch each other, enhancing each others' binding to the DNA. In this case, Equation 9.3 must be modified: Its right-hand side needs an additional factor of (v/V). Following a similar derivation as above, you can now show that the binding and unbinding rates for this reaction have the form $O + 2R \underset{\beta_\text{off}}{\overset{k_\text{on}c^2}{\rightleftharpoons}} ORR$,

and

$$\mathcal{P}(\text{unbound}) = \frac{1}{1 + \bar{c}^2}. \quad \textbf{cooperative binding curve with } n = 2 \quad (9.9)$$

---

[12]We say that, "the binding curve is hyperbolic." Some authors use the adjective "Michaelan" to denote this particular algebraic form, because it's the same function appearing in the Michaelis–Menten formula from enzyme kinetics. Others call it the "Langmuir function."

Your Turn 9C

Adapt the logic of Equations 9.2–9.8 to obtain Equation 9.9. You'll need to define an appropriate equilibrium constant K_d and $\bar{c} = c/K_d$. The definition of K_d isn't quite the same as it was in the noncooperative case. What is it, in terms of k_{on} and β_{off}?

Equation 9.9, and the related expression for $\mathcal{P}(\text{bound})$, are often called **Hill functions**, after the physiologist A. V. Hill. If n repressors must bind cooperatively, then the exponent 2 appearing in Equation 9.9 is replaced by n, the **cooperativity parameter** or **Hill coefficient**. The constants n and K_d characterize a Hill function. Intermediate scenarios are possible too, in which the binding of one repressor merely assists the binding of the second; then n need not be an integer. In practice, both n and K_d are usually taken as phenomenological parameters to be fit to experimental data, characterizing a binding reaction with an unknown or incompletely understood mechanism.

About Hill functions

Let's take a moment to note two qualitative aspects of functions like the ones in Equations 9.8–9.9. First notice that the binding curve may or may not have an **inflection point**, depending on the value of n.

Example Find the condition for there to be an inflection point.

Solution An inflection point is a place where the curvature of a function's graph switches, or equivalently where the function's second derivative crosses through zero. So to investigate it, we must calculate the second derivative of $(1 + \bar{c}^{\,n})^{-1}$ with respect to concentration. First work out

$$\frac{d}{d\bar{c}}(1 + \bar{c}^{\,n})^{-1} = -(1 + \bar{c}^{\,n})^{-2}(n\bar{c}^{\,n-1}),$$

which is always less than or equal to zero: Hill functions for $\mathcal{P}(\text{unbound})$ are strictly decreasing. Next, the second derivative is

$$-(1 + \bar{c}^{\,n})^{-3}\left[-2n\bar{c}^{\,n-1}n\bar{c}^{\,n-1} + (1 + \bar{c}^{\,n})n(n - 1)\bar{c}^{\,n-2}\right].$$

This expression is zero when the factor in large brackets is zero, or in other words at $\bar{c} = \bar{c}_*$, where

$$0 = -2n(\bar{c}_*)^{2n-2} + (n - 1)(1 + (\bar{c}_*)^n)(\bar{c}_*)^{n-2}.$$

Solving for $\bar{c}_*$, we find that the second derivative vanishes at

$$\bar{c}_* = \left(\frac{n - 1}{n + 1}\right)^{1/n}.$$

The corresponding point on the binding curve will not be an inflection point, however, if $\bar{c}_*$ lies at the extreme value of concentration, that is, if $\bar{c}_* = 0$. In fact this does happen when $n = 1$, so we get an inflection point only when $n > 1$.

Your Turn 9D

Confirm this result by getting a computer to plot the function $(1 + \bar{c}^{\,n})^{-1}$ over an interesting range of $\bar{c}$, for various values of n.

A second qualitative property of Hill functions is reminiscent of something we saw when studying power-law probability distribution functions (see Figure 5.6). At low concentration, $c \ll K_d$, $\mathcal{P}(\text{unbound})$ approaches the constant 1. At high concentration, $c \gg K_d$, the function $\mathcal{P}(\text{unbound})$ becomes a power law, which looks like a straight line on a log-log plot. Thus, we can assess the cooperativity of a binding curve simply by plotting it in this way and noting whether the slope of the right-hand part of the graph is -1, or some value more negative than that.

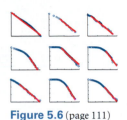

Figure 5.6 (page 111)

Figure 9.6 makes these ideas more concrete by showing curves for two proteins, representing the probability for each one to be bound to a small molecule (a ligand). One dataset shows noncooperative binding; the other is well fit by the cooperative Hill function with $n = 3.1$. One of these proteins is myoglobin, which has a single binding site for oxygen; the other is hemoglobin, which has four interacting binding sites for oxygen.

Your Turn 9E

Figure out which curve represents myoglobin and which represents hemoglobin in Figure 9.6.

9.4.4 The gene regulation function quantifies the response of a gene to a transcription factor

We wish to apply the binding curve idea to gene regulation. We will always make the following simplifying assumptions:

- The cell, or at least some subvolume of its interior, is "well mixed." That is, the transcription factors, and their effectors, if any, are uniformly spread throughout this volume with some average concentration. (The volume can be effectively smaller than the cell's total volume because of crowding and spatial confinement of the molecular actors, but we'll take it to be a constant fraction of the total.)

- The overall rate of production of each gene product, which we will call its gene regulation function (GRF, or just f in equations), equals its maximum value Γ times the fraction of time that its regulatory sequence is *not* occupied by a repressor, that is, $\Gamma \mathcal{P}(\text{unbound})$.

These assumptions are at best qualitatively correct, but they are a starting point. The first is reasonable in the tiny world of a bacterium, where molecular diffusion can be effective at mixing everything rapidly.

The second point above neglects the random ("noisy") character of regulator binding, transcription, and translation. This simplification is sometimes justified because binding and unbinding are rapid, occurring many times during the time scale of interest (typically the cell's division time). Combined with the continuous, deterministic approximation discussed in Chapter 8, it lets us write simple equations for cellular dynamics. A more realistic physical model of gene regulation might keep track of separate inventories for a protein and for its corresponding messenger RNA, and assign each one separate synthesis and clearance kinetics, but we will not attempt this.

With these idealizations, Equations 9.8–9.9 give the **gene regulation function** f:

$$f(c) = \frac{\Gamma}{1 + (c/K_d)^n}. \quad \text{GRF of simplified operon} \tag{9.10}$$

Note that f and Γ have dimensions $\mathbb{T}^{-1}$, whereas c and K_d have dimensions $\mathbb{L}^{-3}$. As before, we'll introduce the dimensionless concentration variable $\bar{c} = c/K_d$.

$\boxed{T_2}$ *Section 9.4.4′ (page 236) discusses some modifications to the simplified picture of gene regulation described above.*

9.4.5 Dilution and clearance oppose gene transcription

The Hill functions derived above can be used to describe the binding and unbinding of a repressor, but they're not the whole story. If all we had was production, then the concentration of gene products would always increase and we'd never get any steady states.

There are at least two other factors that affect the concentration of proteins: clearance (degradation) and changes in cell volume (dilution). For the first, a reasonable guess for the rate of clearance is that it is proportional to concentration, as we assumed when modeling HIV infection.[13] Then

$$\frac{dc}{dt} = -k_\emptyset c, \quad \text{contribution from clearance} \tag{9.11}$$

where the constant $k_\emptyset$ is called the **clearance rate constant**. $1/k_\emptyset$ has dimensions $\mathbb{T}$, so it is sometimes called the "clearance time constant."

To account for changes in cell volume, we'll simplify by assuming that $V(t)$ increases uniformly, doubling after a fixed "doubling time" τ_d. Thus, we have $V(t) = 2^{t/\tau_d} V_0$. We'll rephrase this relation by introducing the **e-folding time**[14] $\tau_e = \tau_d/(\ln 2)$; then

$$V(t) = V_0 \exp(t/\tau_e). \tag{9.12}$$

Taking the reciprocal of each side of Equation 9.12, and multiplying each by the number of molecules, gives an equation for the change of the concentration due to dilution. Its derivative has the same form as Equation 9.11, so the two formulas can be combined into a single equation representing all processes that reduce the concentration of the protein of interest, in terms of a single **sink parameter** $\tau_{tot} = \left(k_\emptyset + \frac{1}{\tau_e}\right)^{-1}$:

$$\frac{dc}{dt} = -\frac{c}{\tau_{tot}}. \quad \text{dilution and clearance} \tag{9.13}$$

The same equation also holds for the dimensionless quantity $\bar{c}$.

---

[13] See Equations 1.1 and 1.2 (page 14), and the discussion of the birth-death process in Section 8.3.1 (page 182). However, this is an approximation—for example, some degradation mechanisms instead saturate at high concentration.

[14] The relation between τ_d and $1/\tau_e$ is similar to that between half-life and clearance rate constant; see Problem 1.3 (page 23). Some authors use "generation time" as a synonym for e-folding time, but this usage can lead to confusion, as others use the same words to mean the doubling time.

We have now arrived at some formulas for regulated production, and loss, of a molecular species in a cell. It's time to get predictions about the behavior of regulatory networks from those equations.

9.5 Synthetic Biology

9.5.1 Network diagrams

We can represent cellular control mechanisms in a rough way by drawing **network diagrams**.[15] Imagining the cell as a small reaction vessel, we draw a box to represent the inventory for each relevant molecular species. Then we use the following graphical conventions to represent interconnections:

- *An incoming solid line represents production of a species, for example, via expression of the corresponding gene.*

- *Outgoing solid lines represent loss mechanisms.*

- *If a process transforms one species to another, and both are of interest, then we draw a solid line joining the two species' boxes. But if a species' precursor is not of interest to us, for example, because its inventory is maintained constant by some other mechanism, we can replace it by the symbol $\varnothing$, and similarly when the destruction of a particular species creates something not of interest to us.* $\quad$ (9.14)

- *To describe the effect of transcription factor population on another gene's transcription, we draw a dashed "influence line" from the former to the latter's incoming line, terminating with a symbol: A blunt end, -----I, indicates repression, whereas an open arrowhead, -----▷, indicates activation.*

- *Other kinds of influence line can instead modulate the loss rate of a species; these lines terminate on an outgoing arrow.*

- *To describe the influence of an effector on a transcription factor, we draw a dashed influence line from the former that impinges on the latter's dashed line.*

Figures 9.7 and 9.10 below illustrate these conventions. To reduce clutter, we do not explicitly draw an influence line indicating that the rate of clearance of a species depends on its population.

The first point above is a simplification: It lumps together the processes of transcription and translation into a single arrow (we do not draw separate boxes for the messenger RNA and its gene product). Although this leads to compact diagrams, nevertheless for some purposes it is necessary to discuss these processes separately. The last point is also a simplification of the real situation: It does not keep separate track of the inventories for transcription factors with and without bound effector (for example, by giving each state its own box). Instead, the amount of active transcription factor is simply assumed to depend on the effector concentration. Also, more complex binding schemes may require additional kinds of elements. Despite these limitations, the simple network diagrams we'll draw are a flexible graphical language that unifies many concepts.

$\boxed{T_2}$ *Section 9.5.1′a (page 236) describes another approximation implicit in Idea 9.14. Many variants of network diagrams exist in the literature; see Section 9.5.1′b.*

---

[15]Network diagrams were already introduced informally in Figures 8.2b (page 182), 8.8b (page 192), and 9.5b (page 212).

Figure 9.7 [Network diagrams.] **Feedback in gene expression.** (a) An unregulated gene creates a birth-death process. This is the same diagram as Figure 8.2b (page 182), except that the influence of protein population on its own clearance is understood tacitly. (b) The genetic governor circuit constructed in *E. coli* by Becskei and Serrano (2000). A gene's product represses its own expression.

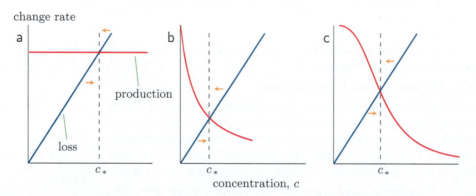

Figure 9.8 [Sketch graphs.] **Graphical understanding of the appearance of stable fixed points in genetic control circuits.** (a) In the birth-death process, the production rate of a protein is constant (*red line*), whereas the loss is proportional to the concentration (*blue line*). The lines intersect just once, at the steady-state value c_*. If the concentration fluctuates upward from that value, loss outstrips production, pushing the concentration back toward c_* (*arrow*), and similarly for downward fluctuations. Thus, the steady-state value c_* is a stable fixed point of the system. (b) With noncooperative feedback control, the production rate depends on concentration (see Figure 9.6a, page 215), but the result is similar: Again there is one stable fixed point. (c) Even with cooperative regulation, we have the same qualitative behavior, as long as loss depends linearly on concentration.

9.5.2 Negative feedback can stabilize a molecule inventory, mitigating cellular randomness

We can now start to apply the general ideas about feedback at the start of this chapter to cellular processes more complex than the birth-death process (Figure 9.7a). Figure 9.8 shows graphically why we expect stable fixed-point behavior to arise in an unregulated gene [panel (a)], and in genes with additional feedback, either noncooperative [panel (b)] or cooperative [panel (c)].

We might expect that a cell with self-regulating genes could achieve better control over inventories than is possible by relying on clearance alone. Indeed,

- In the unregulated case, if there is a fluctuation above the setpoint value c_* then the gene's product is *still being produced* at the usual rate, opposing clearance and dilution; in the self-regulated case, the production rate falls.

- Similarly, if there is a fluctuation below c_*, production ramps up in the self-regulated case, again speeding return of the inventory to the setpoint.

Better reversion to the setpoint implies that cells can better correct the inevitable random fluctuations in the processes that maintain protein levels. We will soon see that these

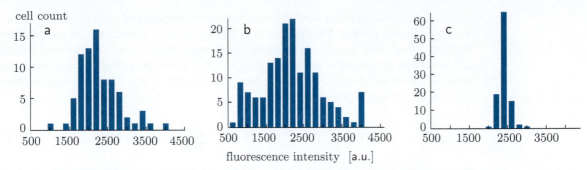

Figure 9.9 [Experimental data.] **Variation in protein content across many cells containing a synthetic control circuit.** (a,b) Feedback regulation in *E. coli* was disabled in two different ways (see text); the cells had a broad distribution in GFP content. (c) The cells' GFP content was much more tightly controlled when feedback regulation was implemented (Figure 9.7b). The maximum production rate and dilution rate were the same as in (a,b). [Data from Becskei & Serrano, 2000.]

expectations are borne out mathematically, but already we can note that cells do make extensive use of negative feedback as a common **network motif**: For example, a large fraction of the known transcription factors in *E. coli* regulate themselves in this way.

A. Becskei and L. Serrano tested these ideas by adding a synthetic gene to *E. coli* (see Figure 9.7b). The gene expressed a fusion protein: One part was the ***tet* repressor**[16] (abbreviated TetR). The other protein domain was a green fluorescent protein, which was used to monitor the amount present in individual cells. The gene's promoter was controlled by an operator that binds TetR. For comparison, the experimenters also created organisms with either a mutated gene, creating a protein similar to TetR but that did not bind its operator, or a mutated operator that did not bind the native TetR molecule. These organisms effectively had no transcriptional feedback, so they created birth-death processes. Figure 9.9 shows that the bacteria with autoregulation maintained much better control over protein population than the others.

9.5.3 A quantitative comparison of regulated- and unregulated-gene homeostasis

We can go beyond the qualitative ideas in the previous section by solving the equation for protein concentration in the cell:

$$\frac{dc}{dt} = f(c) - \frac{c}{\tau_{\text{tot}}}. \qquad \text{production, dilution, and clearance} \qquad (9.15)$$

In this formula, f is the gene regulation function (Equation 9.10) and τ_{tot} is the sink parameter (Equation 9.13). If repression is noncooperative, we may take $n = 1$ in Equation 9.10, obtaining

$$\frac{dc}{dt} = \frac{\Gamma}{1 + (c/K_{\text{d}})} - \frac{c}{\tau_{\text{tot}}}. \qquad (9.16)$$

Here Γ is the maximum production rate, and K_{d} is the dissociation constant for binding of the repressor to its operator.

---

[16]So named because it was first found in studies of bacterial resistance to the antibiotic tetracycline.

Let's solve this equation with the condition that at some initial time, called $t = 0$, there are no molecules present. We could ask a computer to do it numerically, or use advanced methods of calculus, but an approximate solution is more instructive. Suppose that the concentration of nutrient is large enough that $c \gg K_d$ (we will justify this approximation later). Then we may neglect the 1 in the denominator of the GRF. Changing variables from c to $y = c^2$ converts Equation 9.16 to

$$\frac{dy}{dt} = 2\left(\Gamma K_d - \frac{y}{\tau_{tot}}\right). \tag{9.17}$$

This equation is mathematically similar to the one for the *unregulated* gene (birth-death process). Defining $y_* = \tau_{tot}\Gamma K_d$ gives

$$\frac{d(y - y_*)}{dt} = -\frac{2}{\tau_{tot}}(y - y_*), \text{ or}$$

$$y - y_* = Ae^{-2t/\tau_{tot}},$$

where A is a constant. Choosing $A = -\tau_{tot}\Gamma K_d$ enforces the initial condition. We can now substitute $y = c^2$, obtaining our prediction for the time course of the normalized concentration of protein:

$$\frac{c(t)}{c_*} = \sqrt{1 - e^{-2t/\tau_{tot}}}. \quad \text{noncooperatively regulated gene} \tag{9.18}$$

Let's compare Equation 9.18 to the prediction for the unregulated gene:[17]

$$\frac{c(t)}{c_*} = 1 - e^{-t/\tau_{tot}}. \quad \text{unregulated gene} \tag{9.19}$$

To facilitate comparison, consider only the late-time behavior, or equivalently the return to the setpoint after a small deviation. In this case, Equation 9.18 becomes $\approx 1 - \frac{1}{2}e^{-2t/\tau_{tot}}$; *the rate is twice as great* for the self-regulating case, compared to an unregulated gene. A similar analysis for larger values of the Hill coefficient n shows that they correct even faster than this.

N. Rosenfeld and coauthors tested the quantitative predictions in Equations 9.18–9.19 by making a modified form of Becskei and Serrano's construction. They used the fact that the *tet* repressor, TetR, is sensitive to the presence of the antibiotic tetracycline (whose anhydrous form is abbreviated aTc). Binding the effector aTc allosterically modifies TetR, preventing it from binding to its operator (Figure 9.3e). The cellular circuit represented by Figure 9.10a constantly generates enough TetR to reduce GFP production to a low level. When aTc is suddenly added to the growth medium, however, the repression is lifted, and the system becomes effectively unregulated. In the conditions of the experiment, degradation of GFP

Figure 9.3e (page 209)

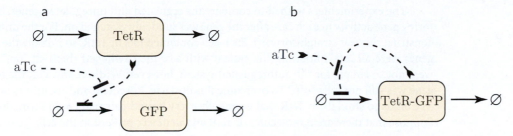

Figure 9.10 [Network diagrams.] **Synthetic gene circuits without and with feedback control.** (a) A modified form of Figure 9.7a, allowing expression of GFP to be switched by the effector aTc. The *tet* repressor is always present, but its action on the fluorescent reporter's promoter can be turned off by adding aTc to the growth medium. The two negative control lines jointly amount to a positive control: Adding aTc turns on GFP production. (b) A modified form of Figure 9.7b, allowing feedback control to be disabled by the addition of aTc.

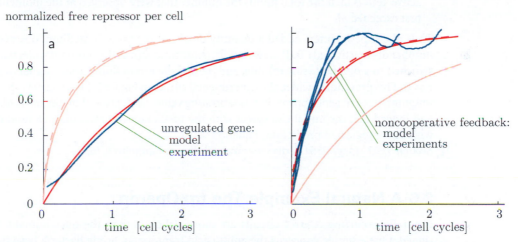

Figure 9.11 [Experimental data with fits.] **Theory and experiment for kinetics of gene autoregulation.** (a) *Lower red curve:* Model prediction of GFP production from an unregulated promoter (Equation 9.19). *Solid blue curve:* Experimental measurement from bacteria with the circuit shown in Figure 9.10a, averaged over many individuals. (The upper curves are the same as in (b) for comparison.) (b) *Dashed red curve:* Approximate solution to the equation for a regulated promoter (Equation 9.18). *Solid red curve:* Exact solution from Section 9.5.3' (page 236). *Blue curves:* Experimental results for the autoregulated gene system, in three trials. The data agree with the model that the initial approach to the setpoint is much faster than in the unregulated case. (The analysis in this chapter does not explain the observed overshoot; see Chapter 11.) [Data from Rosenfeld et al., 2002; see Dataset 14.]

was slow, so the experimenters expected that the sink parameter τ_{tot} would be dominated by dilution from cell growth. Measurement of the time course of fluorescence per cell indeed confirmed the prediction of Equation 9.19 with this value of τ_{tot} (see Figure 9.11a). The experimenters normalized the fluorescence signal by dividing by the total number of bacteria, then normalized again by dividing by the saturating value of the signal per bacterium.

To study the regulated case, the experimenters constructed a modified organism with the network shown in Figure 9.10b. To analyze its behavior, they first noted that TetR binding is effectively noncooperative. Previous biochemical estimates suggested that $\Gamma\tau_{\mathrm{tot}} \approx 4\,\mu\mathrm{M}$ and $K_{\mathrm{d}} \approx 10\,\mathrm{nM}$. Thus, when the system is near its setpoint it satisfies $c/K_{\mathrm{d}} \approx c_{*}/K_{\mathrm{d}} \approx \sqrt{\tau_{\mathrm{tot}}\Gamma/K_{\mathrm{d}}} \approx 20$, justifying the approximation we made earlier, that this quantity was $\gg 1$.

The experimenters wished to compare the regulated and unregulated genes, by monitoring gene activity in each case after the gene was suddenly switched on. For the unregulated construct this was straightforward. Each cell contained lots of TetR, so initially the reporter genes were all "off"; flooding the system with aTc then switched them all "on." Matters were not so simple for the autoregulated system, however: Without any aTc, the regulated gene was still not fully "off." To overcome this obstacle, Rosenfeld and coauthors noted that aTc binds so strongly to TetR that essentially every bound pair that could form, does form. Suppose that there are n molecules of TetR and m of aTc present in the cell:

- If $n < m$, then essentially all of the TetR molecules are inactivated, the gene is effectively unregulated, and n rises. The rise is visible because the GFP domain of the fusion protein fluoresces regardless of whether its TetR domain is active.
- Once $n = m$, then every aTc molecule is bound to a TetR. From this moment, the system switches to regulated mode, starting from zero active TetR molecules. The number of active TetR is then the total minus the number that were observed at the moment when n first exceeded m.

Thus, the experimenters added a quantity of aTc to the growth medium, observed as the number of (inactivated) TetR grew rapidly, chose time zero as the moment when that growth switched to a different (slower) rate, and plotted the excess fluorescence signal over that at $t = 0$. This procedure yielded the experimental curves in Figure 9.11b. The data showed that the autoregulated gene approaches its saturating value much faster than the unregulated one, and indeed that the initial rise agrees with the prediction of Equation 9.18 (and disagrees with the unregulated prediction, Equation 9.19).

$\boxed{T_2}$ *Section 9.5.3′ (page 236) derives an exact solution to Equation 9.16.*

9.6 A Natural Example: The *trp* Operon

Naturally occurring control circuits are more elaborate than the ones humans have designed. One example concerns the amino acid tryptophan. Some bacteria have the ability to synthesize it, if it is missing from their food supply. For these bacteria, it's desirable to maintain a stable inventory of the molecule, mitigating fluctuations in its availability (and in their own demand).

The enzymes needed to synthesize tryptophan belong to an operon, controlled by an operator that binds a repressor. But unlike the *lac* repressor system, which must respond positively to the presence of a food molecule, synthetic pathways must *shut down* when they sense an adequate inventory of their product. Accordingly, the *trp* repressor *binds* DNA when it has bound a molecule of its effector tryptophan (see Figure 9.12). Additional negative feedbacks also improve the system's performance. For example, tryptophan can also bind directly to one of its production enzymes, allosterically blocking its action. A third feedback mechanism, called "attenuation," prematurely terminates transcription of the *trp* operon when sufficient tryptophan is present.

9.7 Some Systems Overshoot on Their Way to Their Stable Fixed Point

Imagine that you are in a room with a manually controlled heater. Initially the room is too cold, so you switch the heat on "high." Later, the room arrives at a comfortable temperature.

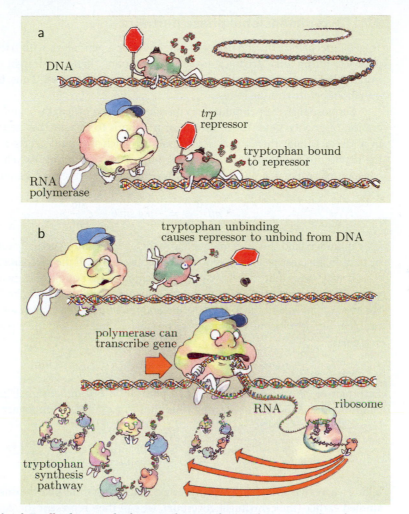

Figure 9.12 [Metaphor.] **Feedback control of tryptophan synthesis.** The genes coding for enzymes needed to synthesize tryptophan are contained in a repressible operon. The operon's repressor, TrpR, is in turn modulated by an effector, which is tryptophan itself. (a) Unlike the *lac* repressor system, TrpR binding *permits* the repressor to bind to its operator, creating a negative feedback loop. Thus, when the tryptophan inventory is adequate, the repressor turns off the operon. (Not shown in this cartoon is the fact that *two* molecules of tryptophan must bind to the repressor in order to turn on the gene.) (b) When inventory falls, the repressor unbinds, allowing expression of the genes for tryptophan synthesis enzymes. [From *The way life works* © Judith Hauck and Bert Dodson.]

If you monitor the temperature continuously, and adjust the power to your heater by smaller increments as you approach the desired temperature, then you can gracefully bring the room temperature to that setpoint and maintain it there, much as the centrifugal governor maintains an engine's speed. But suppose that you are reading a good book. You may fail to notice that the temperature passed your desired setpoint until the room is much too hot. When you do notice, you may readjust the heater to be too low, again lose interest, and end up too cold! In other words, negative feedback with a delay can give a behavior called **overshoot**, even if ultimately the temperature does arrive at your desired setpoint.

9.7.1 Two-dimensional phase portraits

To get a mechanical analog of overshoot, consider a pendulum with friction. Gravity supplies a restoring force driving it toward its "setpoint" (hanging straight down), but inertia implies that its response to disturbance is not instantly corrected. Thus, the pendulum may overshoot before coming to rest.

Suppose that the pendulum is a mass m on a rod of length L, and let g be the acceleration of gravity. One way to introduce friction is to imagine the pendulum submerged in a viscous fluid; then there is a regime in which the fluid exerts a force opposite to the pendulum's motion, with magnitude proportional to minus its velocity. If we call the constant of proportionality ζ, then Newton's law governs the mass's angular position θ:

$$mL\frac{d^2\theta}{dt^2} = -mg\sin\theta - \zeta\frac{d\theta}{dt}. \qquad (9.20)$$

We can understand this system's qualitative behavior, without actually solving the equation of motion, by creating a phase portrait, just as we did with the mechanical governor.

It may seem that the phase portrait technique isn't applicable, because knowing the value of θ at a given time is not enough information to determine its later development: We need to know *both* the initial position and the initial velocity. But we can overcome this problem by introducing a second state variable ω, and writing the equation of motion in two parts:

$$\frac{d\theta}{dt} = \omega; \qquad \frac{d\omega}{dt} = -(mg\sin\theta + \zeta\omega)/(mL). \qquad (9.21)$$

Each pair of (θ, ω) values determines a point on a *two*-dimensional "phase *plane*." If we choose one such point as a starting condition (for example, **P** in Figure 9.13), then Equation 9.21 gives us a vector

$$\boldsymbol{W}(\boldsymbol{P}) = \left(\frac{d\theta}{dt}\Big|_{\boldsymbol{P}}, \frac{d\omega}{dt}\Big|_{\boldsymbol{P}}\right)$$

telling where that point will move at the next moment of time. Like the one-dimensional case studied earlier (the centrifugal governor), we can then represent the equations of motion for our system as a set of arrows, but this time on the phase plane.

Figure 9.13b shows the resulting phase portrait appropriate for Equation 9.21. The straight orange line shows the locus of points for which $d\theta/dt = 0$; thus, the blue arrows are purely vertical there. The curved orange line shows the corresponding locus of points for which $d\omega/dt = 0$ (the arrows are purely horizontal). These lines are called the **nullclines**; their intersections are the points where $\boldsymbol{W} = (0, 0)$, that is, the fixed points. One fixed point is unsurprising: When $\theta = \omega = 0$, the pendulum hangs straight down motionless. The arrows all lead eventually to this stable fixed point, though not directly. The other fixed point may be unexpected: When $\theta = \pm\pi$ and $\omega = 0$, the pendulum is motionless at the *top* of its swing. Although in principle it could sit in this state indefinitely, in practice a small deviation from purely vertical will bring it down, eventually to land at the stable fixed point. One of the black curves underscores this point: It depicts a solution of the equations of motion that starts very close to the **unstable fixed point**, but nevertheless ends at the stable one.[18]

The black curves in Figure 9.13b show that the pendulum can indeed display overshoot: If we release it from a positive value of θ, it crosses $\theta = 0$ and swings over to negative θ

---

[18]This example explains why an unstable fixed point is also called a "tipping point."

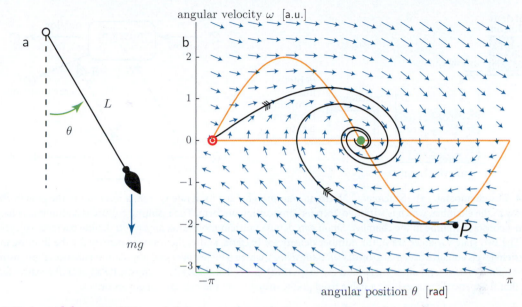

Figure 9.13 A pendulum. (a) [Schematic.] A mass m is attached to a pivot by a rod (whose own mass is negligible). The diagram shows a state with $\theta > 0$; if the mass is released from rest at this position, its angular velocity will initially be zero, but its angular acceleration will be negative. In addition to gravity, a frictional force acts in the direction opposite to the angular velocity ω. (b) [Two-dimensional phase portrait.] *Blue arrows* depict the vector field $W(\theta, \omega)$ as a function of angular position θ and velocity ω. The *straight orange line* is the locus of points for which $d\theta/dt = 0$. The *curved orange line* is the corresponding locus for $d\omega/dt = 0$. These lines are the system's nullclines. Two representative trajectories are shown (*black curves*). One of them starts at an arbitrary point P in the plane; the other starts near the unstable fixed point (*red bull's eye*). Both are attracted to the stable fixed point (*green dot*), but both overshoot before arriving there. The figure shows the case in which $gL(m/\zeta)^2 = 4$; see Problem 9.8 for other behaviors.

before turning back, only to cross back to positive θ, and so on, before finally coming to rest. Such behavior is not possible on a one-dimensional phase portrait like the one we drew for the centrifugal governor (Figure 9.1b). Even in two dimensions, it will only happen if the friction constant is sufficiently small; otherwise we say that the pendulum is **overdamped**.[19]

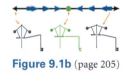

Figure 9.1b (page 205)

In short, the pendulum uses a simple form of negative feedback (the restoring force of gravity) to create a stable fixed point. But the details of its parameters matter if we want to get beyond the most qualitative discussion, for example, to address the question of overshoot. *The phase portrait gives a level of description intermediate between a verbal description and the full dynamical equations,* enabling us to answer such questions without having to solve the equations of motion explicitly.

$\boxed{T_2}$ *Section 9.7.1′ (page 237) describes the taxonomy of fixed points in greater detail.*

9.7.2 The chemostat

It is sometimes desirable to maintain a colony with a fixed number of bacteria. A strategy to accomplish this involves a simple feedback device invented by A. Novick and L. Szilard, who called it the **chemostat**.[20] Figure 9.14a shows the idea of the device. Bacteria need various

---

[19] See Problem 9.8.
[20] Chapter 10 will give an application of the chemostat.

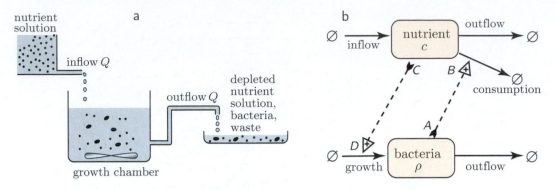

Figure 9.14 The chemostat. (a) [Schematic.] Nutrient solution is fed at a fixed rate Q from a reservoir to the growth chamber. Bacteria grow in the chamber, which is stirred to keep its composition uniform. The resulting culture is continuously harvested, to maintain a constant volume in the chamber. *Small dots* denote nutrient molecules; *larger dots* denote bacteria. (b) [Network diagram.] The two state variables are ρ, the number density for bacteria, and c, the number density for the limiting nutrient. Higher nutrient levels enhance bacterial growth (*left dashed line*), whereas higher bacterial population enhances the consumption of nutrient (*right dashed line*). Following around the loop *ABCD* shows that these effects constitute an overall negative feedback, suggesting that this network may exhibit a stable fixed point, analogous to the mechanical governor.

nutrients to grow, for example, sugar, water, and in some cases oxygen. But they also need a source of *nitrogen*, for example, to make their proteins. Novick and Szilard limited the nitrogen supply by including only one source (ammonia) in their growth medium, at a fixed concentration (number density) c_{in}.

The chemostat consists of a growth chamber of volume V, which is continuously stirred to keep the densities of nutrients and bacteria spatially uniform. Growth medium (nutrient solution) is continuously added to the chamber at some rate Q (volume per unit time). In order to keep the fluid volume in the chamber constant, its contents are "harvested" at the same rate Q, for example, by an overflow pipe. These contents differ in composition from the incoming medium: Nutrient has been depleted down to some concentration c. Also, the removed fluid contains bacteria at some density ρ. Both ρ and c have dimensions of inverse volume; each must be nonnegative.

We might hope that the system would settle down to a steady state, in which bacteria reproduce at exactly the same rate that they are washed out of the chamber. But some less desirable options must be explored: Perhaps all bacteria will eventually leave. Or perhaps the bacteria will reproduce uncontrollably, or oscillate, or something else. We need some analysis to find the time development of both ρ and c.

The analysis to follow involves a number of quantities, so we first list them here:

V	volume of chamber (constant)
Q	inflow and outflow rate (constant)
ρ	number density (concentration) of bacteria; $\bar{\rho}$, its dimensionless form
c	number density (concentration) of nutrient; $\bar{c}$, its dimensionless form
k_g	bacterial growth rate (depends on c); $k_{g,max}$, maximum value
ν	number of nutrient molecules needed to make one bacterium (constant)
K	nutrient concentration for half-maximal growth rate (constant)
t	time; $\bar{t}$, its dimensionless form
$T = V/Q$	a time scale (constant)
$\gamma = k_{g,max} T$	a dimensionless combination of parameters (constant)

Neither c nor ρ is under the direct control of the experimenter; both are unknown functions of time. To understand their behavior, let's first think qualitatively. If the population momentarily dips below its setpoint, then there's more food available per individual and growth speeds up, restoring the setpoint, and conversely for the opposite fluctuation. But there won't be any stable value of population if the timescale of bacterial growth is too large compared to the timescale of filling the chamber; otherwise, the bacteria will be washed away faster than they can replenish themselves, leading to population equal to zero.

Now we can do some analysis to see if the preceding expectations are correct, and to get the detailed criterion for a stable fixed point. First, consider how ρ must change with time. In any short interval dt, a volume $Q\,dt$ leaves the chamber, carrying with it $\rho(t)(Q\,dt)$ bacteria. At the same time, the bacteria in the chamber grow and divide at some rate k_g per bacterium, creating $\rho(t)Vk_g\,dt$ new individuals.[21] One feedback in the system arises because the growth rate k_g depends on the availability of the limiting nutrient. If no other nutrient is growth limiting, then k_g will depend only on c. It must equal zero when there is no nutrient, but it saturates (reaches a maximum value) when there is plenty of nutrient ($k_g \to k_{g,\mathrm{max}}$ at large c). A reasonable guess for a function with those properties is a Hill function $k_g(c) = k_{g,\mathrm{max}}c/(K+c)$. This function is determined by the maximum rate $k_{g,\mathrm{max}}$ and by another constant, K, that expresses how much nutrient is needed to attain half-maximal growth rate. The rate constant $k_{g,\mathrm{max}}$ has dimensions $\mathbb{T}^{-1}$; K has dimensions appropriate for a concentration, that is, $\mathbb{L}^{-3}$.

We can summarize the preceding paragraph in a formula for the net rate of change of the bacterial population with time:

$$\frac{d(\rho V)}{dt} = \frac{k_{g,\mathrm{max}}c}{K+c}\rho V - Q\rho. \tag{9.22}$$

We would like to solve this equation for $\rho(t)$, but it involves the nutrient concentration c. So we also need to find an equation for that quantity and solve it simultaneously with Equation 9.22.

To get the second equation, note that, in a short interval dt, nutrient solution flows into the chamber with concentration c_{in} and volume $Q\,dt$. At the same time, medium flows *out* with concentration $c(t)$ and the same volume. There is also some loss from bacteria consuming the nutrient. A reasonable proposal for that loss rate is to suppose that it is proportional to the growth rate of the bacteria, because they are incorporating the nutrient into their own structure, and every individual is similar to every other. We may therefore write

$$\frac{d(cV)}{dt} = Qc_{\mathrm{in}} - Qc - \nu\frac{k_{g,\mathrm{max}}c}{K+c}\rho V, \tag{9.23}$$

where the constant ν represents the number of nutrient molecules needed to create one bacterium.

Figure 9.14b gives a highly schematic representation of the chemostat—its network diagram. The diagram representation brings out the character of the feedback in the chemostat: One of the dashed lines enhances production of bacteria. The other enhances loss of

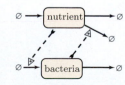

Figure 9.14b (page 228)

---

[21]These statements neglect the fact that bacteria are discrete; actually, the total number in the chamber must always be an integer. Typical values of ρV are so large that this discreteness is negligible, justifying the continuous, deterministic approximation used here.

nutrient, a negative influence. Working around the loop in the diagram therefore shows an *overall negative feedback* in the system, because $(+) \times (-) = (-)$. This observation suggests that the system may have a stable fixed point. We can now look in greater detail to see whether that conclusion is ever true, and if so, when.

Equations 9.22–9.23 are difficult to solve explicitly. But they define a phase portrait in the c-ρ plane, just as Equation 9.21 did for the pendulum, and that portrait gives us the qualitative insight we need. Before drawing this portrait, however, we should pause to recast the equations in the simplest form possible. The simplification involves four steps:

1. First, find a combination of the parameters that defines a natural time scale for the problem. That scale is the time needed for the inflow to fill the chamber, had it been empty initially: $T = V/Q$. Next, we define the dimensionless quantity $\bar{t} = t/T$, and everywhere substitute $\bar{t}T$ for t in the formulas. This "nondimensionalizing procedure" generally simplifies equations, by eliminating explicit mention of some parameters.

2. Similarly, the nutrient concentration c can be expressed as a dimensionless variable times the half-maximal concentration K: Let $c = \bar{c}K$.

3. Once the variables have been recast in terms of dimensionless quantities, the coefficients appearing in the equations must also enter only in dimensionless combinations. One such combination is $k_{g,\max}T$, which can be abbreviated to the single symbol γ.

4. We could also express ρ as a multiple of K, but the equations get even a bit nicer if instead we define $\rho = \bar{\rho}K/\nu$.

Your Turn 9F

Carry out the nondimensionalizing procedure just outlined, and obtain the **chemostat equations:**

$$\frac{d\bar{\rho}}{d\bar{t}} = \left(\gamma \frac{\bar{c}}{1 + \bar{c}} - 1 \right) \bar{\rho}; \qquad \frac{d\bar{c}}{d\bar{t}} = \bar{c}_{in} - \bar{c} - \gamma \frac{\bar{c}\bar{\rho}}{1 + \bar{c}}. \qquad (9.24)$$

In these formulas, $\bar{\rho}(\bar{t})$ and $\bar{c}(\bar{t})$ are state variables; γ and $\bar{c}_{in}$ are constant parameter values set by the experimenter.

Reformulating the original equations as Equation 9.24 is a big help when we turn to explore the possible behaviors of the chemostat, because the six parameters $k_{g,\max}$, K, Q, V, c_{in}, and ν only enter through the *two* dimensionless combinations $\bar{c}_{in}$ and γ.

The nullcline curves of the chemostat system in the $\bar{c}$-$\bar{\rho}$ plane can now be found by setting one or the other time derivative equal to zero. This procedure yields the two curves

$$\bar{\rho} = (\bar{c}_{in} - \bar{c})(1 + \bar{c})/(\gamma\bar{c}); \text{ and}$$
$$\text{either } \bar{c} = 1/(\gamma - 1) \text{ or } \bar{\rho} = 0.$$

The intersections of the nullclines give the fixed points (Figure 9.15):

$$\text{First: } \bar{c}_1 = \frac{1}{\gamma - 1}, \ \bar{\rho}_1 = \bar{c}_{in} - \frac{1}{\gamma - 1}. \quad \text{Second: } \bar{c}_2 = \bar{c}_{in}, \ \bar{\rho}_2 = 0. \qquad (9.25)$$

The second fixed point corresponds to the situation where all bacteria disappear from the chamber; the nutrient concentration approaches its level in the inflowing medium and stays there. The first fixed point requires more discussion, because of the requirements that $\bar{c}$ and $\bar{\rho}$ must not be negative.

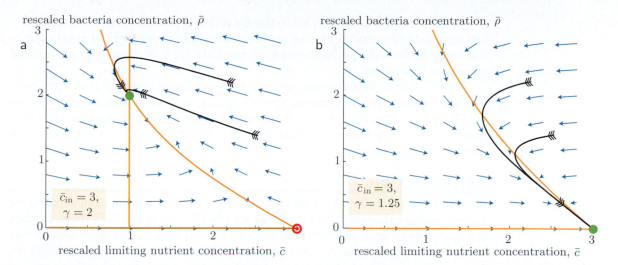

Figure 9.15 [Mathematical functions.] **Phase portraits for the chemostat equations** (Equation 9.24). (a) The case with parameter values $\bar{c}_{in} = 3$ and $\gamma = 2$. There is one stable and one unstable fixed point. Any initial state (point in the $\bar{c}$-$\bar{\rho}$ plane) must ultimately evolve to the steady state corresponding to the *green dot*. Two typical trajectories are shown (*black curves*). In addition, the system has an unstable fixed point (*red bull's eye*). (b) The case $\bar{c}_{in} = 3$ and $\gamma = 1.25$. Now there is only one fixed point, corresponding to an end state with zero bacteria.

Example Find the conditions that the parameters $\bar{c}_{in}$ and γ must satisfy in order for the first fixed point to be valid.

Solution A putative fixed point for which either ρ or c is negative is not physical. This requires that $\gamma > (\bar{c}_{in})^{-1} + 1$ (and $\gamma > 1$, which is redundant).

Figure 9.15a shows a typical phase portrait for the case with $\bar{c}_{in} = 3$ and $\gamma = 2$. The criterion found in the preceding Example is satisfied in this case, and the figure indeed shows two fixed points. Any starting values of $\bar{c}(0)$ and $\bar{\rho}(0)$ will eventually arrive at the stable fixed point as shown. Then *the number density of bacteria, in samples taken from the overflow, will approach a constant,* as desired, although it may overshoot along the way.

Panel (b) shows the corresponding portrait with $\bar{c}_{in} = 3$ and $\gamma = 1.25$. In this case, any initial condition is driven to the second fixed point, with zero bacteria; small $\gamma = k_{g,max} V/Q$ means that the flow rate gives a replacement time T shorter than the minimum needed for bacterial division.

We have obtained the insight that we wanted: Under conditions that we now know, the chemostat will drive itself to the first fixed point, stabilizing the bacterial population in the chamber to the value given in Equation 9.25.

9.7.3 Perspective

Using the phase portrait technique, we can see that, although their physical origins are completely different, the pendulum and chemostat are similar as dynamical systems (compare Figures 9.13b and 9.15a). For example, each can exhibit overshoot before settling down to a steady state. The phase portrait supplies a level of abstraction that unifies mechanical and biochemical dynamics.

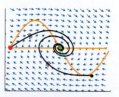

Figures 9.13b (page 227)

We also learned the valuable lesson that the details of parameter values matter; behaviors suggested by the network diagram must be confirmed by calculation. But our calculation could stop far short of explicitly solving the dynamical equations; it was enough to find and classify the fixed points.

Our analysis of the chemostat may seem to rest on the arbitrary choice of a particular function for the growth rate, namely, $k_g(c) = k_{g,max}c/(K + c)$. In fact, growth functions of Hill form are often observed. But the main qualitative result (existence of a stable fixed point under some conditions) is **robust**: It survives if we replace this choice by a variety of other saturating functions.

THE BIG PICTURE

Imagine that you are out walking, and you find an unfamiliar electronic component on the ground (perhaps a transistor). You could take it to the lab and characterize its behavior, then brilliantly intuit that it could be combined with other components to make a "governor" circuit, for example, a constant-current source. You could even simulate that circuit's behavior mathematically, given your characterization of the components, but this is still not enough. You still need to *make* such a circuit and confirm that it works! Maybe your model has left out something important: For example, maybe noise vitiates the analysis.

Synthetic biology has undertaken this program in the context of gene circuits in living cells. At the same time, advances in diagnostic methods like fluorescence imaging of fusion proteins has given us a window onto the internal states of individual cells. Knowing what systems really work, what components are available, and what design principles apply will potentially lead to advances in medicine and biotechnology. The next two chapters will extend these ideas to understand more elaborate behaviors than homeostasis. In each case, feedback control turns out to be key.

KEY FORMULAS

• *Binding:* Suppose that a single molecule O is in a solution containing another molecule R at concentration c. The kinetic scheme $O + R \underset{\beta_{off}}{\overset{k_{on}c}{\rightleftharpoons}} OR$ (the noncooperative binding model) means that

  – The probability to bind in time Δt, if initially unbound, is $k_{on}c\,\Delta t$, and
  – The probability to unbind in time Δt, if initially bound, is $\beta_{off}\,\Delta t$.

After this reaction system comes to equilibrium, we have $\mathcal{P}_{unbound} = (1 + \bar{c})^{-1}$, where $\bar{c} = c/K_d$ and $K_d = \beta_{off}/k_{on}$ is the dissociation equilibrium constant describing how strong the binding is.

More generally, Section 9.4.3 also introduced a cooperative model with

$$\mathcal{P}_{unbound}(c) = (1 + \bar{c}^{\,n})^{-1}. \tag{9.9}$$

Here n is called the cooperativity parameter, or Hill coefficient. If $n > 1$, this function has an inflection point.

• *Gene regulation function (GRF) of simplified operon:* To study genetic switching, we constructed a simplified model that lumped together the separate processes of regulation, transcription, and translation. Thus, we assumed that production of a gene product (protein) by an operon proceeds at a maximal rate Γ in the absence of any repressor. Section 9.4.4 argued that the production rate would be reduced by a factor of the probability that the operator is not bound to a repressor molecule R:

$$f(c_R) = \frac{\Gamma}{1 + \bar{c}^{\,n}}. \tag{9.10}$$

- *Clearance:* We also simplified by lumping together the separate processes of messenger RNA clearance, protein clearance, and dilution by cell growth. Thus, we assumed that the concentration of gene product X decreases at the rate c_X/τ_{tot}, where the "sink parameter" τ_{tot} is a constant with dimensions $\mathbb{T}$. Letting f denote the gene regulation function, steady state then requires that

$$c_X/\tau_{tot} = f(c_X).$$

- *Chemostat:* Let ρ be the number density of bacteria and c the number density of molecules of some limiting nutrient, supplied to a chemostat of volume V at flow rate Q and incoming concentration c_{in}. Let $k_{g,max}$ be the maximum division rate of the bacteria, K the concentration of nutrient corresponding to half-maximal growth, and ν the number of nutrient molecules needed to generate one individual bacterium. Then ρ and c obey

$$\frac{d(\rho V)}{dt} = \frac{k_{g,max}\, c}{K + c}\rho V - Q\rho \tag{9.22}$$

$$\frac{d(cV)}{dt} = Qc_{in} - Qc - \nu \frac{k_{g,max}\, c}{K + c}\rho V. \tag{9.23}$$

The system always has a fixed point with $\rho = 0$ and $c = c_{in}$; depending on parameter values, it may also have another fixed point.

FURTHER READING

Semipopular:
Echols, 2001.

Intermediate:
Cell and molecular biology: Alberts et al., 2014; Karp, 2013; Lodish et al., 2012.
Autoregulation: Alon, 2006; Myers, 2010; Wilkinson, 2006.
Feedback, phase portraits: Bechhoefer, 2005; Ingalls, 2013; Otto & Day, 2007; Strogatz, 2014.
Binding of transcription factors: Bialek, 2012, §2.3; Dill & Bromberg, 2010, chapt. 28; Marks et al., 2009; Phillips et al., 2012, chapt. 19.
trp operon: Keener & Sneyd, 2009, chapt. 10.
Feedback control in metabolism and morphogenesis: Bloomfield, 2009, chapts. 9–10; Klipp et al., 2009, chapts. 2–3 and 8.

Technical:
Allostery in *lac* repressor: Lewis, 2005.
Synthetic governor circuit: Becskei & Serrano, 2000; Rosenfeld et al., 2002.

T_2 | **Track 2**

9.3.1′a Contrast to electronic circuits

Electronic circuits use only a single currency, the electron. They must therefore *insulate* each wire so that their state variables don't all jumble together, creating **crosstalk**. Insulation permits electrons to spread only in authorized ways. Electronic circuit diagrams implicitly assume that any two points not joined by a line are insulated from each other.

Cells use a very different strategy to solve the same problem. They use *many* distinct molecular species as counters, and only a limited amount of spatial partitioning. In fact, we will neglect spatial partitioning in cells, even though it's important, particularly in eukaryotes. Cells control crosstalk via the **specificity** of molecular recognition. That is, all the molecular actors that are intended to interact with a particular molecule *recognize* it; they only act on (or are acted on by) that type molecule. Specificity is not perfect, however. For example, an enzyme or other control element can be spoofed by molecules, such as TMG or IPTG, that partially resemble the one that it normally binds (see Sections 9.3.4 and 10.2.3).

9.3.1′b Permeability

The main text mentioned that macromolecules generally cannot spontaneously cross a cell's outer membrane. Cells do have specialized pores and motors to "translocate" impermeable molecules, but such mechanisms, when present, will be explicitly noted on our network diagrams.

Some small molecules are more freely permeable, so their internal concentrations reflect the environment and are not independent state variables. But size isn't everything: Other small molecules, for example, ions, are blocked by the cell membrane. Even tiny single-atom ions, such as Ca^{++}, can only cross the membrane under the cell's control.

The main text noted that some cells use electric potential across the membrane as a state variable. Even potential, however, is tightly coupled to chemical inventories: It is determined by the populations of ions on either side of the membrane.

T_2 | **Track 2**

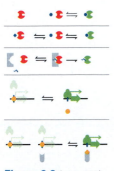

Figure 9.3 (page 209)

9.3.3′ Other control mechanisms

Figure 9.3 shows a few implementations of control used in cells, but there are many others, such as these examples:

- Panels (a,b) describe activation in response to binding an effector, but some enzymes are *in*activated instead.
- There may be both activating and inactivating effectors present, and either (but not both) may bind to the same active site on the enzyme; then the enzyme's state will depend on the competition between them.
- In addition to the allosteric (indirect) mechanisms shown, some enzymes are directly blocked by an inhibitor that occupies their active site in preference to the substrate. Or an inhibitor may bind to free substrate, interfering with its binding to the enzyme.
- Enzymes and their substrates may be confined ("sequestered") into distinct compartments in a cell.
- Enzyme activity may be modulated by the general chemical environment (temperature, pH, and so on).

- Panel (c) imagines a second enzyme that covalently attaches a small group, like a phosphate, and a third that cuts off that group. A protein's functions can also be controlled by cutting ("cleaving") its amino acid chain, removing a part that otherwise would prevent its functioning.

- Bacteria contain DNA sequences that are transcribed into small RNA molecules (about 50–250 base pairs), but whose transcripts ("sRNAs") are not translated into proteins (they are "noncoding"). One known function of sRNAs is that they can bind via base pairing to target messenger RNAs, inhibiting their translation. sRNAs can also bind to proteins, altering their activity.

- Plants, animals, and some viruses encode even shorter "micro" RNAs ("miRNAs," about 22 base pairs long). These bind to a complex of proteins, notably including one called Argonaute, which can then bind to a target mRNA, preventing translation and also marking it for clearance ("RNA interference"). Alterations in cell function from RNA interference can persist long after the causative condition (for example, starvation) has stopped, and even into multiple succeeding generations (Rechavi et al., 2014).

- Some messenger RNAs act as their own regulators: They contain a noncoding segment (a "riboswitch") that folds, creating a binding site for an effector. When an effector molecule binds, the result is a change in production of the protein(s) encoded by the mRNA.

T_2 | **Track 2**

9.3.4′a More about transcription in bacteria

The main text discussed RNA polymerase binding to DNA as the first step in transcription. Actually, although the polymerase does bind directly to DNA, this binding is weak and nonspecific; it must be assisted by an auxiliary molecule called a "sigma factor." For example, the most commonly used sigma factor, σ^{70}, binds specifically both to polymerase and to a promoter sequence, helping bring them together. The sigma factor also helps the polymerase with the initial separation of the strands of DNA to be transcribed. After initiation, the sigma factor is discarded and transcription proceeds without it.

9.3.4′b More about activators

Some activators work by touching the polymerase, increasing its energy payoff for binding, and hence increasing its probability of binding. Others exert an allosteric interaction on a bound polymerase, increasing the probability per unit time that it will open the DNA double helix and begin to transcribe it.

T_2 | **Track 2**

9.3.5′ Gene regulation in eukaryotes

Transcription in eukaryotic cells requires the formation of a complex of many proteins, in addition to a polymerase, all bound to the cell's DNA. In some cases, however, the mechanism of control is reminiscent of the simpler bacterial setup. For example, hormone molecules such as steroids can enter cells from the outside, whereupon they act as effector ligands binding to "nuclear hormone receptors." Each receptor molecule has a DNA-binding domain that recognizes a specific sequence in the cell's genome.

Unlike the case in bacterial repression, however, the receptor binds to DNA regardless of whether it has also bound its ligand. Nor does the bound receptor directly obstruct the binding of polymerase to the DNA, as in the bacterial case. Instead, the presence of ligand controls whether another molecule (the "coactivator") will bind to the receptor. The receptor's job, then, is to sense the presence of its ligand and bring coactivator to a particular place on the genome (or not, as appropriate for its function). When bound, the coactivator then enables transcription of the controlled gene(s).

T_2 | **Track 2**

9.4.4′a More general gene regulation functions

The main text assumed that transcription rate was a constant Γ times the fraction of time that an operator site is not occupied by a repressor. For activators, Section 10.4.1′ (page 266) will make the analogous assumption that production rate is a constant times the fraction of time the activator is *present* at its binding site.

9.4.4′b Cell cycle effects

If there are m copies of a regulated gene in a cell, we may take Γ to be the maximal rate of any one, times m. But there is a moment prior to cell division at which each gene has been duplicated, though the cell has not yet divided. After that moment, the rate of production is double its previous value, a "cell cycle effect"; equivalently, m is not constant. We will not model such effects, instead supposing that they are partly compensated by cell volume growth and can be accounted for by using an effective, averaged value for the gene regulation function.

T_2 | **Track 2**

9.5.1′a Simplifying approximations

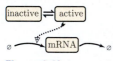

Figure 8.8b (page 192)

The main text described an approximate formulation of gene regulation, in which the activity of a transcription factor is taken to be some function of its effector's concentration. Similarly, we will lump any other state transitions between active and inactive genes into an averaged transcription rate; thus, we do not work at the level of detail shown in Figure 8.8b.

9.5.1′b The Systems Biology Graphical Notation

Many different graphical schemes are used to describe reaction networks. The one used in this book is a simplified version of the Systems Biology Graphical Notation (Le Novère et al., 2009).

T_2 | **Track 2**

9.5.3′ Exact solution

Section 9.5.3 discussed the dynamics of a noncooperatively regulated gene and found an approximate solution in the limit $c \gg K_d$. But certainly the approximation isn't fully valid for the situation of interest to us, where c starts out equal to zero.

To find an exact solution (but still in the continuous, deterministic approximation), begin by defining dimensionless time $\bar{t} = t/\tau_{tot}$, concentration $\bar{c} = c/K_d$, and a parameter $S = \Gamma \tau_{tot}/K_d$. Then Equation 9.16 (page 221) becomes

$$\frac{d\bar{c}}{d\bar{t}} = \frac{S}{1+\bar{c}} - \bar{c}.$$

We can solve this equation by separation of variables: Write

$$d\bar{t} = d\bar{c} \left[S(1+\bar{c})^{-1} - \bar{c} \right]^{-1},$$

and integrate both sides from the initial to the final values. We are interested in solutions for which $\bar{c}(0) = 0$.

We can do the integral by the method of partial fractions. First, though, note that the condition for a steady state is $S = \bar{c}(1+\bar{c})$, whose roots are $\bar{c} = x_\pm$, where

$$x_\pm = \tfrac{1}{2}(-1 \pm \sqrt{1+4S}).$$

Only x_+ is physical, because x_- is negative; nevertheless x_- is a useful abbreviation. Then the integrated form of the equation says

$$\int_0^{\bar{t}} d\bar{t} = -\int_0^{\bar{c}} d\bar{c} \left[\frac{P}{\bar{c} - x_+} + \frac{Q}{\bar{c} - x_-} \right],$$

where

$$P = (1+x_+)/(x_+ - x_-), \qquad Q = 1 - P.$$

So

$$\bar{t} = -P \ln\left(1 - \bar{c}/x_+\right) - Q \ln\left(1 - \bar{c}/x_-\right).$$

You can now compute $\bar{t}$ for a range of $\bar{c}$ values, convert back to t and $c/c_* = \bar{c}/x_+$, and plot the results. This procedure led to the solid red curve in Figure 9.11b.

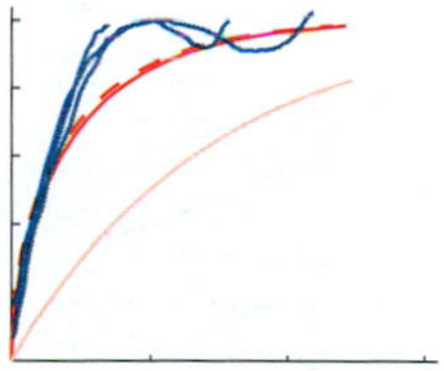

Figure 9.11b (page 223)

Track 2

9.7.1′ Taxonomy of fixed points

The pendulum in Section 9.7.1 (page 226) has two isolated fixed points in its phase portrait. One of them, at $\theta = 0$ and $\omega = 0$, was stable against any sort of small perturbation. Such a point is also sometimes called a "stable spiral." The other fixed point, at $\theta = \pi$ and $\omega = 0$, was called "unstable," indicating that there is at least one small perturbation from it that grows with time. More precisely, this point is a "saddle," which indicates that not *every* perturbation grows: We can displace the pendulum from vertical, and give it a carefully chosen initial angular velocity that brings it to rest exactly at $\theta = \pi$.

A third sort of unstable fixed point is unstable to *any* perturbation; such points are called "unstable nodes."

PROBLEMS

9.1 Viral dynamics

Use the graphical conventions in Idea 9.14 (page 219) to sketch a network diagram appropriate for the model of HIV dynamics after administration of an antiviral drug described in Section 1.2.3 (page 12).

9.2 Cooperative autoregulation

a. Repeat the derivation that led to Equation 9.18 (page 222), but this time assume that the repressor binds cooperatively to its operator with Hill coefficient $n > 1$. Make the same approximation as was used in the main text.

b. How does your answer in (a) behave at late times, $t \gg \tau_{\text{tot}}$?

9.3 Jackrabbit start

a. Find the initial rate of increase in the solution to the unregulated gene problem with initial concentration zero. [*Hint:* Start from Equation 9.19.]

b. Repeat for the approximate solution to the regulated gene problem (Equation 9.18), and comment.

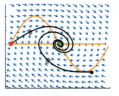

Figure 9.13b (page 227)

9.4 Pendulum phase portrait

a. Use a computer to create a phase portrait for the pendulum, similar to the one in Figure 9.13b. Include the vector field (Equation 9.21, page 226) and the two nullclines. Try using illustrative parameter values $g/L = 1\,\text{s}^{-2}$ and $\zeta/(mL) = 1/(3\,\text{s})$.

b. Trace the arrows with your finger to argue that the system exhibits overshoot.

9.5 Numerical solutions of phase-portrait equations

Continue Problem 9.4 by using a computer (not your finger) to follow the vector field and find some typical trajectories. Such curves, obtained by following a vector field, are often called its **streamlines**.

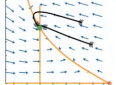

Figure 9.15a (page 231)

9.6 Fixed points of chemostat equations

a. Obtain Equations 9.25 (page 230) from Equation 9.24.

b. Get a computer to draw the nullclines for the case with $\bar{c}_{\text{in}} = 3$ and $\gamma = 1.35$.

c. Comment on the difference between panels (a) and (b) in Figure 9.15.[22]

Figure 9.15b (page 231)

9.7 $\boxed{T_2}$ Exact solution

Section 9.5.3 studied the dynamics of a noncooperatively regulated gene, in the continuous, deterministic approximation, and showed that in the limit $c \gg K_d$ the approximate solution Equation 9.18 (page 222) could be used. The approximate solution had the remarkable property that $c(t)/c_*$ depended only on τ_{tot}, not on the values of K_d and Γ.

a. Section 9.5.3′ (page 236) gave an exact solution for the same equation. Plot the approximate solution as a function of $\bar{t} = t/\tau_{\text{tot}}$. Superimpose plots of the exact solution, for various values of the dimensionless parameter $S = \Gamma\tau_{\text{tot}}/K_d$, and see how well they agree with the approximate solution when S is large.

b. Obtain Dataset 14, and compare your plots with experiment.

---

[22]Chapter 10 will introduce the concept of "bifurcation" for such qualitative changes as a parameter is changed.

9.8 $\boxed{T_2}$ **Over- and underdamping**

a. Nondimensionalize the pendulum equation (Equation 9.21, page 226) by introducing the dimensionless variable $\bar{t} = t/T$, where T is a conveniently chosen time scale constructed out of the parameters m, g, and ζ. For small angular excursions, the equation simplifies: You can substitute θ in place of $\sin\theta$. Then the equation becomes linear in θ with constant coefficients, so you know that its solution is some sort of exponential in time. Solve the equation of motion using this approximation.

b. Examine your solution. Whether or not it exhibits overshoot depends on the value of a dimensionless combination of the parameters. If it does overshoot, the system is called **underdamped**; otherwise it is overdamped. Find the criterion separating these two regimes.[23]

c. Make a graph like the one you made in Problem 9.4, but illustrating the overdamped case.

---

[23] In the underdamped case, the stable fixed point is an example of a stable spiral (Section 9.7.1′, page 237); in the overdamped case, it's an example of another class of fixed points called "stable nodes."

Genetic Switches in Cells

Perhaps we can dimly foresee a day when the hallowed subject of logic will be recognised as an idealisation of physiological processes that have evolved to serve a useful purpose.
—Horace Barlow, 1990

10.1 Signpost

One way that living organisms respond to their environment is by evolutionary change, for example, the acquisition of resistance to drugs or other toxic chemicals. But evolution is slow, requiring many generations; sudden environmental changes can wipe out a species long before it has adapted. Thus, cells with faster response mechanisms have a fitness advantage over those lacking them. Cells belonging to multicellular organisms also need to *specialize*, or commit to a variety of very different forms and functions, despite all having the same genome. Even more dramatically, a cell sometimes needs to engage a pathway of programmed death, or **apoptosis**, for example, as a stage in normal embryonic development, or in response to internal or external signals that indicate the cell is seriously compromised in some way.

Each of these situations illustrates a need for cells to implement *switch-like* behavior among a discrete menu of options. This chapter will study some illustrative examples in bacteria. As with Chapter 9, we will then introduce a mechanical analogy, develop some graphical ideas for how to study the phenomena, see how the ideas have been implemented artificially, and finally return to natural systems.

This chapter's Focus Question is

Biological question: How can you make decisions without a brain?

Physical idea: Cellular elements can implement logic circuitry and remember the answers by using bistability.

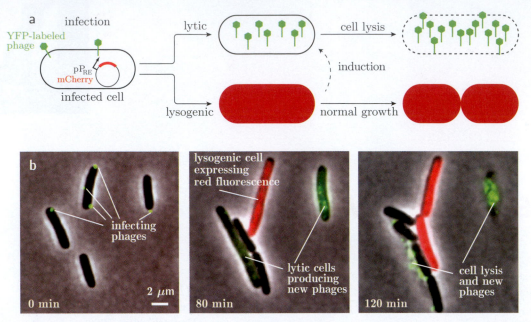

Figure 10.1 Cell-fate decision. Following infection of *E. coli* by phage *lambda*, the virus can either replicate and kill the host cell (lysis), or it can integrate into the bacterial chromosome, where it replicates as part of the host genome (lysogeny). (a) [Cartoon.] A schematic description of a cell-fate assay. One or more fluorescently labeled virions (*green*) simultaneously infect a cell. If the infected cell chooses the lytic program, this will be seen via production of new fluorescent virions, followed by cell lysis. If the cell chooses the lysogenic program, this will be seen via production of red fluorescence from the P_{RE} promoter, followed by resumed growth and cell division. (b) [Optical micrographs.] Frames from a time-lapse movie showing the infection events sketched in (a). At time $t = 0$ (*left*), two cells are each infected by a single phage (*green spots*), and one cell is infected by three phages. At $t = 80$ min (*middle*), the two cells infected by single phages have each entered the lytic program, as indicated by the intracellular production of new phages (*green*). The cell infected by three phages has entered the lysogenic state, as indicated by the red fluorescence from P_{RE}. At $t = 120$ min (*right*), the lytic cells have burst, whereas the lysogenic cell has divided normally. [Photos courtesy Ido Golding from Zeng et al., 2010, Figures 1b–c, © Elsevier. See also Media 14.]

10.2 Bacteria Have Behavior

10.2.1 Cells can sense their internal state and generate switch-like responses

Even bacteria can get sick: A class of viruses called bacteriophage attack bacteria such as *Escherichia coli*.[1] One of the first to be studied was dubbed enterobacteria phage *lambda* (or more simply **phage *lambda***). Like other viruses, phage *lambda* injects its genetic material into the host, where it integrates into the host's genome. From this moment, the host bacterium is essentially a new organism: It now has a modified genome, which implements a new agenda (Figure 10.1).

Some infected cells proceed with the classic virus behavior, called the **lytic program**: They redirect their resources to producing new virions (virus particles), then **lyse** (burst) to release them. In other cells, however, the integrated viral genome (or **provirus**) remains inactive. These cells behave as ordinary *E. coli*. This cell state is called **lysogenic**.[2] We can interpret the virus's "strategy" to preserve a subpopulation of the infected cells as promoting

---

[1]Bacteriophages were also mentioned in Section 4.4.
[2]See Media 14.

survival, because a virus that completely destroyed its host population would itself be unable to survive.

When infected bacteria divide, both resulting daughter cells inherit the provirus, which can remain inactive for many generations. But if an infected cell receives a life-threatening stress, for example, DNA damage from ultraviolet light, the cell can rapidly exit its inactive lysogenic state and switch to the lytic program, a process called **induction**. That is, the virus opts to destroy its host when it "decides" that the host is doomed anyway.

One interesting aspect of this story is that the infected cell needs to be decisive: There would be no point to engaging the lytic program partway. In short,

> *A bacterium infected with phage* lambda *contains a switch that initially commits it to one of two discrete programs. If the lysogenic program is chosen, then the infected cell waits until it is under stress, and only then re-commits irreversibly to the lytic program.* (10.1)

The behavior just described is called the ***lambda* switch**. Similar switches are found in eukaryotes, not just bacteria: For example, after HIV integrates its genome into a cell, its provirus can lie dormant until it is triggered.

10.2.2 Cells can sense their external environment and integrate it with internal state information

Bacteria have other decision-making elements, which, unlike the *lambda* switch, operate without any phage infection. The first of these to be discovered involves metabolism. A bacterium can obtain energy from a variety of simple sugar molecules. Given an ample supply of glucose, for example, each *E. coli* bacterium divides in about half an hour, leading to an exponentially growing population. When the food supply runs out, the population stabilizes, or even decreases as the starving bacteria die.

J. Monod noticed a strange variation on this story in 1941, while doing his PhD research on *E. coli* and *Bacillus subtilis*. Monod knew that the bacteria could live on glucose, but also on other sugars, for example, lactose. He prepared a growth medium containing *two* sugars, then inoculated it with a small number of identical bacteria. Initially the population grew exponentially, then leveled off or fell, as expected. But to Monod's surprise, after a delay the population spontaneously began once again to grow exponentially (Figure 10.2). Monod coined the word **diauxie** to describe this two-phase growth. He eventually interpreted it as indicating that his cells were initially unable to metabolize lactose, but somehow gained that ability after the supply of glucose was exhausted.

Similar behaviors occur in other contexts as well. For example, some bacteria can become resistant to the antibiotic tetracycline by producing a molecular pump that exports molecules of the drug from the cell. In the absence of antibiotic, the cells don't bother to produce the pump; they switch on production only when threatened.

10.2.3 Novick and Weiner characterized induction at the single-cell level

10.2.3.1 The all-or-none hypothesis

Monod's surprising observation led to an intensive hunt for the mechanism underlying diauxie, which eventually yielded insights touching on every aspect of cell biology. One of the key steps involved ingenious experiments by A. Novick and M. Weiner in 1957. By this

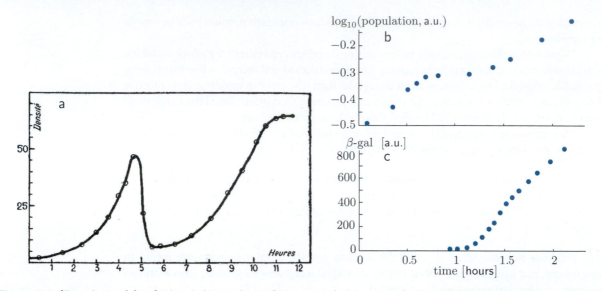

Figure 10.2 [Experimental data.] **Diauxic (2-stage) population growth.** (a) Some of J. Monod's historic original data, showing growth of a culture of *B. subtilis* on a synthetic medium containing equal amounts of sucrose and dextrin. The horizontal axis gives time after the initial inoculation, in hours. The vertical axis shows the amount of light scattering by the culture, related to the number of bacteria present, in arbitrary units. [From Monod, 1942.] (b) Diauxic growth of *E. coli* fed a mixture of glucose and lactose. The two exponential growth phases appear on this semilog plot as roughly straight-line segments. (c) The measured number of β-gal molecules per cell, in the same experiment as in (b). The number is small throughout the first growth phase, during which the bacteria ignore the supplied lactose. After about half an hour of starvation, the cells finally begin to create β-gal. [(b,c): Data from Epstein et al., 1966.]

time, it was known that every potential food source requires a different, specific cellular machinery (enzymes) to import it into the bacterial cell, oxidize it, and capture its chemical energy. For example, *E. coli* requires a set of enzymes to metabolize lactose; these include **beta-galactosidase** (or "β-gal"), which splits lactose molecules into simpler sugars. There would be considerable overhead cost in producing all these enzymes at all times. A bacterium cannot afford to carry such unnecessary baggage, which would slow down its main business of reproducing faster than its competitors, so *E. coli* normally only implements the pathway needed to metabolize its favorite food, glucose. It maintains a repertoire of latent skills, however, in its genome. Thus, when glucose is exhausted, the bacteria sense that (*i*) glucose is not available and (*ii*) another sugar is present; only in this case do they synthesize the enzymatic machinery needed to eat the other sugar, a process again called **induction**. A fully induced cell typically contains several thousand β-gal molecules, in contrast to fewer than ten in the uninduced state. The response to this combination of conditions takes time to mount, however, and so the population pauses before resuming exponential growth.

Novick and Weiner realized that the study of induction is complicated by the dual role of lactose: It is a signal to the cell (an inducer), triggering the production of β-gal, but also a food source, consumed by the cell. The experimenters knew, however, that earlier work had uncovered a class of related molecules, similar enough to lactose that they also trigger induction, but different enough that the cell's enzymes cannot metabolize them. Two such "gratuitous inducer" molecules are called TMG and IPTG.[3] Novick and Weiner

---

[3] See Section 9.3.4 (page 208).

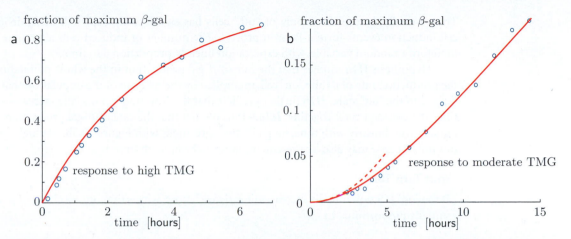

Figure 10.3 [Experimental data with fits.] **Novick and Weiner's data on induction in *E. coli*.** The horizontal axis represents time after adding the gratuitous inducer TMG to a bacterial culture. The vertical axis represents the measured number of β-gal molecules per cell, divided by its maximal observed value. (a) The rise in β-gal activity following addition of 500 μM TMG to a bacterial culture in a chemostat. (b) The same, but with 7 μM TMG. The *solid curves* in each panel are fits, discussed in Section 10.2.3.2 (page 246) and Problem 10.2. The *dashed curve* in (b) is the quadratic function that approximates the short-time behavior of the full solution (see the Example on page 247). [Data from Novick & Weiner, 1957; see Dataset 15.]

grew *E. coli* in a medium without any inducer, then suddenly introduced TMG at various concentrations. Next, they sampled the culture periodically and measured its β-gal content. Because they grew their cultures in a chemostat, they knew the concentration of bacteria,[4] and hence the total number in their sample; this allowed them to express their results as the average number of β-gal molecules per *individual bacterium*.

At high inducer concentration, the experimenters found that the β-gal level initially rose linearly with time, then leveled off (Figure 10.3a). This result had a straightforward interpretation: When TMG was added, all cells switched on their β-gal production apparatus and produced the enzyme at their maximum rate. Eventually this birth-death process reached a steady state, in which the production rate (reduced by the continual growth and division of the cells) balanced the normal clearance and dilution of any protein in a cell. At lower TMG concentrations, however, a surprising feature emerged (Figure 10.3b): In contrast to the high-TMG case, the initial rise of β-gal was not linear in time. That is, when exposed to low TMG levels the cell culture did not immediately begin producing β-gal at its maximum rate; instead, the production rate increased *gradually* with time.

Novick and Weiner realized that their measurement of β-gal was ambiguous, because it involved samples containing many bacteria: Measuring the overall level of β-gal production did not specify whether every individual was producing the enzyme at the same rate or not. In fact, the experimenters suspected that there were wide discrepancies between individuals in the partially induced case. They formulated the most extreme form of this hypothesis:

> **H1a:** *Each individual cell is always either fully "on" (producing β-gal at its maximum rate) or else fully "off" (producing β-gal at a negligible rate); and*
> **H1b:** *When suddenly presented with inducer, individual cells begin to flip randomly* (10.2) *from "off" to "on," with a probability per unit time of flipping that depends on the applied inducer concentration.*

---
[4]See Section 9.7.2 (page 227).

Thus, initially, before the supply of "off" cells has significantly depleted, the individual cell induction events form a Poisson process. The number of induced cells after time t is therefore a random variable, with expectation linearly proportional to time.[5]

Hypothesis **H1a** implies that the rate of β-gal production in the whole population is the production rate of a fully "on" cell, multiplied by the fraction of the population that has flipped to the "on" state. **H1b** further predicts that the "on" fraction initially increases with time at a constant rate. Together, **H1a,b** thus predict that the rate of β-gal production also initially rises linearly with time, in qualitative agreement with Figure 10.3b. (At high TMG, this initial phase may also be present, but it is too brief to observe.)

Your Turn 10A

If the "on" fraction initially increases linearly with time, why is the concentration of β-gal initially proportional to t^2?

$\boxed{T_2}$ *Section 10.2.3.1′ (page 266) gives some further details about Novick and Weiner's experiments.*

10.2.3.2 Quantitative prediction for Novick-Weiner experiment

We could oppose **H1** to an alternative, "commonsense," hypothesis **H2**, which states that a population of genetically identical bacteria, grown together in a well-stirred, uniform environment, must each behave in the same way. This hypothesis requires an additional assumption, however, that each individual bacterium gradually turns on its β-gal activity with the particular time dependence seen in Figure 10.3b. Such behavior arises naturally under **H1**. Let's make this claim (**H1**) precise.

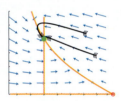

Figure 9.15a (page 231)

Novick and Weiner obtained their data (Figure 10.3) by growing a culture in a chemostat with volume V and flow rate Q, and letting it reach steady state prior to introducing the gratuitous inducer TMG. In this way, they ensured a fixed density of bacteria, ρ_* (Figure 9.15a), and hence also fixed population size $N_* = \rho_* V$. They then made successful quantitative predictions of the detailed time courses seen in Figure 10.3, supporting their hypothesis about induction. We can understand those predictions by writing and solving an equation for the number of β-gal molecules in the chemostat.

Let $z(t)$ be the total number of β-gal molecules in the chemostat at time t and $S(t)$ the average rate at which each individual bacterium creates new enzyme molecules. No molecules of β-gal flow into the chamber, but in every time interval dt a fraction $(Q dt)/V$ of all the molecules flow out. Combining this loss with the continuous creation by bacteria yields $dz/dt = N_* S - (Q/V)z$. To see through the math, it's again useful to apply a nondimensionalizing procedure like the one that led to Equation 9.24 (page 230). Thus, we again express time as a dimensionless variable $\bar{t}$ times the natural scale V/Q. Substituting $(V/Q)\bar{t}$ everywhere for t then gives

$$\frac{dz}{d\bar{t}} = \frac{V}{Q} N_* S - z. \tag{10.3}$$

We cannot solve this equation, however, until we specify how S depends on time.

H1b says that, under high inducer conditions, every bacterium rapidly switches "on"; thus, each begins generating enzyme at its maximal rate, so $S(\bar{t})$ is a *constant* after time zero.

---

[5]See Idea 7.6 (page 160).

Your Turn 10B

In this situation, the z dynamics follows a birth-death process. So apply the strategy of the Example on page 206 to show that the function

$$z(\bar{t}) = (VN_*S/Q)(1 - e^{-\bar{t}})$$

solves Equation 10.3.

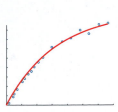

The solution says that z rises from zero until its creation rate matches the loss rate. The curve in Figure 10.3a shows that a function of this form fits the data well.

Figure 10.3a (page 245)

At lower inducer levels, Section 10.2.3.1 argued that the fraction of induced cells will initially increase linearly with time. The average production rate per cell, S, is then that fraction times a constant, so $S(\bar{t})$ should also be a linear function: $S(\bar{t}) = \alpha\bar{t}$, where α is a constant. After substituting this expression into Equation 10.3, we can simplify still further by rescaling z, obtaining

$$\frac{d\bar{z}}{d\bar{t}} = \bar{t} - \bar{z}. \qquad (10.4)$$

Example Find the appropriate rescaled variable $\bar{z}$ that converts Equation 10.3 to Equation 10.4. Then solve the equation and graph its solution with a suitable initial condition. [*Hint:* You may be able to guess a trial solution after inspecting Figure 10.3b.]

Solution Following the nondimensionalizing procedure, let $z = P\bar{z}$, where P is some unknown combination of the constants appearing in the equation. Substitute $P\bar{z}$ for z in the equation, and notice that choosing $P = VN_*\alpha/Q$ simplifies Equation 10.3 to the form Equation 10.4. The initial condition is no β-gal, $\bar{z}(0) = 0$.

You can solve the equation by using mathematical software. Alternatively, notice that if we can find any *one* solution to the equation, then we can get others by adding any solution of the associated linear problem $du/d\bar{t} = -u$, which we know all take the form $u = Ae^{-\bar{t}}$.

To guess one particular solution, notice that the data in Figure 10.3b become linear in t at long times. Indeed, the linear function $\bar{z} = \bar{t} - 1$ does solve the differential equation, though it does not satisfy the initial condition.

To finish, then, we consider the combination $\bar{z} = \bar{t} - 1 + Ae^{-\bar{t}}$ and adjust the free parameter A to satisfy the initial condition. This gives $\bar{z}(\bar{t}) = e^{-\bar{t}} - 1 + \bar{t}$.

The curve in Figure 10.3b shows that a function of this form also fits the initial induction data well. At time $\bar{t} \ll 1$, it takes the limiting form $\bar{z} \to \bar{t}^2$, as seen in the experimental data.[6]

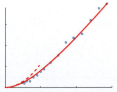

Figure 10.3b (page 245)

Your Turn 10C

Although initially the β-gal production rate increases with time, it cannot increase without limit. Discuss qualitatively why not, and how Equation 10.4 and its solution must be modified at long times.

---

[6]See page 19.

10.2.3.3 Direct evidence for the all-or-none hypothesis

The analysis in the preceding section doesn't prove the "all-or-none" hypothesis *H1*, but it does show that it is compatible with observation. For a direct test, Novick and Weiner used another known fact about induction. When a culture of "off" (uninduced) cells is placed in a high-TMG medium, they all turn "on." Similarly, when "on" (induced) cells are placed in a medium with no TMG (or an extremely low concentration), eventually they all turn "off." But there is a range of intermediate inducer concentrations in which each cell of a culture maintains whichever state it had prior to the change in medium. That is, the fraction of cells expressing β-gal does not change, even though the inducer concentration does change. This phenomenon is called **maintenance**; a growth medium with the intermediate inducer concentration is a **maintenance medium**.

Novick and Weiner interpreted the phenomenon of maintenance by introducing another hypothesis:

> *H1c: In a maintenance medium, individual bacteria are **bistable**. That is, they can persist indefinitely in either the induced or uninduced state. Even when a cell divides,* (10.5) *both daughters inherit its state.*

The memory of cell state across division is generically called **epigenetic inheritance**; it's not simply "genetic," because both induced and uninduced cells have exactly the same genome. Epigenetic inheritance underlies the ability of your body's organs to grow properly: For example, skin cells beget only skin cells, despite having the same genome as nerve cells. Similarly, in the *lambda* switch system, after division the daughter cells retain the lysogenic state of their parent.[7]

The experimenters realized that they could grow an individual cell up to a large population, every one of whose members had the same induction state, by using the phenomenon of maintenance. Thus, to determine the tiny quantity of β-gal in one cell, it sufficed to grow a culture from that cell in maintenance medium, then assay the entire culture, all of whose cells would be in the same state. In this way, Novick and Weiner anticipated by decades the development of today's single-cell technologies. To obtain single-cell samples, they diluted a culture with maintenance medium to the point where a single sample was unlikely to contain *any* bacteria. More precisely, each sample from their diluted culture had a bacterial population drawn from a Poisson distribution with expectation 0.1 bacterium per sample. When the very dilute samples were then used to create new cultures, as expected about 90% of them contained no cells. But the remaining 10% of the samples were very likely to contain no more than one bacterium.[8]

Novick and Weiner took a partially induced culture with β-gal at 30% of the level found in maximum induction, made many single-cell samples, and grew them under the same maintenance conditions. They then measured the β-gal content of each new culture. As predicted by *H1a,c*, they found that each culture was either making β-gal at the maximum rate or else at a much lower rate, with no intermediate cases. Moreover, the *number* of cultures synthesizing β-gal at the maximum rate was just 30% of the total, explaining the initial bulk measurement. Repeating the experiment with other levels of partial induction gave similar results, falsifying *H2*.

Today it is possible to check *H1* quite directly. For example, E. Ozbudak and coauthors modified the genome of *E. coli* to make it synthesize green fluorescent protein RNA each time it made a β-gal transcript. Looking at individual cells then showed that each was either

---

[7] See Section 10.2.1.
[8] See Equation 4.6 (page 76).

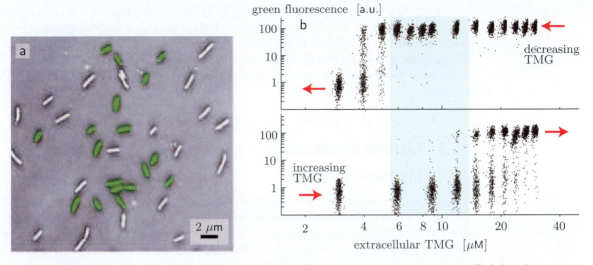

Figure 10.4 Individual *E. coli* cells expressing a green fluorescent protein gene controlled by *lac* repressor. (a) [Micrographs.] Overlaid green fluorescence and inverted phase-contrast images of cells that are initially uninduced for LacI expression, then grown for 20 hours with the gratuitous inducer TMG at concentration 18 μM. The cells showed a bi-modal distribution of expression levels, with induced cells having over one hundred times the fluorescence of uninduced cells. (b) [Experimental data.] Behavior of a series of cell populations. Each culture was initially fully induced (upper panel) or fully uninduced (lower panel), then grown in media containing various amounts of TMG. For each value of TMG concentration, a cloud representation is given of the distribution of measured fluorescence values, for about 1000 cells in each sample. *Arrows* indicate the initial and final states of the cell populations in each panel. The TMG concentration must be increased above 30 μM to turn on all initially uninduced cells, whereas it must be lowered below 3 μM to turn off all initially induced cells. The *pale blue* region shows the range of hysteretic (maintenance) behavior, under the conditions of this experiment. [From Ozbudak et al., 2004, Figures 2a–b, pg. 737.]

"on" or "off," with few intermediates (Figure 10.4a). The data also confirm bistability in a range of TMG concentrations corresponding to maintenance (Figure 10.4b, blue region).[9] That is, when the experimenters raised the level gradually from zero, some bacteria remained uninduced until inducer concentration exceeded about 30 μM. But when they *lowered* TMG gradually from a *high* value, some bacteria remained induced, until TMG concentration fell below about 3 μM. That is, the level of induction in the maintenance region depends on the *history* of a population of bacteria, a property called **hysteresis**.

T_2 *Section 10.2.3.3′ (page 266) discusses other recently discovered genetic mechanisms.*

10.2.3.4 Summary

In short, Novick and Weiner's experiments documented a second classic example of cellular control, the *lac* **switch:**

> E. coli *contains a bistable switch. Under normal conditions, the switch is "off," and individual bacteria don't produce the enzymatic machinery needed to metabolize lactose, including β-gal. If the bacterium senses that inducer is present above a threshold, however, then it can transition to a new "on" state, with high production of β-gal.* (10.6)

---

[9] T_2 The precise limits of the maintenance regime depend on the experimental conditions and the bacterial strain used. Thus, for example, 7 μM lies in the maintenance regime in Figure 10.4, although this same inducer concentration was high enough to induce cells slowly in Novick and Weiner's experiments (Figure 10.3b).

With these results in hand, it became urgent to discover the detailed cellular machinery implementing the *lac* switch. The physicist L. Szilard proposed that the system involved **negative control**: Some mechanism *prevents* synthesis of β-gal, but induction somehow disables that system. Section 10.5.1 will explain how Szilard's intuition was confirmed. To prepare for that discussion, however, we first explore a mechanical analogy, as we did in Chapter 9.

10.3 Positive Feedback Can Lead to Bistability

10.3.1 Mechanical toggle

To replace a flat tire on a car, you need to lift the car upward a few centimeters. A lever can give you the mechanical advantage needed to do this. But a simple lever will let the car back down immediately, as soon as you release it to begin your repairs! A device like the **toggle** shown in Figure 10.5a would be more helpful.

In the left panel of Figure 10.5a, the downward force from the heavy load is converted into a clockwise torque on the handle. If we overcome that torque by lifting the handle, it moves up, and so does the load. After we pass a critical position for the handle, however, the load starts to exert a *counterclockwise* torque on it, locking it into its upward position even after we release it. We can reset the toggle (return it to the lower stable state), but only by pushing downward on the handle. In other words, if we start at the critical position, then whichever way we move leads to a torque tending to push the handle further in that same direction—a **positive feedback** making the critical position an unstable fixed point.

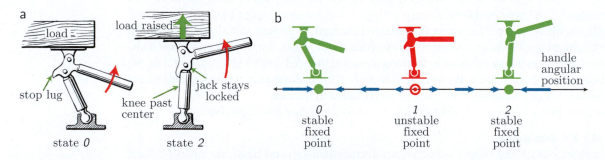

Figure 10.5 The concept of a toggle. (a) [Schematic.] The "knee jack," a mechanical example. Rotating the handle counterclockwise lifts the load. After a critical angle is reached, the handle snaps to its upper, locked position and remains there without requiring any continuing external torque. (b) [Phase portrait.] The corresponding phase portrait is a line representing angular position of the handle. *Arrows* represent the torque on the handle created by the load. The arrows shrink to zero length at the unstable fixed point indicated by the *red bull's eye*. When the angle gets too large, the stop lug exerts a clockwise torque, indicated by the large arrow to the right of *2*; thus, the *green dot 2* is a stable fixed point. A similar mechanism creates another stable fixed point, the other *green dot* labeled *0*. In the absence of external torques, the system moves to one of its stable fixed points and stays there. [(a) From Walton, 1968, Figure 22. Used with permission of Popular Science. Copyright ©2014. All rights reserved.]

Figure 10.5b represents these ideas with a one-dimensional phase portrait. In contrast to the governor (Figure 9.1b), this time the system's final state depends on where we release it. For any initial position below *1*, the system is driven to point *0*; otherwise it lands at point *2*. If we make a small excursion away from one of the stable fixed points, the arrows in the phase portrait show that there is negative feedback. But near the unstable fixed point *1*, the arrows point away. Because there are just two stable fixed points, the toggle is called "bistable" (it exhibits bistability).

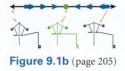

Figure 9.1b (page 205)

In short, the toggle system has *state memory*, thanks to its bistability, which in turn arises from the feedbacks shown in Figure 10.5b. In computer jargon, it can "remember" one binary digit, or **bit**. We can read out the state of a mechanical toggle by linking it to an electrical switch, forming a **toggle switch**. Unlike, say, a doorbell button ("momentary switch"), a toggle switch remembers its state history. It also has a decisive action: When given a subthreshold push, it returns to its undisturbed state. In this way, the toggle filters out noise, for example, from vibrations.

Example a. Consider a slightly simpler dynamical system:[10] Suppose that a system's state variable x follows the vector field $W(x) = -x^2 + 1$. What kinds of fixed points are present?
b. Find the time course $x(t)$ of this system for a few interesting choices of starting position x_0 and comment.

Solution a. The phase portrait reveals two values of x where W equals zero. Looking at W near these two fixed points shows that $x = 1$ is stable, because at nearby values of x, W points toward it. Similarly we find that $x = -1$ is unstable. There are no other fixed points, so if we start anywhere below $x = -1$, the solution for $x(t)$ will never settle down; it "runs away" to $x \to -\infty$.
b. The equation $dx/dt = -x^2 + 1$ is separable. That is, it can be rewritten as

$$\frac{dx}{1 - x^2} = dt.$$

Integrate both sides to find the solution $x(t) = (1 + Ae^{-2t})/(1 - Ae^{-2t})$, where A is any constant. Evaluate $x_0 = x(0)$, solve for A in terms of x_0, and substitute your result into the solution.

In Problem 10.1, you'll graph the function in the Example for various x_0 values and see the expected behavior: A soft landing at $x = +1$ when $x_0 > -1$, or a runaway solution for $x_0 < -1$. For an everyday example of this sort of system, imagine tilting backward on the back legs of a chair. If you tilt by a small angle and then let go, gravity returns you to the usual, stable fixed point. But if you tilt beyond a critical angle, your angular velocity increases from zero in a runaway solution, with an uncomfortable conclusion.

Although the preceding Example had an explicit analytic solution, most dynamical systems do not.[11] As in our previous studies, however, the phase portrait can give us a qualitative guide to the system's behavior, without requiring an explicit solution.

---

[10] See Section 9.2 (page 204).
[11] In Problem 9.5, you'll get a computer to solve differential equations numerically.

10.3.2 Electrical toggles

10.3.2.1 Positive feedback leads to neural excitability

Neurons carry information to, from, and within our brains. Section 4.3.4 (page 78) mentioned one part of their mechanism: A neuron's "input terminal" (or dendrite) contains ion channels that open or close in response to the local concentration of a neurotransmitter molecule. Opening and closing of channels in turn affects the electric potential across the membrane, by controlling whether ions are allowed to cross it.

But a neuron must do more than just accept input; it must also *transmit* signals to some distant location. Most neurons have a specialized structure for this purpose, a long, thin tube called the axon that projects from the cell body. Instead of chemosensitive ion channels, the axon's membrane is studded with **voltage-sensitive ion channels**. A burst of neurotransmitter on the dendrite opens some channels, locally decreasing the electric potential drop across it. This decrease in turn affects nearby voltage-sensitive channels at the start of the axon, causing them to open and extending the local depolarization further. That is, the axon's resting state is a stable fixed point of its dynamical equations, but for disturbances beyond a threshold there is a positive feedback that creates a chain reaction of channel opening and depolarization. This change propagates along the axon, transmitting the information that the dendrite has been stimulated to the distant axon terminal. (Other processes "reset" the axon later, creating a second, trailing wave of repolarization that returns it to its resting state.)

10.3.2.2 The latch circuit

The mechanical toggle does not require any energy input to maintain its state indefinitely. But it inspired a *dynamic* two-state system, an electronic circuit called the "latch" that contributed to a transformation of human civilization starting in the mid-20th century. In its simplest form, the circuit consists of two current amplifiers (transistors). A battery attempts to push current through both transistors. But each one's output is being fed to the other one's input, in such a way that when *1* is conducting, its output turns off *2* and vice versa. That is, each transistor *inhibits the other one*, a double-negative feedback loop that renders the whole system bistable. We could add external wires that, when energized, overcome the internal feedback and thus can be used to reset the latch to a desired new state. When the reset signal is removed, the system's bistability maintains the set state indefinitely.

Thus, like the mechanical toggle, the latch circuit acts as a 1-bit memory. Unlike the mechanical version, however, the latch can be switched extremely fast, using extremely small amounts of energy to reset its state. This circuit (and related successors) formed part of the basis for the computer revolution.

10.3.3 A 2D phase portrait can be partitioned by a separatrix

We previously discussed the pendulum, a system with a single stable fixed point (Figure 9.13). Figure 10.6 shows a modified version of this system. If we give it a small push, from a small starting angle, then as before it ends up motionless in the down position. But if given a big enough initial push, the pendulum can arrive at a magnet, which then holds it—another example of bistability. As with the toggle, there are two different stable fixed points. Instead of an isolated unstable fixed point separating them as in Figure 10.5b, however, there is now an entire **separatrix** curve in the

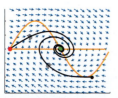

Figure 9.13b (page 227)

Figure 10.5b (page 250)

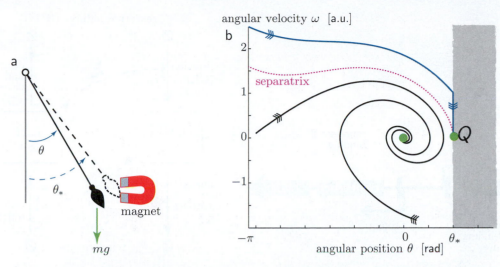

Figure 10.6 Mechanical system with a separatrix. (a) [Schematic.] A pendulum with a stop. The fixed magnet has little effect until the pendulum's angular position reaches a critical value θ_*; then the magnet grabs and holds the pendulum bob. (b) [Phase portrait.] The phase portrait now has two stable fixed points (*green dots*). A separatrix (*magenta curve*) separates nearby initial states that will end up at the origin (*black curves*) or at *Q* (*blue curve*). Any trajectory starting out above the separatrix will end at *Q*; those below it will end at the origin.

θ-ω plane, dividing it into "basins of attraction" corresponding to the two possible final states.

10.4 A Synthetic Toggle Switch Network in *E. coli*

10.4.1 Two mutually repressing genes can create a toggle

The ideas behind the mechanical and electronic toggles can yield bistable behavior in cells, reminiscent of the *lac* and *lambda* switches.[12]

T. Gardner and coauthors wanted to create a generic circuit design that would yield bistable behavior. Their idea was simple in principle: Similarly to the electronic toggle, imagine two genes arranged so that each one's product is a transcription factor repressing the *other* one. For their demonstration, Gardner and coauthors chose to combine the *lac* repressor (LacI) and its operator with another repressor/operator pair known to play a role in the *lambda* switch (Section 10.2.1). The *lambda* repressor protein is denoted by the abbreviated name cI.[13] The network diagram in Figure 10.7b shows the overall scheme.

It's not enough to draw a network diagram, however. Before attempting the experiment, Gardner and coauthors asked, "Will this idea always *work*? Or never? Or only if certain conditions are met?" Answering such questions requires a phase-portrait analysis. The situation is similar to the chemostat story, where we found that the qualitative behavior we sought depended on specific relations between the system's parameters (Figure 9.15).

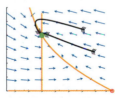

Figure 9.15a (page 231)

Figure 9.15b (page 231)

---

[12]See Ideas 10.1, 10.2, and 10.5.

[13]The letter "I" in this name is a Roman numeral, so it is pronounced "see-one," not "see-eye."

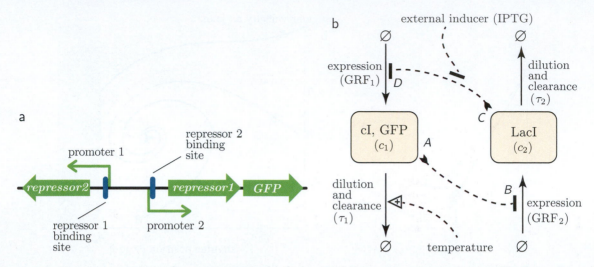

Figure 10.7 Two representations of a genetic toggle. (a) [Schematic.] Two genes repress each other. The figure shows the placement of elements on a short DNA molecule in the cell (a "plasmid"). *Wide arrows* denote genes; *hooked arrows* denote promoters. Repressor 1 inhibits transcription from promoter 1. Repressor 2 inhibits transcription from promoter 2. The operon on the right contains, in addition to the *repressor 1* gene, a "reporter gene" coding for green fluorescent protein. [After Gardner et al., 2000.] (b) [Network diagram.] The same circuit, drawn as a network diagram. The loop *ABCD* has overall positive feedback (two negative feedbacks), suggesting that this network may exhibit an unstable fixed point analogous to the mechanical or electronic toggles. To provide external "command" input, one of the transcription factors (LacI) can be rendered inactive by externally supplying the inducer IPTG. The other one (cI) is a temperature-sensitive mutant; its clearance rate increases as temperature is raised.

The discussion will involve several quantities:

c_1, c_2	concentrations of repressor; $\bar{c}_i$, their dimensionless forms
n_1, n_2	Hill coefficients
$K_{d,1}, K_{d,2}$	dissociation equilibrium constants
Γ_1, Γ_2	maximal production rates of repressors; $\bar{\Gamma}_i$, their dimensionless forms
V	volume of cell
τ_1, τ_2	sink parameters; see Equation 9.13 (page 218)

Section 9.4.4 obtained a formula for the gene regulation function of a simple operon (Equation 9.10, page 218). It contains three parameters: the maximal production rate Γ, dissociation equilibrium constant K_d, and cooperativity parameter n. Duplicating Equation 9.10, with the connections indicated on the network diagram Figure 10.7, yields[14]

$$\frac{dc_1}{dt} = -\frac{c_1}{\tau_1} + \frac{\Gamma_1/V}{1 + (c_2/K_{d,2})^{n_1}}$$

$$\frac{dc_2}{dt} = -\frac{c_2}{\tau_2} + \frac{\Gamma_2/V}{1 + (c_1/K_{d,1})^{n_2}}.$$

(10.7)

To simplify the analysis, suppose that both operons' cooperativity parameters are equal, $n_1 = n_2 = n$. Because we are interested in possible switch-like (bistable) behavior of this

---

[14]The external controls indicated on the network diagram will be discussed in Section 10.4.2.

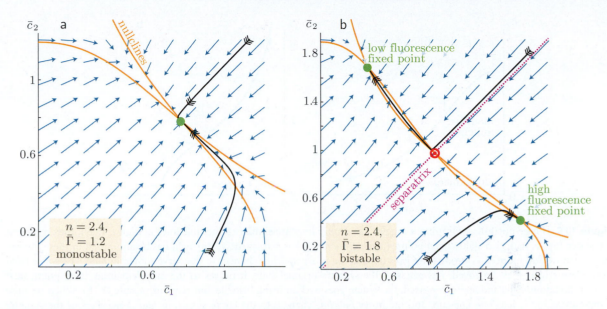

Figure 10.8 [Mathematical functions.] **Phase portraits of the idealized toggle system.** (a) The nullclines of the two-gene toggle system (*orange curves*) intersect in a single, stable fixed point (Equation 10.7 with $n = 2.4$ and $\bar{\Gamma}_1 = \bar{\Gamma}_2 = 1.2$). Two solutions to the dynamical equations are shown (*black curves*); both evolve to the stable fixed point (*green dot*). (b) Changing to $\bar{\Gamma}_1 = \bar{\Gamma}_2 = 1.8$, with the same Hill coefficient, instead yields bistable (toggle) behavior. The same initial conditions used in (a) now evolve to very different final points. One of these (*upper black curve*) was chosen to begin very close to the separatrix (*dotted magenta line*); nevertheless, it ends up at one of the stable fixed points. (Compare to the bistable behavior in Figure 10.6b on page 253.)

system, we first investigate the steady states by setting the derivatives on the left-hand sides of Equation 10.7 equal to zero. The equations are a bit messy—they have many parameters—so we follow the nondimensionalizing procedure used earlier. Define

$$\bar{c}_i = c_i / K_{\mathrm{d},i} \quad \text{and} \quad \bar{\Gamma}_i = \Gamma_i \tau_i / (K_{\mathrm{d},i} V).$$

The nullclines are the loci of points at which one or the other concentration is not changing in time. They are curves representing solutions of

$$\bar{c}_1 = \bar{\Gamma}_1 / (1 + (\bar{c}_2)^n); \qquad \bar{c}_2 = \bar{\Gamma}_2 / (1 + (\bar{c}_1)^n).$$

We can understand the conditions for steady state, that is, for a simultaneous solution of these two equations, graphically: Draw the first one as a curve on the $\bar{c}_1$-$\bar{c}_2$ plane. Then draw the second as another curve on the same plane and find the intersection(s) (see Figure 10.8).[15]

The case of noncooperative binding ($n = 1$) never generates toggle behavior. But when $n > 1$, the nullcline curves have inflection points,[16] and more interesting phenomena can occur. Figures 10.8a,b show two possibilities corresponding to different choices of the parameters in our model (that is, two sets of $\bar{\Gamma}_i$ values). Panel (b) shows three fixed points, two of which turn out to be stable, similar to the mechanical toggle. That is, as we

---

[15]This same procedure was used earlier for the pendulum and chemostat; see Figures 9.13 (page 227) and 9.15 (page 231).
[16]See Section 9.4.3.

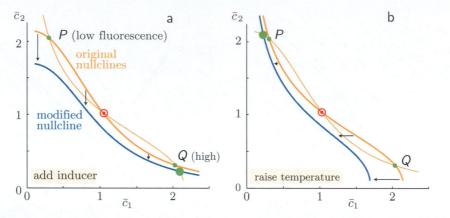

Figure 10.9 [Phase portraits.] **Flipping a toggle.** The *orange lines* are similar to the nullclines in Figure 10.8b; they represent a bistable toggle, with two stable fixed points (*small green dots*). For clarity, the vector field has been omitted in these figures. (a) When we add inducer, $\bar{c}_1$ production goes up, and the *heavy nullcline* changes to the curve shown in *blue*. The stable fixed point formerly at *P* is destroyed; regardless of the system's original state, it moves to the only remaining fixed point, near *Q*—a bifurcation. After inducer is removed, the system goes back to being bistable, but remains in state *Q*. (b) When we raise temperature, the $\bar{c}_1$ loss rate goes up, and the *heavy nullcline* changes to the curve shown in *blue*. Regardless of the system's original state, it moves to the fixed point near *P*. After temperature is restored, the system will remain in state *P*.

adjust the system's parameters there is a qualitative change in the behavior of the solutions, from monostable to bistable. Such jumps in system behavior as a control parameter is continuously changed are called **bifurcations**.

$\boxed{T_2}$ *Section 10.4.1′ (page 266) describes a synthetic genetic switch involving only a single gene.*

10.4.2 The toggle can be reset by pushing it through a bifurcation

Figure 10.8 shows how a single stable fixed point can split into three (two stable and one unstable), as the values of $\bar{\Gamma}_i$ are varied.[17] The figure showed situations where $\bar{\Gamma}_1$ and $\bar{\Gamma}_2$ are kept equal. A more relevant case, however, is when one of these values is changed while holding the other fixed.

For example, in the genetic toggle (Figure 10.7b), temporarily adding external inducer neutralizes some of the LacI molecules. This lifts the repression of cI. Production of cI then represses production of LacI and hence flips the toggle to its stable state with high fluorescence. If the cell was originally in that state, then its state is unchanged. But if it was initially in the state with low fluorescence, then it changes to the "high" state and stays there, even after inducer is removed. Mathematically, we can model the intervention as increasing $K_{d,2}$ (and hence reducing $\bar{\Gamma}_2$). The phase portrait in Figure 10.9a shows that in this situation, one nullcline is unchanged while the other moves. The unstable fixed point and one of the two stable ones then move together, merge, and annihilate each other; then the system has no choice but to move to the other stable fixed point, which itself has not moved much. In short, *raising inducer flips the toggle* to the state with high concentration of cI (and hence high fluorescence).

The other outside "control line" shown in the network diagram involves raising the temperature, which raises the rate of clearance of the temperature-sensitive mutant cI. We

---

[17] $\boxed{T_2}$ More precisely, the unstable fixed point is a saddle; see Section 9.7.1′ (page 237).

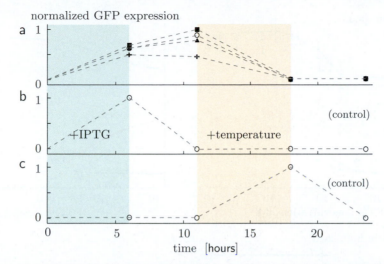

normalized GFP expression

Figure 10.10 [Experimental data.] **Bistable behavior of the two-gene toggle.** In these graphs, *symbols* represent measurements of green fluorescent protein expression, in each of several constructs. Different symbols refer to variant forms of the ribosome binding sites used in each gene. *Gray dashed lines* just join the symbols; they are neither measurements nor predictions. (a) After adding external inducer (*left colored region*), the toggle switches "on" and stays that way indefinitely after inducer is returned to normal. Raising the temperature (*right colored region*) switches the toggle "off," where it can also remain indefinitely, even after the temperature is reduced to normal. Thus, both *white regions* represent the same growth conditions, but the bacteria behave differently based on past history—the system displays hysteresis (memory). (b) A control, in which the *cI* gene was missing. In this incomplete system, GFP expression is inducible as usual, but bistability is lost. (c) Another control, this time lacking LacI. For this experiment, the GFP gene was controlled by cI, in order to check that repressor's temperature sensitivity. [Data courtesy Timothy Gardner; see also Gardner et al., 2000.]

model this by lowering the value of the clearance time constant τ_1 (and hence reducing $\bar{\Gamma}_1$). Figure 10.9b shows that the result is to flip the toggle to the state with low concentration of cI.

After either of these interventions, we can return the system to normal conditions (low inducer and normal temperature); the two lost fixed points then reappear, but the system remains stuck in the state to which we forced it. Figure 10.10 shows that the biological toggle switch indeed displayed this bistable behavior; Figure 10.11 shows a more detailed view for the case of raising/lowering inducer concentration. Figure 10.12 demonstrates that the system is bistable, by showing two separated peaks in the histogram of individual cell fluorescence when inducer is maintained right at the bifurcation point.

T_2 *Section 10.4.2′ (page 272) discusses time-varying parameter values more thoroughly.*

10.4.3 Perspective

The artificial genetic toggle system studied above displays behavior strongly reminiscent of the *lac* switch, including bistability and hysteresis. But the verbal descriptions seem already to explain the observed behavior, without all the math. Was the effort really worthwhile?

One reply is that physical models help us to imagine what behaviors are possible and catalog the potential mechanisms that could implement them. When we try to understand an existing system, the first mechanism that comes to mind may not be the one chosen by Nature, so it is good to know the list of options.[18]

---

[18] T_2 Section 10.4.1′ (page 266) discusses another implementation of a genetic toggle.

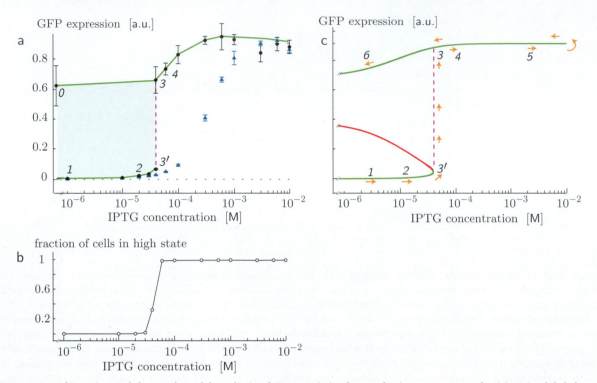

Figure 10.11 [Experimental data and model prediction.] **Hysteresis in the synthetic two-gene toggle.** (a) *Point labeled 0:* cI production when the toggle is stuck in its "on" state with no IPTG inducer present. *Other black points, left to right:* Production when the toggle is initially prepared in its "off" state and IPTG level is gradually increased. At a critical value, the cell population divides into subpopulations consisting of those cells that are entirely "on" or "off"; their expression levels are indicated separately as *3* and *3'*, respectively. At still higher inducer levels, all cells in the population are in the same state. The *shaded region* is the region of bistability. The hysteretic behavior shown is reminiscent of the naturally occurring *lac* switch (Figure 10.4, page 249). For comparison, the control strain lacking cI gave the *blue triangles*, showing gradual loss of repression as inducer is added (Equation 9.10, page 218). That system shows no hysteresis. (b) Fraction of bacteria in the "on" state for various inducer concentrations. (c) Predicted system behavior using a physical model similar to Figure 10.7. As IPTG is decreased, a stable fixed point (*green*) bifurcates into one unstable (*red*) and two stable fixed points. From lower left, *arrows* depict a sequence of steps in which the initially "off" state is driven "on," then stays that way when returned to zero IPTG. [Reprinted by permission of Macmillan Publishers, Ltd.: *Nature*, 403(6767), Gardner et al., Figure 5a,b, pg. 341. ©2000.]

Figure 10.10a (page 257)

Physical modeling also reminds us that a plausible cartoon does not guarantee the desired behavior, which generally emerges only for a particular range of parameter values (if at all). When designing an artificial system, modeling can give us qualitative guides to which combinations of parameters control the behavior and how to tweak them to improve the chances of success. Even when success is achieved, we may *still* wish to tweak the system to improve its stability, make the stable states more distinct, and so on. For example, Gardner and coauthors noted that cooperative repression was needed in order to obtain bistability and that larger values of the Hill coefficients increased the size of the bistable regime; such considerations can guide the choice of which repressors to use. They also noted that the rates of synthesis of the repressors $\bar{\Gamma}_{1,2}$ should be roughly equal; this observation led them to try several variants of their construct, by modifying one of the gene's ribosome binding sites. Some of these variants indeed performed better than others (Figure 10.10a).

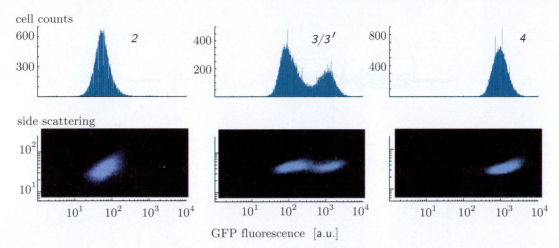

Figure 10.12 [Experimental data.] **Bistability in the synthetic two-gene toggle.** The data show that, in the bistable region, genetically identical cells belong to one of two distinct populations. The *top* part of each subpanel is a histogram of the fluorescence from a reporter gene that tracks cI concentration by synthesizing green fluorescent protein. The *bottom* part of each subpanel shows a cloud representation of the joint distribution of GFP fluorescence (*horizontal axis*) and a second observed variable correlated to cell size, which helps to distinguish the peaks of the histogram. [Data courtesy Timothy Gardner; see also Gardner et al., 2000, Figure 5, pg. 341.]

10.5 Natural Examples of Switches

This chapter opened with two examples of mysterious behavior in single-cell organisms. We then saw how positive feedback can lead to bistability in mechanical systems and how experimenters used those insights to create a synthetic toggle switch circuit in cells. Although the synthetic system is appealing in its simplicity, now it is time to revisit the systems actually invented by evolution.

10.5.1 The *lac* switch

Section 10.2.2 (page 243) discussed how *E. coli* recognizes the presence of the sugar lactose in its environment and in response synthesizes the metabolic machinery needed to eat that sugar. It would be wasteful, however, to mount the entire response if only a few molecules happen to arrive. Thus, the cell needs to set a threshold level of inducer, ignoring small transients. The positive-feedback design can create such sharp responses.

To implement this behavior, *E. coli* has an operon containing three genes (the **lac operon**, Figure 10.13). One gene, called *lacZ*, codes for the enzyme beta-galactosidase, whose job is to begin the metabolism of lactose. The next gene, called *lacY*, codes for a **permease** enzyme. This enzyme does not perform any chemical transformation; rather, it embeds itself in the cell membrane and actively pulls any lactose molecules that bump into it inside the cell. Thus, a cell expressing permease can maintain an interior concentration of inducer that exceeds the level outside.[19]

A separate gene continuously creates molecules of the *lac* repressor (LacI). Figure 10.13a sketches how, in the absence of lactose, LacI represses production of the enzymes. Panel (b)

---

[19] $\boxed{T_2}$ A third gene in the operon, called *lacA*, codes for another enzyme called beta-galactoside transacetylase, which is needed when the cell is in lactose-eating mode.

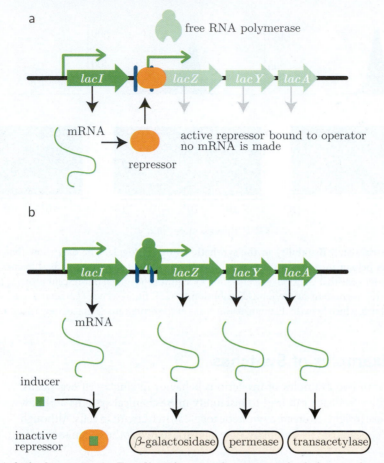

Figure 10.13 [Schematic.] **The *lac* operon in *E. coli.*** *Wide arrows* denote genes; *hooked arrows* denote promoters. (a) In the absence of inducer, the three genes in the operon are turned off. (b) Inducer inactivates the *lac* repressor, allowing transcription of the genes.

sketches the response to an externally supplied inducer, as in the experiment of Novick and Weiner: The inducer inhibits binding of LacI to the promoter, increasing the fraction of time in which the gene is available for transcription.

Figure 10.14b shows how the requirement of switch-like response is met. Even if inducer is present outside an uninduced cell, little will enter because of the low baseline production of permease. Nevertheless, if exposure to high levels of inducer persists, eventually enough can enter the cell to trigger production of more permease, which pulls in more inducer, creating positive feedback. The bistability of this loop explains Novick and Weiner's observation of all-or-nothing induction.

Actually, repression is not absolute, because of randomness in the cell: Even without inducer, LacI occasionally unbinds from its operator, leading to a low baseline level of β-gal and permease production (Figure 10.15a). However, a few permeases are not enough to bring the intracellular inducer level up much, if its exterior concentration is low. Even in the maintenance regime, where an alternative stable fixed point exists, this source of stochastic noise is not sufficient to push the cell's state through its separatrix, so an uninduced cell stays near its original state.

a

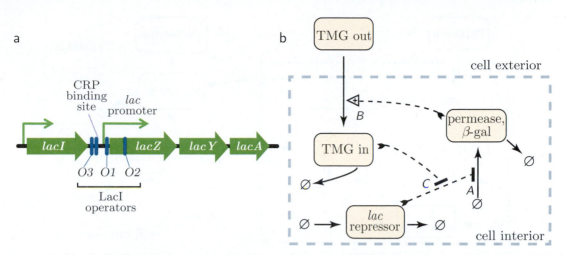

b

Figure 10.14 **Positive feedback in the *lac* system.** (a) [Schematic.] A more detailed schematic than Figure 10.13. *Vertical bars* are binding sites for transcription factors: three for LacI and one for CRP, to be discussed later. (b) [Network diagram.] Another representation for this system, via its network diagram. Increasing concentration of the inducer molecule, here TMG, inside the cell inhibits the negative effect of the *lac* repressor on the production of permease. The net effect of the double negative is that the feedback loop *ABC* is overall positive, leading to switch-like behavior. This simplified diagram neglects complications arising from having three operators for LacI, and from the fact that the repressor must form a tetramer before it binds to DNA.

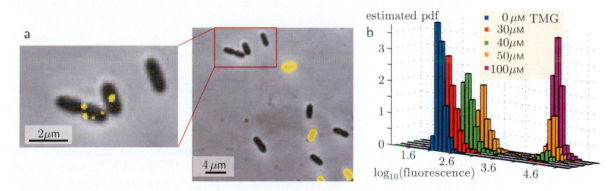

Figure 10.15 **Induction in the *lac* system.** (a) [Micrographs.] A strain of *E. coli* expressing a fusion of permease (LacY) and yellow fluorescent protein. *Right:* There is a clear distinction between induced and uninduced cells. *Inset:* Uninduced cells nevertheless contain a few copies of the permease. (b) [Experimental data.] Measuring the fluorescence from many cells confirms that the population is bimodal at intermediate inducer concentration (in this case, 40–50 μM TMG). [(a) Courtesy Paul Choi; see also Choi et al., 2008, and Media 15. Reprinted with permission from AAAS. (b) Data courtesy Paul Choi.]

Matters change at higher levels of external inducer. Novick and Weiner's hypothesis was that, in this situation, initially uninduced cells have a fixed probability per unit time of making a transition to the induced state.[20] Section 10.2.3.2 showed that this hypothesis could explain the data on induction at inducer levels slightly above the maintenance regime, but at the time there was no known molecular mechanism on which to base it. Novick and Weiner speculated that the synthesis of a single permease molecule could be the event triggering

---

[20]See Idea 10.2 (page 245).

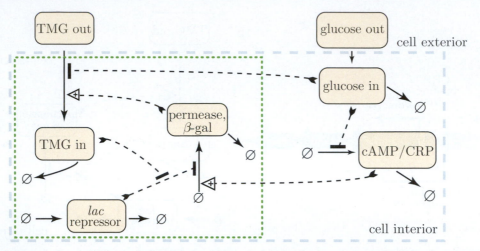

Figure 10.16 [Network diagram.] **Extended decision-making circuit in the** *lac* **system.** The *inner box* contains the same decision module shown in Figure 10.14b. Additional elements outside that box override the signal of external inducer, when glucose is also available.

induction, but Figure 10.15a shows that this is not the case. Instead, P. Choi and coauthors found that permease transcription displays two different kinds of bursting activity:

- Small bursts, creating a few permeases, maintain the baseline level but do not trigger induction.
- Rarer large bursts, creating hundreds of permeases, can push the cell's control network past its separatrix and into the induced state, if the external inducer level is sufficiently high (Figure 10.15b).

T_2 *Section 10.5.1′ (page 273) discusses the distinction between the large and small bursts just mentioned.*

Logical cells

Like all of us, *E. coli* prefers some foods to others; for example, Section 10.2.3 mentioned that it will not fully activate the *lac* operon if glucose is available, even when lactose is present. Why bother to split lactose into glucose, if glucose itself is available? Thus, remarkably, *E. coli* can compute a *logical operation*:

Turn on the operon when (lactose is present) **and** *(glucose is absent).*   (10.8)

Figures 10.2b,c (page 244)

Such a control scheme was already implicit when we examined diauxic growth (Figures 10.2b,c).

Figure 10.16 shows how *E. coli* implements the logical computation in Idea 10.8. The circuit discussed earlier is embedded in a larger network. When glucose is unavailable, the cell turns on production of an internal signaling molecule called cyclic adenosine monophosphate, or **cAMP**. cAMP in turn is an effector for a transcription factor called cAMP-binding receptor protein, or CRP, which acts as a necessary activator for the *lac* operon.[21] Thus, even if lactose is present, glucose has the effect of keeping the production of β-gal, and the rest of the lactose metabolic apparatus, turned off.

---

[21] CRP's binding site is shown in Figure 10.14a.

A second control mechanism is known that supplements the one just described: The presence of glucose also inhibits the action of the permease, an example of a more general concept called **inducer exclusion**.

10.5.2 The *lambda* switch

Section 10.2.1 described how a bacterium infected with phage *lambda* can persist in a latent state for a long time (up to millions of generations), then suddenly switch to the self-destructive lytic program. The network diagram in Figure 10.17 outlines a greatly simplified version of how this works. The diagram contains one motif that is already familiar to us: an autoregulating gene for a transcription factor (called Cro). A second autoregulating gene, coding for the *lambda* repressor, is slightly more complicated. At low concentration, this gene's product activates its own production. However, a second operator is also present, which binds cI less strongly than the first. When bound to this site, cI acts as a repressor, decreasing transcription whenever its concentration gets too large and thus preventing wasteful overproduction.

The two autoregulated genes also repress each other, leading to bistable behavior. External stress, for example, DNA damage from ultraviolet radiation, triggers a cascade of reactions called the **SOS response**, involving a protein called RecA. Although *E. coli* evolved the SOS response for its own purposes (initiating DNA repair), phage *lambda* coopts it as the trigger for switching to the lytic program: It eliminates the cI dimers that hold Cro, and hence the lytic program, in check.

T_2 *Section 10.5.2' (page 273) discusses the role of cellular randomness in control networks. Section 10.4.2' (page 272) gives more information about the operators appearing in the* lambda *switch.*

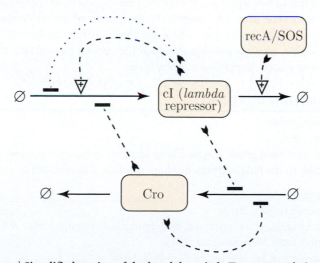

Figure 10.17 [Network diagram.] **Simplified version of the *lambda* switch.** Two autoregulating genes repress each other in an overall positive feedback loop, leading to bistability. The two kinds of influence lines coming from the *lambda* repressor box are discussed in the text. The lytic state corresponds to high Cro and low repressor (cI) levels; the lysogeny state is the reverse. DNA damage flips the switch by initiating the SOS response (*upper right*). The diagram omits the effects of another set of operators present in the genome, located far from the promoters for the *cI* and *cro* genes. Also omitted are dimerization reactions; individual cI and Cro molecules must associate in pairs before they can bind to their operators (Section 10.4.1', page 266).

THE BIG PICTURE

Physicists originally developed phase portrait analysis for a variety of problems describable by sets of coupled differential equations. The systematic classification of fixed points by their stability properties, and their appearance (and disappearance) via bifurcations, form part of a discipline called "dynamical systems." We have seen how these ideas can give us insight into system behaviors, even without detailed solutions to the original equations. Along the way, we also developed another powerful simplifying tool, the reduction of a system to nondimensionalized form.

These ideas are applicable to a much broader class of biological control than the strictly genetic networks we have studied so far. For example, the next chapter will introduce biological *oscillators,* implemented via circuits involving the repeated activation and inactivation of populations of enzymes, without any need to switch genes on and off.

KEY FORMULAS

- *Novick/Weiner:* After a stable bacterial population with density ρ is prepared in a chemostat with volume V and inflow Q, the bacteria can be induced by adding an inducer molecule to the feedstock. Let $S(t)$ be the average rate at which each individual bacterium creates new copies of β-gal. Letting $z(t)$ be the number of those molecules, then

$$\frac{\mathrm{d}z}{\mathrm{d}\bar{t}} = \frac{V}{Q}N_*S - z, \tag{10.3}$$

where $\bar{t} = tQ/V$ and $N_* = \rho_*V$. This equation has different solutions depending on the assumed dependence of S on time.

- *Switch:* If protein *1* acts as a repressor for the production of protein *2*, and vice versa, then we found two equations for the time development of the two unknown concentrations c_1 and c_2 (Equation 10.7, page 254). We examined the stable states implied by setting both time derivatives to zero, and used phase-portrait analysis to find when the system was bistable. As we continuously adjust the control parameters, the system's steady state can jump discontinuously—a bifurcation.

FURTHER READING

Semipopular:
Bray, 2009.
Dormant viral genes, inserted long ago into *in our own genome,* can be reactivated in a way similar to the lytic pathway in phage *lambda*: Zimmer, 2011.

Intermediate:
General: Ingalls, 2013.
Fixed points, phase portraits, and dynamical systems: Ellner & Guckenheimer, 2006; Otto & Day, 2007; Strogatz, 2014.
Discovery of *lac* operon: Müller-Hill, 1996.
lac switch: Keener & Sneyd, 2009, chapt. 10; Wilkinson, 2006.
lambda switch: Myers, 2010; Ptashne, 2004; Sneppen & Zocchi, 2005.
Switches in general: Cherry & Adler, 2000; Ellner & Guckenheimer, 2006; Murray, 2002, chapt. 6; Tyson et al., 2003.
Another switch, involving chemotaxis: Alon, 2006, chapt. 8; Berg, 2004.

Transmission of information by neurons: Dayan & Abbott, 2000; Nelson, 2014, chapt. 12; Phillips et al., 2012, chapt. 17.

Technical:
Historic: Monod, 1949; Novick & Weiner, 1957.
Artificial switch: Gardner et al., 2000; Tyson & Novák, 2013.
The Systems Biology Markup Language, used to code network specifications for computer simulation packages: http://sbml.org/; Wilkinson, 2006.
lac switch: Santillán et al., 2007; Savageau, 2011.
lambda switch: Little & Arkin, 2012.
Cell fate switches: Ferrell, 2008.

T_2 | **Track 2**

10.2.3.1′ More details about the Novick-Weiner experiments

- It's an oversimplification to say that TMG and IPTG molecules mimic lactose. More precisely, they mimic allolactose, a modified form of lactose, produced in an early step of its metabolism; allolactose is the actual effector for the *lac* repressor. (These gratuitous inducers do mimic lactose when they fool the permease enzyme that normally imports it into the cell.)

- To measure the amount of active β-gal, Novick and Weiner used an optical technique. They exposed a sample of culture to yet another lactose imitator, a molecule that changes color when attacked by β-gal. Measuring the optical absorption at a particular wavelength then allowed them to deduce the amount of β-gal present.

T_2 | **Track 2**

10.2.3.3′a Epigenetic effects

Some authors use the term "epigenetic" in a more limited sense than was used in the chapter; they restrict it to only those mechanisms involving covalent modification of a cell's DNA, for example, methylation or histone modification.

10.2.3.3′b Mosaicism

The main text asserted that all the somatic (non-germ) cells in an individual have the same genome ("skin cells have the same genome as nerve cells"), or in other words that all genetic variation is over for any individual once a fertilized egg has formed, so that any further variation in cell fate is epigenetic. We now know that this statement, while generally useful, is not strictly true. Heritable mutations can arise in somatic cells, and can be passed down to their progeny, leading to genome variation within a single individual generically called "mosaicism."

For example, gross abnormalities of this sort are present in cancer cells. A more subtle example involves "jumping genes," or LINE-1 retrotransposons, regions of mobile DNA that duplicate themselves by reverse transcription of their transcript back to DNA, followed by its reinsertion into the genome. The insertion of a genetic element at a random point in a genome can disrupt another gene, or even upregulate it. This activity seems to be particularly prevalent in fetal brain tissue (Coufal et al., 2009). Finally, our immune system generates a large repertoire of antibodies by genetic reshuffling occurring in white blood cells. A small fraction of these cells are selected by the body and preserved, forming our immune "memory."

More broadly still, recent research has documented the extent to which your microbiome, whose independent genetic trajectory interacts with that of your somatic cells, must be considered as an integral part of "you."

T_2 | **Track 2**

10.4.1′a A compound operator can implement more complex logic

Section 10.4.1 showed one way to obtain toggle behavior, by using a feedback loop consisting of two mutually repressing genes. F. Isaacs and coauthors were able to construct a different

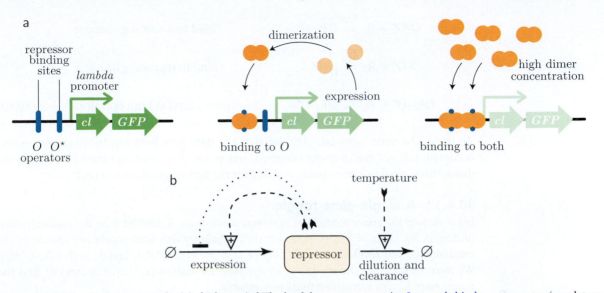

Figure 10.18 **The single-gene toggle.** (a) [Schematic.] The *lambda* repressor protein cI can only bind to an operator (regulatory sequence) after it has formed a dimer. It activates its own production if it binds to operator O, but represses its own production if it binds to $O^\star$. (A third operator $O^\dagger$ was present in the experiment, but not in our simplified analysis.) (b) [Network diagram.] To provide external "command" input, a temperature-sensitive variant of cI was used: Raising the temperature destabilizes the protein, contributing to its clearance as in the two-gene toggle. Compare this diagram to Figure 10.7b (page 254).

sort of artificial genetic toggle in *E. coli* using just *one* gene (Isaacs et al., 2003). Their method relied on a remarkable aspect of the *lambda* repressor protein (cI), mentioned in Section 10.5.2 (page 263): One of the repressor binding sites, which we'll call $O^\star$, is situated so that bound cI physically obstructs the RNA polymerase, so as usual it prevents transcription. Another operator, which we'll call O, is *adjacent to* the promoter; here cI acts as an *activator*, via an allosteric interaction on the polymerase when it binds.[22] Thus, controlling the *cI* gene by operators binding cI itself implements both positive and negative feedback (Figure 10.18).

Isaacs and coauthors mimicked *E. coli*'s natural arrangement, by using both of the mechanisms in the previous paragraph. Operator O created positive feedback; $O^\star$ gave autorepression (Figure 10.18b).[23] The two regulatory sequences differed slightly: $O^\star$ had a weaker affinity for cI than O. In this way, as the cI level rose, first the positive feedback would set in, but then at higher concentration, transcription shut off. Qualitatively it may seem reasonable that such a scheme could create bistability, but some detailed analysis is needed before we can say that it will really work.

In this situation, the gene regulation function is more complicated than that discussed in Section 9.4.4. Rather than a single binding reaction $O + R \underset{\beta_{\mathrm{off}}}{\overset{k_{\mathrm{on}} c}{\rightleftharpoons}} OR$, there are *four* relevant reactions. Again denoting the repressor by the generic symbol R, they are

$$2R \overset{K_{\mathrm{d},1}}{\rightleftharpoons} R_2 \qquad\qquad \text{form dimer} \qquad (10.9)$$

---

[22] The operators O and $O^\star$ are more traditionally called $O_R 2$ and $O_R 3$, respectively.

[23] Isaacs and coauthors actually used *three* regulatory sequences, which is the situation in natural phage *lambda*. Our analysis will make the simplifying assumption that only two were present.

$$O\text{-}O^\star + R_2 \overset{K_{d,2}}{\rightleftharpoons} OR_2\text{-}O^\star \qquad\qquad \text{bind to activating operator}$$

$$O\text{-}O^\star + R_2 \overset{K_{d,3}}{\rightleftharpoons} O\text{-}O^\star R_2 \qquad\qquad \text{bind to repressing operator}$$

$$OR_2\text{-}O^\star + R_2 \overset{K_{d,4}}{\rightleftharpoons} OR_2\text{-}O^\star R_2. \qquad\qquad \text{bind to both operators} \qquad (10.10)$$

In these schematic formulas, $O\text{-}O^\star$ refers to the state with both regulatory sequences unoccupied, $OR_2\text{-}O^\star$ has only one occupied, and so on. The notation means that $K_{d,1}$ is the dissociation equilibrium constant β_{off}/k_{on} for the first reaction, and so on.[24]

10.4.1′b A single-gene toggle

Let x denote the concentration of repressor monomers R, and let y be the concentration of dimers R_2. Let $\alpha = \mathcal{P}(O\text{-}O^\star)$ denote the probability that both regulatory sequences are unoccupied, and similarly $\zeta = \mathcal{P}(OR_2\text{-}O^\star)$, $\gamma = \mathcal{P}(O\text{-}O^\star R_2)$, and $\delta = \mathcal{P}(OR_2\text{-}O^\star R_2)$. We must now reduce the six unknown dynamical variables (x, y, α, ζ, γ, and δ), and the parameters $K_{d,i}$, to something more manageable.

The logic of Section 9.4.4 gives that, in equilibrium,

$$x^2 = K_{d,1}y, \qquad y\alpha = K_{d,2}\zeta, \qquad y\alpha = K_{d,3}\gamma, \qquad y\zeta = K_{d,4}\delta. \qquad (10.11)$$

Your Turn 10D

A fifth reaction should be added to Equation 10.10, in which $O\text{-}O^\star R_2$ binds a second repressor dimer. Confirm that this reaction leads to an equilibrium formula that is redundant with the ones written in Equation 10.10, and hence is not needed for our analysis.

It's helpful to express the binding of R_2 to $O^\star$ in terms of its binding to O, by introducing the quantity $p = K_{d,2}/K_{d,3}$. Similarly, let q represent the amount of encouragement that cI already bound to O gives to a second cI binding to $O^\star$, by writing $K_{d,4} = K_{d,2}/q$. Following Equation 9.6, next note that the regulatory sequences must be in one of the four occupancy states listed, so that $\alpha + \zeta + \gamma + \delta = 1$:

$$1 = \alpha + \frac{\alpha x^2}{K_{d,1}K_{d,2}} + \frac{\alpha x^2 p}{K_{d,1}K_{d,2}} + \frac{\alpha x^2}{K_{d,1}K_{d,2}}\frac{x^2 q}{K_{d,1}K_{d,2}},$$

or

$$\alpha = \left(1 + \frac{x^2(1+p)}{K_{d,1}K_{d,2}} + \frac{x^4 q}{(K_{d,1}K_{d,2})^2}\right)^{-1}. \qquad (10.12)$$

We can now write the gene regulation function by supposing that all the fast reactions in Equation 10.10 are nearly in equilibrium, and that the average rate of cI production is

---

[24]See Section 9.4.4 (page 217).

a constant Γ times the fraction of time ζ that O is occupied (but $O^\star$ is unoccupied). That assumption yields[25]

$$\mathrm{d}x/\mathrm{d}t = \Gamma\zeta + \Gamma_{\mathrm{leak}} - x/\tau_{\mathrm{tot}}. \qquad (10.13)$$

In this formula, Γ again determines the maximum (that is, activated) rate of production. There will be some production even without activation; to account for this approximately, a constant "leak" production rate Γ_{leak} was added to Equation 10.13. Finally, there is loss of repressor concentration from dilution and clearance as usual.[26]

Equation 10.13 involves two unknown functions of time, x and ζ. But Equation 10.11 gives us ζ in terms of α and x, and Equation 10.12 gives α in terms of x. Taken together, these formulas therefore amount to a dynamical system in just one unknown x, whose behavior we can map by drawing a 1D phase portrait. The equations have seven parameters: $\Gamma, \Gamma_{\mathrm{leak}}$, $\tau_{\mathrm{tot}}, K_{\mathrm{d},1}, K_{\mathrm{d},2}, p$, and q. Biochemical measurements give $\Gamma/\Gamma_{\mathrm{leak}} \approx 50, p \approx 1$, and $q \approx 5$, so we can substitute those values (Hasty et al., 2000). As in our discussion of the chemostat,[27] we can also simplify by using the nondimensionalizing procedure. Letting $\bar{x} = x/\sqrt{K_{\mathrm{d},1}K_{\mathrm{d},2}}$ and $\bar{t} = \Gamma_{\mathrm{leak}}t/\sqrt{K_{\mathrm{d},1}K_{\mathrm{d},2}}$ gives

$$\frac{\mathrm{d}\bar{x}}{\mathrm{d}\bar{t}} = \frac{50\bar{x}^2}{1 + 2\bar{x}^2 + 5\bar{x}^4} + 1 - M\bar{x}, \qquad (10.14)$$

where $M = \sqrt{K_{\mathrm{d},1}K_{\mathrm{d},2}}/(\tau_{\mathrm{tot}}\Gamma_{\mathrm{leak}})$. That is, the remaining parameters enter only in this one combination.

Our earlier experience has taught us that, to learn the qualitative behavior of Equation 10.14, we need to start by mapping out the fixed points. It may not seem easy to solve the quintic equation obtained by setting Equation 10.14 equal to zero. But it's straightforward to *graph* the first two terms, which don't depend on M. If we then superimpose a line with slope M on the graph, then its intersection(s) with the curve will give the desired fixed point(s).[28] We can also find the system's behavior as M is changed, by changing the slope of the straight line (pivoting it about the origin), as shown in Figure 10.19a.

For large M (for example, if leak production is slow), the figure shows that there is only one fixed point, at very low repressor concentration. At small M, there is also only one fixed point, at high repressor. But for intermediate M, there are *three* fixed points: Two stable ones flank an unstable one, much like the mechanical toggle shown in Figure 10.5a. The system is bistable in this range of parameter values.

Figure 10.5a (page 250)

The foregoing analysis was involved, but one key conclusion is familiar from other systems we have studied: Although at the verbal/cartoon level it seemed clear that the system would be bistable, *in fact this property depends on the details of parameter values.*

---

[25] Recall that O has higher binding affinity than $O^\star$. Equation 10.12 relies on the same idealizations as Equation 9.10 (page 218): the continuous, deterministic approximation and repressor binding/unbinding/dimerization that is fast compared to other time scales in the problem.

[26] See Equation 9.13 (page 218).

[27] See Section 9.7.2 (page 227).

[28] Figure 9.8 (page 220) introduced a similar graphical solution to a set of equations.

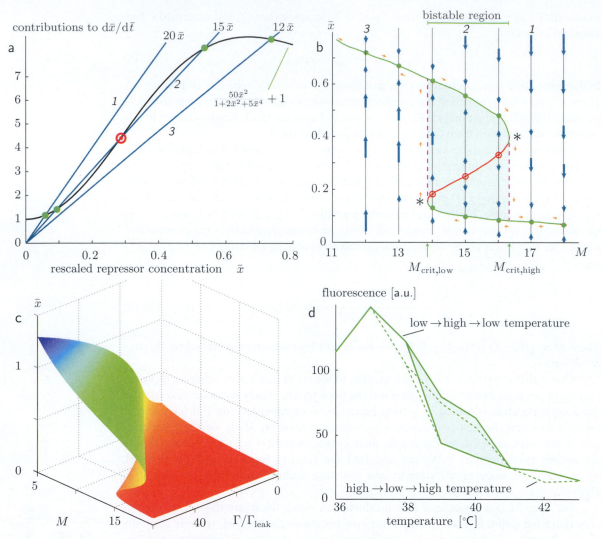

Figure 10.19 Analysis of the single-gene toggle system. (a) [Mathematical functions.] *Black curve:* the first two terms on the right side of Equation 10.14. *Colored lines:* minus the last term of Equation 10.14 with (top to bottom) $M = 20$, 15, and 12. The top line labeled *1* has M slightly above the upper critical value $M_{\text{crit,high}}$; line *3* is slightly below $M_{\text{crit,low}}$. On each line, *colored dots* highlight the fixed points (intersections with the black curve). (b) [Bifurcation diagram.] In this panel, each vertical line is a 1D phase portrait for $\bar{x}$ (analogous to Figure 9.1b on page 205), with a particular fixed value of M. *Green dots* and *red bull's eyes* denote stable and unstable fixed points, respectively. The loci of all these points form the *green and red curves*. *Tiny orange arrows* depict the system's behavior as M is gradually changed: As M is slowly increased from a low initial value, the system's steady state tracks the upper part of the green curve, until at $M_{\text{crit,high}}$ (*asterisk*) it suddenly plunges to the lower part of the curve (*right dashed line*). When driven in the reverse direction, the system tracks the lower part of the curve until M falls below $M_{\text{crit,low}}$ (*asterisk*), then jumps to the upper curve (*left dashed line*): Thus, the system displays hysteresis. (c) [Extended bifurcation diagram.] Many graphs like (b), for various values of $\Gamma/\Gamma_{\text{leak}}$, have been stacked; the loci of fixed points then become the surface shown. The "pleated" form of the surface for large $\Gamma/\Gamma_{\text{leak}}$ indicates bistability as M is varied. As $\Gamma/\Gamma_{\text{leak}}$ is decreased, the surface "unfolds" and there is no bistability for any value of M. (d) [Experimental data.] Measurements displaying hysteresis, as ambient temperature is used to control M. The plots qualitatively resemble (b). [(d): Data from Isaacs et al., 2003.]

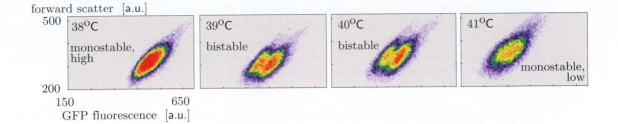

Figure 10.20 [Experimental data.] **Bistability in the single-gene toggle.** Cultures containing the temperature-sensitive variant of cI were bistable at temperatures between 39–40°C. Each panel is a 2D histogram, with larger observed frequencies indicated as redder colors. The horizontal axis shows the fluorescence from a reporter gene that tracks cI concentration by synthesizing green fluorescent protein. The vertical axis is a second observed variable correlated to cell size, which helps to separate the peaks of the histogram (see also Figure 10.11d on page 258). Cultures containing the autoregulatory system with the natural, temperature-insensitive cI protein never displayed bistability (*not shown*). [Data courtesy Farren Isaacs; from Isaacs et al., 2003, Figure 1c, pg. 7715. ©2003 National Academy of Sciences, USA.]

Your Turn 10E

M increases as we raise $K_{d,1}$ or $K_{d,2}$, pushing the system toward monostable behavior at low $\bar{x}$. It decreases as we raise τ_{tot} and Γ_{leak}, pushing the system toward monostable behavior at high $\bar{x}$. Discuss why these dependences are reasonable.

Up to now, we have regarded the system parameters as fixed, and used our graphical analysis to catalog the possible system behaviors. But in an experiment, it is possible to adjust some of the parameters as specified functions of time (Figure 10.19b,c). For example, suppose that we start with high M. Then there is only one possible outcome, regardless of the initial value of $\bar{x}$: $\bar{x}$ ends up at the only fixed point, which is low. As we gradually lower M through a critical value $M_{crit,high}$, suddenly an alternative stable fixed point appears; the system undergoes a bifurcation, giving it a new possible final state. Nevertheless, having started in the lower state the system will remain there until M decreases past a second critical value $M_{crit,low}$. At this point the intermediate, unstable fixed point *merges* with lower stable fixed point and both *disappear*; they annihilate each other.[29] The system then has no choice but to move up to the remaining, high-concentration fixed point.

If we play the story of the previous paragraph in reverse, *increasing* M from a low level, then eventually the system's state pops from the high concentration back to the low. However, this time the transition occurs at $M_{crit,high}$: The system exhibits hysteresis. Isaacs and coauthors observed this phenomenon in their system. They used a mutant cI protein that was temperature sensitive, becoming more unstable as the surroundings became warmer. Thus, changing the ambient temperature allowed them to control τ_{tot}, and hence the parameter M (see the definition below Equation 10.14). Figures 10.19d and 10.20 show that, indeed, the artificial genetic circuit functioned as a toggle in a certain temperature range; it displayed hysteresis.

In short, the single-gene toggle uses temperature as its command input. Pushing the input to either end of a range of values destroys bistability and "sets the bit" of information to be remembered; when we bring the command input to a neutral value in the hysteretic

---

[29] See also Figure 10.9 (page 256).

region, each cell remembers that bit indefinitely. This result is reminiscent of the behavior of the *lac* switch as a gratuitous inducer is dialed above, below, or into the maintenance range (Section 10.2.3). It also reminds us of a nonliving system: The ability of a tiny magnetic domain on a computer hard drive to remember one bit also rests on the hysteresis inherent in some magnetic materials.

$\boxed{T_2}$ | **Track 2**

10.4.2′ Adiabatic approximation

A dynamical system has "dynamical variables" and "parameters." Parameters are imposed from outside the system. We do not solve any equation to find them. Usually we suppose that they are constant in time. Dynamical variables are under the system's control. We imagine giving them initial values, then stepping back and watching them evolve according to their equations of motion and parameter values.

The main text considered two closely related forms of discontinuity in systems whose equations are strictly continuous:

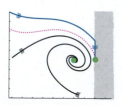

Figure 10.6b (page 253)

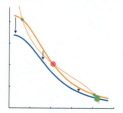

Figure 10.9a (page 256)

- We may hold parameters fixed, but consider a family of initial conditions. Normally, making a small change in initial conditions gives rise to a small change in final state. If, on the contrary, the outcome changes abruptly after an infinitesimal change of initial conditions, one possible reason is that our family of initial conditions straddled a separatrix (Figure 10.6b).[30] That is, for any fixed set of parameters, phase space can be divided into "basins of attraction" for each fixed point, separated by separatrices.

- We may consider a family of systems, each with slightly different parameter values (but each set of parameter values is unchanging in time). Normally, making a small change in parameter values gives rise to a small change in the arrangement of fixed points, limit cycles, and so on. If, on the contrary, this arrangement changes abruptly after an infinitesimal change of parameter values, we say the system displayed a bifurcation (Figure 10.9a). That is, parameter space itself can be partitioned into regions. Within each region we get qualitatively similar dynamics. The regions are separated by bifurcations.

Sections 10.4.2 and 10.4.1′ implicitly generalized this framework to a slightly different situation. What happens if we impose time-dependent parameter values? That is, an experimenter can "turn a knob" during the course of an observation. The parameters are then still externally imposed, but they are changing over time. Mathematically this is a completely new problem, but there is a limiting case where we can formulate some expectations: Suppose that we vary the parameters slowly, compared to the characteristic time of the dynamics. Such variation is called "adiabatic." In that case, we expect the system to evolve to a fixed point appropriate to the starting parameter values, then track that fixed point as it continuously, and slowly, shifts about. But if the externally imposed parameter values cross a bifurcation, then the fixed point may disappear, and the corresponding time evolution can show an abrupt shift, as discussed in Sections 10.4.2 and Section 10.4.1′b.

---

[30] In a system with more than two variables, another possible behavior for such a family is deterministic chaos; see Section 11.4′b (page 291).

T_2 | **Track 2**

10.5.1′ DNA looping

The *lac* operon has another remarkable feature. Molecules of LacI self-assemble into tetramers (sets of four identical molecules of LacI), each with *two* binding sites for the *lac* operator. Also, *E. coli*'s genome contains *three* sequences binding LacI (Figure 10.14a). A repressor tetramer can bind directly to the operator *O1* that obstructs transcription of the gene. But there is an alternate pathway available as well: First, one binding site of a tetramer can bind to one of the other two operators. This binding then holds the repressor tethered in the immediate neighborhood, increasing the probability for its second binding site to stick to *O1*. In other words, the presence of the additional operators effectively raises the concentration of repressor in the neighborhood of the main one; they "recruit" repressors, modifying the binding curve.

Binding a single LacI tetramer to two different points on the cell's DNA creates a *loop* in the DNA.[31] When a repressor tetramer momentarily falls off of one of its two bound operators, the other one can keep it in the vicinity, increasing its probability to rebind quickly, giving only a short burst of transcription. Such events are responsible for the baseline level of LacY and β-gal. If repressor ever unbinds at *both* of its operators, however, it can wander away, become distracted with nonspecific binding elsewhere, and so leave the operon unrepressed for a long time. Moreover, unbinding from DNA increases the affinity of the repressor for its ligand, the inducer; binding inducer then reciprocally reduces the repressor's affinity to rebind DNA ("sequestering" it).

P. Choi and coauthors made the hypothesis that complete-unbinding events were responsible for the large bursts that they observed in permease production (Section 10.5.1, page 259), and that a single long, unrepressed episode generated in this way could create enough permease molecules to commit the cell to switching its state. To test the hypothesis, they created a new strain of *E. coli* lacking the two auxiliary operators for LacI. The modified organisms were therefore unable to use the DNA looping mechanism; they generated the same large transcriptional bursts as the original, but the small bursts were eliminated (Choi et al., 2008).

Thus, induction does require a cell to wait for a rare, single-molecule event, as Novick and Weiner correctly guessed from their indirect measurements (Figure 10.3b), even though the nature of this event is not exactly what they suggested.

Figure 10.14a (page 261)

Figure 10.3b (page 245)

T_2 | **Track 2**

10.5.2′ Randomness in cellular networks

In this chapter, we have neglected the fact that cellular processes are partly random (see Chapter 8). It may seem that randomness is always undesirable in decision making. For example, when we make financial decisions, we want to make optimal use of every known fact; even when our facts are incomplete, the optimal decision *based on those facts* should be uniquely defined. Indeed, we have seen how cells can use feedback to mitigate the effects of noise (Chapter 9).

But when a lot of independent actors all make decisions that appear optimal to each individual, the result can be catastrophic for the population. For example, a long stretch

---

[31] DNA looping has also been discovered in the *lambda* operon (Ptashne, 2004), and others as well.

of favorable environmental factors could lead to the best estimate that this situation will continue, leading every individual to opt for maximal growth rather than food storage. If that estimate proves wrong, then the entire population is unprepared for the downturn in conditions. It appears that even single-cell organisms "understand" this principle, allowing enough randomness in some decisions so that even in a community of genetically identical individuals, there is some diversity in behavior (Acar et al., 2008; Eldar & Elowitz, 2010; Raj & van Oudenaarden, 2008).

PROBLEMS

10.1 My little runaway

Complete the solution to the Example on page 251. Then graph some illustrative solutions showing the possible behaviors for various initial conditions.

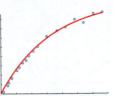

10.2 Novick-Weiner data

Obtain Dataset 15, which contains the data displayed in Figure 10.3.

a. Plot the data contained in the file novickA. Find a function of the form $\bar{z}(t) = 1 - e^{-t/T}$ **Figure 10.3a** (page 245) that fits the data, and display its graph on the same plot. You don't need to do a full maximum-likelihood fit. Just find a value of T that looks good.

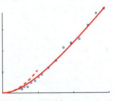

b. Plot the data in novickB. Find a function of the form $z(t) = z_0 \left(-1 + t/T + e^{-t/T} \right)$ that fits the data for $t < 15\,\mathrm{hr}$, and graph it. [*Hint:* To estimate the fitting parameters, exploit the fact that the function becomes linear for large t. Fit a straight line to the data for $10\,\mathrm{hr} \le t \le 15\,\mathrm{hr}$, and relate the slope and intercept of this fit to z_0 and T. Then adjust around these values to identify a good fit to the full function. (Your Turn 10C, page 247, explains why we do not try to fit the region $t \ge 15\,\mathrm{hr}$.)]

Figure 10.3b (page 245)

10.3 Analysis of the two-gene toggle

Section 10.4.1 arrived at a model of how the concentrations of two mutually repressing transcription factors will change in time, summarized in Equation 10.7 (page 254). In this problem, assume for simplicity that $n_1 = n_2 = 2.2$ and $\tau_1 = \tau_2 = 10\,\mathrm{min}$.

a. Nondimensionalize these equations, using the method in Section 10.4.1 (page 253). For concreteness, assume that $\bar{\Gamma}_1 = \bar{\Gamma}_2 = 1.2$.

b. Setting the left sides of these two differential equations equal to zero gives two algebraic equations. One of them gives $\bar{c}_1$ in terms of $\bar{c}_2$. Plot it over the range from $\bar{c}_2 = 0.1\bar{\Gamma}$ to $1.1\bar{\Gamma}$. On the same axes, plot the solution of the other equation, which tells us $\bar{c}_2$ in terms of $\bar{c}_1$. The two curves you have found will intersect in one point, which represents the only fixed point of the system.

c. To investigate whether the fixed point is *stable*, get a computer to plot a vector field in the $\bar{c}_1$-$\bar{c}_2$ plane in the range you used in (b). That is, at each point on a grid of points in this plane, evaluate the two quantities in your answer to (a), and represent them by a little arrow at that point.

d. Combine (overlay) your plots in (b,c). What does your answer to (c) tell you concerning the stability of the fixed point in (b)? Also explain geometrically something you will notice about what the arrows are doing along each of the curves drawn in (b).

e. Repeat (b) with other values of $\bar{\Gamma}$ until you find one with three intersections. Then repeat (c,d).

f. Repeat (a–e) with the values $n_1 = n_2 = 1$ (no cooperativity). Discuss qualitatively how the problem behaves in this case.

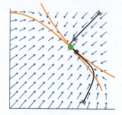

Figure 10.8a (page 255)

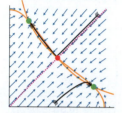

Figure 10.8b (page 255)

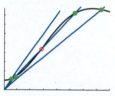

Figure 10.19a (page 270)

10.4 Toggle streamlines

If you haven't done Problem 9.5, do it before starting this one. Figures 10.8a,b show the vector field describing a dynamical system, its nullclines, and also some actual trajectories obtained by following the vector field.

a. First re-express the vector field, Equations 10.7, in terms of the dimensionless variables $\bar{\Gamma}_i$ and $\bar{c}_i$. Consider the case $n_1 = n_2 = 2.2$, $\bar{\Gamma}_1 = \bar{\Gamma}_2 = 1.2$.

b. Choose some interesting initial values of the concentration variables, and plot the corresponding streamlines on your phase portrait.

10.5 $\boxed{T_2}$ Bifurcation in the single-gene toggle

Figure 10.19a shows a graphical solution to the fixed-point equations for the single-gene toggle system. Get a computer to make such pictures. Find the critical values, $M_{\text{crit,high}}$ and $M_{\text{crit,low}}$, where the bifurcations occur.

Cellular Oscillators

We need scarcely add that the contemplation in natural science of a wider domain than the actual leads to a far better understanding of the actual.
—Sir Arthur Eddington

11.1 Signpost

Previous chapters have described two classes of control problems that evolution has solved at the single-cell level: homeostasis and switch-like behavior. This chapter examines a third phenomenon that is ubiquitous throughout the living world, from single, free-living cells all the way up to complex organisms: Everywhere we look, we find *oscillators,* that is, control networks that generate periodic events in time. Biological oscillators range from your once-per-second heartbeat, to the monthly and yearly cycles of the endocrine system, and even the 17-year cycles of certain insects. They are found even in ancient single-cell organisms such as archaea and cyanobacteria. Some periodic behavior depends on external cues, like daylight, but many kinds are **autonomous**—they continue even without such cues.

To begin to understand biological clocks, we will once again investigate a mechanical analogy, and then its synthetic implementation in cells. Finally, we'll look at a natural example. Once again, we will find that feedback is the key design element.

This chapter's Focus Question is

Biological question: How do the cells in a frog embryo know when it's time to divide?

Physical idea: Interlocking positive and negative feedback loops can generate stable, precise oscillations.

11.2 Some Single Cells Have Diurnal or Mitotic Clocks

Our own bodies keep track of time in a way that is partially autonomous: After changing time zones, or work shifts, our sleep cycle remains in its previous pattern for a few days. Remarkably, despite the complexity of our brains, our **diurnal** (daily) clock is based on the periodic activity of individual cells, located in a region of the brain called the suprachiasmatic nucleus. Although we have many of these oscillating cells linked together, they can be separated and grown individually; each one then continues to oscillate with a period of about 24 hours. In fact, *most* living organisms contain single-cell diurnal oscillators.

Not all cells use genetic circuits for their oscillatory behavior. For example, individual human red blood cells display circadian oscillation, despite having no nucleus at all! Also, Section 11.5.2 will describe how individual cells in a growing embryo display intrinsic periodic behavior, implemented by a "protein circuit."

11.3 Synthetic Oscillators in Cells

11.3.1 Negative feedback with delay can give oscillatory behavior

Section 9.7 imagined a control system, consisting of you, a room heater, and a good book: If you respond too slowly, and too drastically, to deviations from your desired setpoint, then the result can be overshoot. In that context, overshoot is generally deemed undesirable. But we can imagine a similar system giving damped, or even sustained, oscillations in other contexts, where they may be useful.

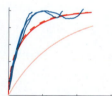

Figure 9.11b (page 223)

In a cell-biology context, then, we now return to negative autoregulation, and examine it more closely as a potential mechanism to generate oscillation. Certainly there is some delay between the transcription of a repressor's gene and subsequent repression: The transcription itself must be followed by translation; then the nascent proteins must fold to their final form; some must then find two or more partners, forming dimers or even tetramers, before they are ready to bind their operators.[1] And indeed, the synthetic network created by Rosenfeld and coauthors did seem to generate some overshoot on its way to its steady state (Figure 9.11b). However, the experiment discussed there looked only at the total production of an entire culture of bacteria. We know from our experience with genetic switches that we need to examine *individual cells* if we wish to see deviations from steady behavior, because each cell can lose synchrony with its neighbors.

J. Stricker and coauthors performed such an experiment in *E. coli*. Again using a fluorescent protein as a reporter for repressor concentration, they indeed found oscillation in a simple negative feedback loop (Figure 11.1). However, the oscillation was not very pronounced, and its amplitude was irregular. Moreover, the period of the oscillator was set by fundamental biochemical processes, and hence could not be adjusted by varying external parameters.

So although the single-loop oscillator is simple, it's not a design of choice when accurate, adjustable beating is needed—for example, in our heart's pacemaker. What other designs could we imagine?

11.3.2 Three repressors in a ring arrangement can also oscillate

One way to adjust timing might be to incorporate more elements into a loop that is still overall negative. In one of the early landmarks of synthetic biology, M. Elowitz and S. Leibler accomplished this feat by using the *lambda*, *lac*, and *tet* repressors, arranging each one's promoter to be controlled by the preceding one (Figure 11.2). As with the one-gene loop, however, oscillation was not robust (many cells did not oscillate at all), and no continuous

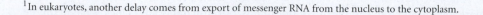

[1] In eukaryotes, another delay comes from export of messenger RNA from the nucleus to the cytoplasm.

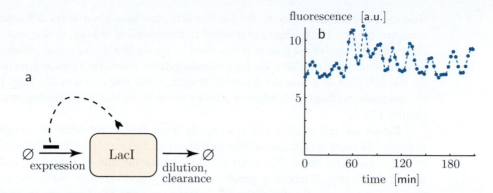

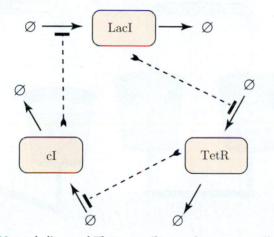

Figure 11.1　A one-gene oscillator circuit. (a) [Network diagram.] A gene controlled by LacI (*not shown*) creates fluorescent protein, allowing visualization of the time course of gene expression. The diagram has the same structure as the governor (Figure 9.10b, page 223), but the actual implementation involved different elements, potentially with different time delays. (b) [Experimental data.] Results from a representative individual *Escherichia coli* cell. The oscillations are somewhat irregular, with a small dynamic range: The peaks are only about 1.3 times the level of the minima. Nevertheless, oscillations are clearly visible. [Data from Stricker et al., 2008; see also Media 16.]

Figure 11.2 [Network diagram.] **The repressilator, a three-gene oscillator circuit.**

control of the period was possible. Despite these limitations in the synthetic realization, the "repressilator" design appears to be used in Nature, for example, in creating the 24-hour clock in some plants.

11.4 Mechanical Clocks and Related Devices Can Also be Represented by Their Phase Portraits

11.4.1 Adding a toggle to a negative feedback loop can improve its performance

Basic mechanical oscillator

To see how to create more decisive, robust, and precise oscillations, as usual we begin by imagining a mechanical device. Figures 11.3a–c depict two buckets, with water flowing in from the top. At any moment, the two buckets may have a net mass difference in their contents, $\Delta m(t)$. The buckets are attached to a mechanical linkage, so that one rises if the other falls; the state of the linkage is described by the angle $\theta(t)$. A second linkage controls a valve depending on θ, so that *the higher bucket gets the water.* We know at least two options for the behavior of a negative-feedback system: Some come to a stable fixed point, like the governor in Chapter 10, whereas others may oscillate, like the room-heater example in Section 11.3.1.

Before we can analyze the system, we must first characterize it completely. For example, we must specify the differential flow rate, $\mathrm{d}(\Delta m)/\mathrm{d}t$, as a function of θ. When $\theta = 0$, Figure 11.3 shows that water should flow equally into each bucket, so $\mathrm{d}(\Delta m)/\mathrm{d}t$ should equal zero. When θ is positive, we want the differential flow rate to be negative, and vice versa. Finally, when $|\theta|$ is large we want the function to saturate; here the valve is directing all the water to one side or the other. A function with all these properties is

$$\mathrm{d}(\Delta m)/\mathrm{d}t = -Q\theta/\sqrt{1+\theta^2},\qquad(11.1)$$

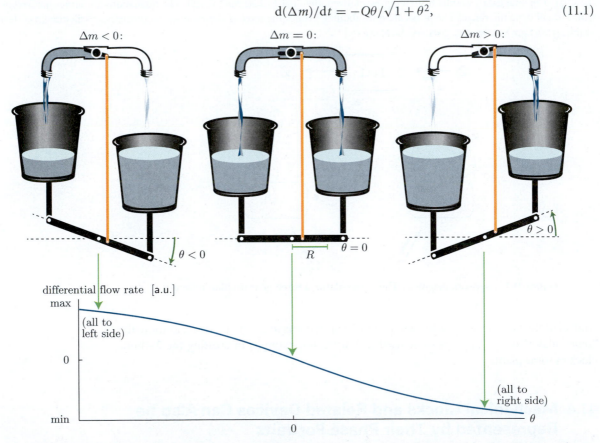

Figure 11.3 Basic mechanical oscillator. *Top:* [Cartoons.] When $\theta = 0$, water flows equally into each bucket. The mass of water in the left bucket, minus that in the right bucket, is called Δm. When an imbalance causes θ to be nonzero, the inlet nozzle changes the flow to correct it. The negative feedback is implemented by a linkage (*orange*) between θ and the water valve. *Bottom:* [Mathematical function.] Thus, the differential flow rate depends on the value of θ.

where the constant Q is the total flow rate of mass into the buckets. We will only be interested in values of θ between about ± 1.5, so the fact that θ is a periodic variable is immaterial. That is, Equation 11.1 is only valid in that limited range.

The buckets respond to an imbalance by moving, with some friction. As with the pendulum, we will assume a viscous-drag type friction, that is, a drag torque proportional to the velocity and in the opposite direction.[2] Thus, θ obeys Newton's law for rotary motion of a rigid body:

$$I d^2\theta/dt^2 = -\zeta\, d\theta/dt + (\Delta m)gR\cos\theta,$$

where I is the moment of inertia and R is shown on the figure. Unlike our discussion of the pendulum, however, this time we will also simplify by supposing that the friction constant ζ is very large. In such a situation, inertia plays no role, so we can neglect the acceleration term in Newton's law:[3]

$$d\theta/dt \approx (\Delta m)\gamma\cos\theta. \qquad \text{basic mechanical oscillator} \qquad (11.2)$$

The constant $\gamma = gR/\zeta$ conveniently lumps together the acceleration of gravity, the length of the lever arms, and the friction constant.

Your Turn 11A

Sketch the nullclines of this system, follow the vector field defined by Equations 11.1–11.2, and describe the motion qualitatively.

Actually, it's also easy to solve the system explicitly, if θ is small: Then $d(\Delta m)/dt \approx -Q\theta$ and $\cos\theta \approx 1$. Taking the derivative of Equation 11.2 and substituting those approximate results gives

$$d^2\theta/dt^2 = -\gamma Q\theta. \qquad (11.3)$$

Your Turn 11B

Show that the system in Figure 11.3 oscillates with frequency $\sqrt{Q\gamma}/2\pi$, and *any* fixed amplitude (as long as θ is small).

You may recognize Equation 11.3 as being mathematically identical to the harmonic oscillators in first-year physics. Because it's the same equation, it has the same behavior. But there is a big physical difference between the two situations. The oscillatory behavior of a pendulum, say, arises from the interplay of inertia and a restoring force that depends only on θ. If friction is present, it makes the system run down and eventually stop at a stable fixed point.

---

[2] See Section 9.7.1 (page 226).

[3] $\boxed{T_2}$ See Problem 11.2. Because we neglect inertia, we also don't need to deal with the fact that I changes in time as the buckets fill up.

In contrast, we assumed that our system had so much friction that inertial effects could be *ignored*. Despite all that friction, the system never runs down, because energy is constantly being added from the outside: Water is falling into the device. The system's behavior reflects

- A torque, proportional to the mass imbalance: Positive Δm (more mass on the left) gives a positive torque (counterclockwise). The angle θ responds by moving in the direction of the torque.
- Inflow: Δm changes with time at a rate proportional to *minus* the angle. Positive θ (counterclockwise displacement) causes water to flow into the right-hand bucket (driving Δm negative).

These two influences always oppose each other (negative feedback), but with a time delay (it takes time for Δm to change). That's a recipe that can, in the right circumstances, give oscillation.

Feedback oscillator with toggle

The mechanical system in Figure 11.3 does oscillate—but it's not very robust. The system does not choose any specific amplitude, and indeed one of the options is amplitude *zero*. We say that the point $\Delta m = 0$, $\theta = 0$ is a **neutral fixed point** (or "center"). The system's trajectories are neither driven toward nor away from it; instead, they orbit it.

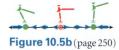
Figure 10.5b (page 250)

For biological applications (and even to design a good mechanical clock), the crucial improvement is to add a *toggle* element to the negative feedback. Suppose that a spring arrangement tries to push θ to either of the values ± 1. That is, suppose that with the water turned off, θ would have a phase portrait like the one in Figure 10.5b, with two stable fixed points flanking an *unstable* fixed point at $\theta = 0$ (see Figure 11.4). You can

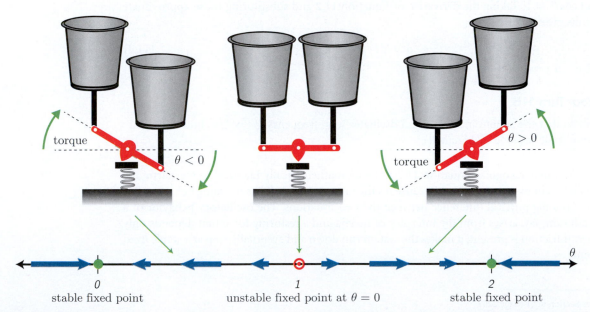

Figure 11.4 [Cartoons; phase portrait.] **Relaxation oscillator.** A toggle element has been added to the device in Figure 11.3; the point $\theta = 0$ is now an unstable fixed point. In the state shown on the *left*, the spring, pushing upward on the eccentric shaft (*red*), exerts a clockwise torque driving θ farther away from 0. When water is added as before, this modification turns the system into a relaxation oscillator. See also Media 17.

probably imagine what will happen: The higher bucket will overbalance the lower one for a while, until $|\Delta m|$ gets large enough to "flip" the toggle; then the system will suddenly snap to its other state and wait for Δm to get sufficiently unbalanced in the other direction.

Let's see how the phase portrait method makes the preceding intuition precise. We can add the toggle to our mathematical model by adding an extra torque due to the spring. The torque should depend on θ in a way that implements a vector field like the one shown at the bottom of Figure 11.4. A suitable choice is to modify Equation 11.2 to

$$d\theta/dt = (\Delta m)\gamma \cos\theta + \alpha(\theta - \theta^3). \quad \text{with toggle element} \quad (11.4)$$

This extra torque is always directed toward one of the points $\theta = \pm 1$; the constant α sets its overall strength. To find the resulting motion, we now draw the nullclines, by solving the equations that we get by setting either of Equations 11.1 or 11.4 equal to zero.

The S-shaped nullcline in Figure 11.5a is particularly significant. Suppose that we shut off the water flow, so that Δm becomes constant. If it's large and positive, then we slice the figure vertically at that value, finding that θ has only one steady state, the one with the left bucket down in Figure 11.4. Similarly, if $\Delta m \ll -2\,\mathrm{g}$, the only steady state solution has the left bucket up. But for intermediate values of Δm, the toggle creates *bistable* behavior: Slicing the phase portrait along a vertical line gives *three* possible steady values of θ, of which two are stable.

Suppose that we initialize our system with small positive values of Δm and θ (point P in Figure 11.5a). Following the arrows, we see that the system arrives at the θ nullcline, then slowly tracks it, moving leftward as $|\Delta m|$ builds up. As it reaches the end of the bistable

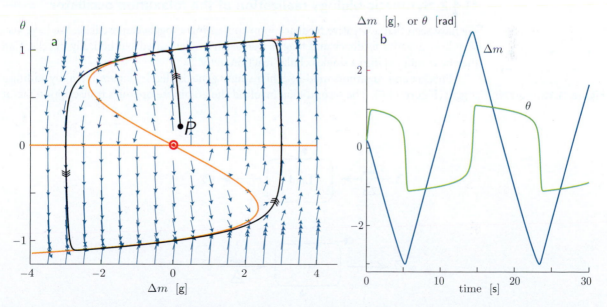

Figure 11.5 Behavior of a relaxation oscillator. (a) [Phase portrait.] The nullclines appear in *orange*. A single unstable fixed point at the origin repels the trajectory shown in *black*, which starts at P but soon begins to trace out the system's limit cycle. (b) [Mathematical functions.] Time dependence of θ and Δm for the trajectory shown in (a). After a brief initial transient, the time course takes a repetitive form. The figures show a solution to Equations 11.1 and 11.4 with parameter values $Q = 1\,\mathrm{g\,s^{-1}}$, $\gamma = 1\,\mathrm{g^{-1}s^{-1}}$, and $\alpha = 5\,\mathrm{s^{-1}}$.

region, it abruptly jumps to the lower branch of the θ nullcline, then begins to follow *that*, moving rightward, until in turn it loses stability and hops back to the upper branch. Thus, after an initial "transient," the system settles down to a **limit cycle**, executing it over and over regardless of its initial condition.

Your Turn 11C

Argue qualitatively from the phase portrait that, had we started from a point *outside* the limit cycle (such as $\Delta m = 4\,\text{g}, \theta = -1/2$), our system would still end up executing the same behavior after a transient.

The mechanism introduced in this section belongs to a class called **relaxation oscillators**. They are very robust: We can make major changes in the dynamics and still achieve oscillation, as long as the θ nullcline has some region of bistability. Moreover, this network architecture confers resistance to noise, because the period is controlled by the time needed for a bucket to fill to a certain point. Thus, it's the *integral* of the inflow over a long time that controls the switching. The integral of a noisy quantity is generally less noisy, because random excursions over and under the average flow rate partially cancel. Finally, a relaxation oscillator is easily tuned, for example, by changing the toggle's thresholds for switching.

T₂ *Section 11.4′ (page 291) introduces the concept of an attractor in the phase portrait, and new phenomena that can occur on higher-dimensional phase portraits. Section 11.4.1′ (page 291) introduces another analysis method to classify system behavior, and also describes a noise-mediated oscillation mechanism.*

11.4.2 Synthetic-biology realization of the relaxation oscillator

We have seen that a negative feedback loop containing a toggle can oscillate. We have also seen how negative feedback and toggles were each implemented synthetically in cells. Can they be combined into a single cellular reaction network?

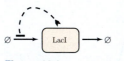

Figure 11.1a (page 279)

Stricker and coauthors modified their single-gene oscillator (Figure 11.1a) by adding a toggle (Figure 11.6). The second loop involved the arabinose operon's transcription factor

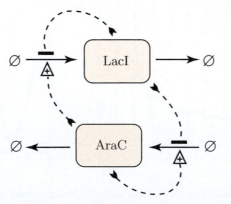

Figure 11.6 [Network diagram.] **A genetic relaxation oscillator.** The central loop involves one repression and one activation line, leading to overall negative feedback with delay. The lower loop is autocatalytic (positive feedback), playing the role of the toggle element in the relaxation oscillator. (The negative feedback loop on the top further improved performance in the experiment.)

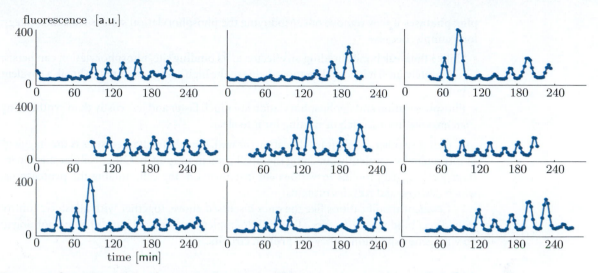

Figure 11.7 [Experimental data.] **Genetic relaxation oscillator.** The units on the vertical axes are arbitrary, but the same for each of nine individual cells. The differences between peaks and valleys are more pronounced than in the one-gene circuit (Figure 11.1b, page 279). [Data from Stricker et al., 2008; see also Media 18.]

AraC, which activated its own expression as well as that of LacI. This architecture achieved robust, high-amplitude oscillations (Figure 11.7). By adjusting the externally supplied inducers arabinose and IPTG, the experimenters also found that they could control the period of oscillations.

11.5 Natural Oscillators

This chapter began with the observation that organisms from cyanobacteria to vertebrates make use of biochemical oscillators to drive repetitive processes and to anticipate periodic environmental events. Now that we have seen some artificial implementations, it's time to return to the natural world, in the context of a system for which powerful experimental techniques allow a fairly detailed analysis.

11.5.1 Protein circuits

So far, our picture for cellular control has been genetic: Repressors and activators bind near the start of a gene, influencing its transcription rate. We also saw that the binding of an effector molecule can influence the behavior of an enzyme or a transcription factor directly, via an allosteric interaction. For example, allolactose or one of its mimics can bind to and modify LacI; tryptophan can bind to and modify one of its production enzymes, and so on.

Many other kinds of control exist. One that is particularly prevalent in cells is the modification of an enzyme by covalently linking a phosphate group to it (**phosphorylation**, Figure 9.3c) or, conversely, by clipping one off (**dephosphorylation**). Enzymes that modify other enzymes in this way are generically called **kinases** if they add phosphate, or

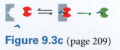

Figure 9.3c (page 209)

phosphatases if they remove one. Modifying the phosphorylation state is a useful strategy, for example, because

- Unlike the weak bonds holding an effector to its binding site, phosphorylation can persist indefinitely until it is actively removed, due to the high energy required to break a covalent bond.
- Phosphorylation and dephosphorylation are much faster and less costly than synthesizing an enzyme from scratch or waiting for it to clear.

Cells use other covalent modifications as well. One that will interest us is the "tagging" of molecules by covalently attaching the small protein **ubiquitin**. Other cellular machinery transports tagged proteins to various compartments in the cell, notably the **proteasome**, which destroys and recycles proteins.

Covalent modifications like the ones discussed above, together with allosteric control from direct binding of effectors, allow cells to implement control circuits having nothing to do with gene expression—they are **protein circuits**.

11.5.2 The mitotic clock in *Xenopus laevis*

Cell division is very complex, even in bacteria. It gets much worse when we raise our sights to single-cell eukaryotes—or even, if we dare, vertebrates. The cycle of eukaryotic cell division (**mitosis**) usually involves numerous "checkpoints," at which the cycle pauses until some critical step (for example, DNA replication) has completed. Researchers are starting to understand these checkpoints in terms of feedback switches, but the whole process is daunting.

So if we wish to think about mitosis, we should look for the simplest possible example. The South African clawed frog *Xenopus laevis* offers such an example. Its fertilized egg undergoes rounds of division, in which all the cells divide in synchrony.[4] Even if we dissociate the embryo into individual cells, those cells continue to initiate mitosis (enter the "M phase") on schedule—each has an autonomous oscillator. Nor are there checkpoints—for example, the embryo does not pause even in the presence of DNA-damaging agents. One can even block cell division altogether, and still find that the clock itself proceeds fairly normally.

There is a big step up in complexity as we pass from synthetic biological systems to natural ones. The tens of thousands of genes in a vertebrate's genome give rise to many actors, with complex interactions that are mostly still not known. Thus, any model that invokes only a few actors is bound to be provisional. Nevertheless, extensive experimental work has identified a module in *Xenopus* that is small and minimally affected by altering molecules other than the few that we will discuss here. J. Tyson and B. Novák proposed a physical model for a cell-cycle clock in the *Xenopus* embryo, in which this module acts as a relaxation oscillator.

The names of those molecules and their complexes can make for cumbersome notation; accordingly we will give the actors in the following discussion one-letter abbreviations:

Q	cyclin-Cdk1 complex
P	APC-Cdc20 complex
R	Wee1
S	Cdc25

Figure 11.8a shows the fundamental negative feedback circuit implicated in the mitotic clock. Mitotic cyclins are synthesized at a constant rate β_Q, and during interphase they are

---

[4]See Media 19.

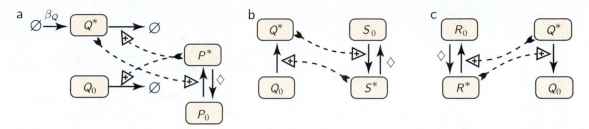

Figure 11.8 [Network diagrams.] **The mitotic clock in the early embryo of $\mathcal{X}enopus$.** For clarity, the network has been divided into three overlapping subsystems. (a) Central negative feedback loop giving oscillations. (b) Positive feedback loop creating a toggle. (c) Another positive feedback loop, reinforcing the toggle behavior. Arrows marked with the $\diamond$ symbol denote constitutive (unregulated) processes not discussed in the text.

stable. Immediately after synthesis, each cyclin binds to a cyclin-dependent kinase called Cdk1 (the "universal M-phase trigger"), forming a complex we will call Q. The Q complex has various phosphorylation states, which we lump into "active" (Q^*) and "inactive" (Q_0). The active form in turn phosphorylates many different targets, initiating a round of mitosis. For our purposes, the important point is that in particular, Q^* activates a complex consisting of the cell division cycle protein Cdc20, bound to the anaphase-promoting complex (APC); we will abbreviate this complex as P. The activated form P^* in turn tags molecules of cyclin, including those in complex Q, for destruction by the cell's proteasome. Thus, the species Q^* and P^* *form an overall negative feedback loop.*

To introduce a toggle into the loop, the cell has two other proteins interacting with Q: Figures 11.8b–c separate these out for easier discussion, but all three panels belong to a single network:

- Panel (b) involves another cell division cycle protein, a phosphatase named Cdc25, which we will call S. Active S^* can activate Q by dephosphorylating it; conversely, Q^* can activate S by phosphorylating it. The result of this double-positive loop is an *overall positive feedback* on Q. Chapter 10 showed that such feedback can create a toggle element.

- Finally, panel (c) shows another loop involving the kinase Wee1, which we will call R. Active R^* can inactivate Q by phosphorylating it; conversely, Q^* can inactivate R by phosphorylation. The result of this double-negative loop is another *overall positive feedback* on Q.

In short, the *Xenopus* system has negative feedback plus toggle elements, the ingredients needed to constitute a relaxation oscillator. To make quantitative predictions, we need the functional forms for all of the influences outlined in words above. Even before these became available, however, some qualitative results gave evidence for the physical model of a relaxation oscillator. For example, J. Pomerening and coauthors simultaneously monitored the level of Q^* and the total of both forms of cyclin (Q_0 and Q^*). Figure 11.9a shows that the two quantities oscillate in step, tracing a loop that is reminiscent of Figure 11.5a. Moreover, when the experimenters intervened to disable one of the positive feedback loops, the time course of Q^* changed from a strong, sharply peaked form, resembling Figure 11.5b, to one that was more sinusoidal in form, and damped (Figures 11.9b,c).

Q. Yang and J. Ferrell characterized the overall negative (oscillator) feedback loop by replacing cyclin B1 by a variant that was impervious to the usual degradation. Although this broke the feedback loop, nevertheless complex P was still able to tag other species for degradation. Following one such species gave the experimenters a rough estimate of how

Figure 11.5a (page 283)

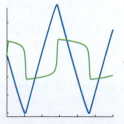

Figure 11.5b (page 283)

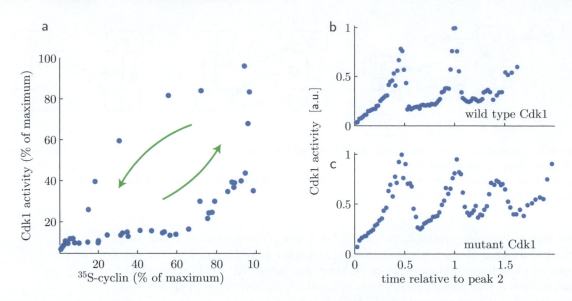

Figure 11.9 [Experimental data.] **Oscillation observed in *Xenopus* egg cell extracts.** (a) Scatter plot of the relationship between cyclin levels and H1 kinase activity, a proxy for Cdk1 activity. The data points lie on a wide loop (*arrows*), which the system traverses repeatedly. (b) The time courses from individual experiments were rescaled to make the second peak of Cdc2 activation occur at $t = 2$ units. (c) In this trial, the positive feedback loop was broken by substituting a mutant form of Cdk1, in which the two inhibitory phosphorylation sites were changed to nonphosphorylatable residues. The resulting time courses show less sharp peaks, and less pronounced difference between maxima and minima, than the wild type cells in (b). [Data courtesy J Pomerening; see also Pomerening et al., 2005.]

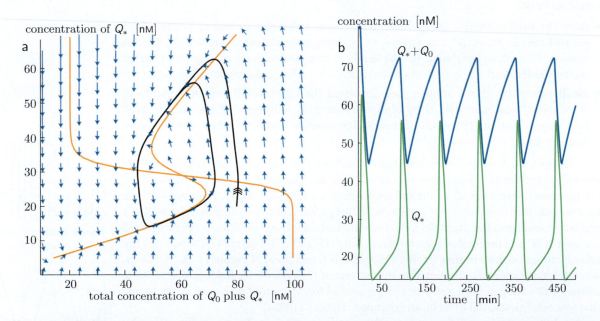

Figure 11.10 **Behavior of a *Xenopus* embryo oscillator model.** (a) [Phase portrait.] This figure is qualitatively similar both to the mechanical relaxation oscillator (Figure 11.5a) and to experimental observations (Figure 11.9a). (b) [Mathematical functions.] Time course of the variables in the model. The *green trace* is similar to the sharply peaked pulses in the experimental observations (Figure 11.9b). Details of the calculation are given in Section 11.5.2′ (page 293).

fast cyclin *would have* been degraded in the *Xenopus* embryo system, as a function of the Q^* population. In fact, they found that the relation had an extremely large Hill coefficient, $n \approx 17$. Combining this result with other biochemists' measurements on the reactions listed above gave them all the ingredients needed to specify the physical model based on the network diagrams in Figure 11.8. In particular, the high Hill coefficient meant that one of the system's nullclines had a broad, nearly level region (Figure 11.10a), similar to the one in our relaxation oscillator (Figure 11.5a).

The resulting physical model indeed displayed oscillations of relaxation type, which qualitatively agreed with experimental observations on the full system (Figure 11.10).

$\boxed{T_2}$ *Section 11.5.2′ (page 293) gives some details of the model.*

THE BIG PICTURE

Chapters 9–11 have studied three exemplary classes of control problems faced by individual cells. Analogous problems, and solutions involving feedback, also arise at the whole-organism level.

Actually, we have barely scratched the surface of biological control. Individual bacteria can also seek out food supplies, or sunlight, and swim toward them; they can detect noxious chemicals, or other environmental dangers, and swim away. Single-celled eukaryotes have more sophisticated behaviors than these, and so on up to vertebrates. Everywhere we look, we see organisms working through their mandate to gather *energy and information* and take appropriate *actions.*[5]

Beyond wishing to understand systems evolved by Nature, however, we may also have technological or therapeutic goals that can be addressed by artificial design. Chapters 9–11 have given some examples of synthetic control systems, embodying simple physical models, that perform lifelike functions to specifications.

Throughout our discussions, we have focused on fairly simple physical models of living systems. Although we seem to have had some successes, one may wonder whether this strategy is really wise. Have we just cherry-picked the few cases that do appear susceptible to this kind of reduction?

One reason to pursue simple, physical models first is that every complex system has evolved from something simpler. For example, at the heart of the enormously complex cell cycle we found the positive-plus-negative feedback motif. We may not always be so lucky, but certainly simple models based on physical ideas supply a set of starting hypotheses that can be investigated before looking farther afield. Second, even complex systems often appear to have a modular structure, in which simpler elements coexist, with limited communication to the others. Third, systems with many participating dynamical variables often mimic simpler ones, because some of the variables evolve rapidly, bringing the system to a lower-dimensional subspace, for example, a limit cycle.[6] Finally, once evolution has found a solution to one problem, it often recycles that solution, modifying it and pressing it into service for other uses.

---

[5] See Section 2.1.

[6] $\boxed{T_2}$ See Section 11.4′ (page 291) and Problem 11.2.

KEY FORMULAS

- *Oscillators:* Simple mechanical oscillator: $d(\Delta m)/dt = -Q\theta/\sqrt{1+\theta^2}$; $d\theta/dt = (\Delta m)\gamma\cos\theta$.
  The mechanical relaxation oscillator adds $\alpha(\theta - \theta^3)$ to the right-hand side of the second equation (Equation 11.4, page 283).

FURTHER READING

Semipopular:
Bray, 2009; Strogatz, 2003.

Intermediate:
Relaxation and other oscillator mechanisms: Keener & Sneyd, 2009;
Murray, 2002, chapts. 6, 10; Strogatz, 2014; Tyson et al., 2003.
Cell cycle: Alberts et al., 2008, chapt. 17; Klipp et al., 2009.
Other oscillators, as dynamical systems: Gerstner et al., 2014; Ingalls, 2013; Winfree, 2001.
Control via phosphorylation: Marks et al., 2009.
$\boxed{T_2}$ Linear stability analysis and linear algebra: Klipp et al., 2009, chapts. 12, 15;
Otto & Day, 2007.

Technical:
Repressilator: Elowitz & Leibler, 2000.
Cell cycle and other oscillators: Novák & Tyson, 1993b; Sha et al., 2003; Tyson & Novák, 2010. Reviews: Novák & Tyson, 2008; Ferrell et al., 2011; Pomerening et al., 2005.
Applications of synthetic biology: Burrill & Silver, 2011; Ro et al., 2006;
Weber & Fussenegger, 2012.

Track 2

11.4′a Attractors in phase space

The concepts of stable fixed point and limit cycle are examples of a bigger idea. An **attractor** in a dynamical system is a subset of its phase space with three properties:

- Any trajectory that begins in the attractor stays there throughout its evolution.
- There is a larger set, with the full dimensionality of the phase space (an "open set"), all of whose points converge onto the attractor under their time development.
- No smaller subset within the attractor has these same properties.

Thus, a stable fixed point is a zero-dimensional attractor; a limit cycle is a one-dimensional attractor (see Strogatz, 2014, chapt. 9).

11.4′b Deterministic chaos

A deterministic dynamical system with phase space dimension larger than two can display another sort of long-time behavior, besides the limit cycles, runaways, and stable fixed points studied in the main text. This behavior is called **deterministic chaos**, or just "chaos." A chaotic system has at least some trajectories that remain bounded but never settle down to steady, or even periodic, behavior. In fact, such a trajectory's behavior is so complex as to appear random, even though mathematically it is completely predictable given its initial condition.

If the chaotic system is dissipative, like the ones we have studied, then its trajectories *can* settle down, but to a very unusual kind of attractor. Such "strange" attractors are fractals, that is, subsets of phase space with noninteger dimension. (Again see Strogatz, 2014, chapt. 9.) Chaotic dynamics can appear in animal populations, and may be relevant for both pathological conditions like cardiac fibrillation and even normal brain function.

Track 2

11.4.1′a Linear stability analysis

Section 11.4.1 began with the system

$$\mathrm{d}(\Delta m)/\mathrm{d}t = Qg(\theta), \quad \mathrm{d}\theta/\mathrm{d}t = (\Delta m)\gamma \cos\theta,$$

where $g(\theta) = -\theta/\sqrt{1+\theta^2}$. Close to the fixed point, $\theta \approx 0$, we can simplify this system by replacing the nonlinear function g by its Taylor series expansion close to $\theta = 0$, truncated after the linear term, and similarly with $\cos\theta$:

$$\frac{\mathrm{d}}{\mathrm{d}t}\begin{bmatrix} \Delta m \\ \theta \end{bmatrix} = \begin{pmatrix} 0 & -Q \\ \gamma & 0 \end{pmatrix} \begin{bmatrix} \Delta m \\ \theta \end{bmatrix}. \tag{11.5}$$

The equations are now easy to solve directly.[7] But for more general problems, recall that first-order, linear differential equations with constant coefficients have exponential solutions.[8]

---

[7] See Equation 11.2 (page 281).
[8] There are exceptional cases; see Problem 1.6 (page 25).

Substitute the trial solution $\begin{bmatrix} \Delta m \\ \theta \end{bmatrix} = e^{\beta t} \begin{bmatrix} x \\ y \end{bmatrix}$ to find that

$$\begin{pmatrix} -\beta & -Q \\ \gamma & -\beta \end{pmatrix} \begin{bmatrix} x \\ y \end{bmatrix} = 0.$$

The only way to get a nonzero solution to this matrix equation is for $\begin{bmatrix} x \\ y \end{bmatrix}$ to be a null eigenvector of the matrix; this in turn requires that the matrix must have a zero eigenvalue, and so must have determinant zero:

$$(-\beta)^2 - (\gamma)(-Q) = 0. \tag{11.6}$$

Making the abbreviation $\bar{\gamma} = \gamma Q$, the solutions to this equation are $\beta = \pm i\sqrt{\bar{\gamma}}$; the corresponding eigenvectors are

$$\begin{bmatrix} x \\ y \end{bmatrix} = \begin{bmatrix} x \\ \mp ix\sqrt{\bar{\gamma}}/Q \end{bmatrix}.$$

Any physical solution must have real values of Δm and θ; we arrange this by combining the two mathematical solutions found so far:

$$\begin{bmatrix} \Delta m(t) \\ \theta(t) \end{bmatrix} = e^{i\sqrt{\bar{\gamma}}t} \begin{bmatrix} x \\ -ix\sqrt{\bar{\gamma}}/Q \end{bmatrix} + e^{-i\sqrt{\bar{\gamma}}t} \begin{bmatrix} x \\ ix\sqrt{\bar{\gamma}}/Q \end{bmatrix} = 2x \begin{bmatrix} \cos(\sqrt{\bar{\gamma}}t) \\ \sqrt{\bar{\gamma}}\sin(\sqrt{\bar{\gamma}}t)/Q \end{bmatrix}. \tag{11.7}$$

The solutions oscillate with frequency $\sqrt{\bar{\gamma}}/2\pi$. The amplitude is arbitrary but constant in time—it neither grows nor decays.

The preceding rigamarole starts to show its value when we move on to less simple systems. The main text next considered adding a toggle element, an extra contribution to $d\theta/dt$ of the form $\alpha(\theta - \theta^3)$, where α is a positive constant. Again expanding near the fixed point $\theta \approx 0$, Equation 11.5 becomes

$$\frac{d}{dt}\begin{bmatrix} \Delta m \\ \theta \end{bmatrix} = \begin{pmatrix} 0 & -Q \\ \gamma & \alpha \end{pmatrix} \begin{bmatrix} \Delta m \\ \theta \end{bmatrix}. \tag{11.8}$$

Then Equation 11.6 becomes

$$(-\beta)(\alpha - \beta) - (\gamma)(-Q) = 0. \tag{11.9}$$

whose solutions are $\beta = \frac{1}{2}(\alpha \pm \sqrt{\alpha^2 - 4\bar{\gamma}})$. We can now notice that both of these solutions have *positive real part*. Thus, both of them lead to system behaviors in which x and y are growing—the fixed point is unstable, driving the state outward, in this case toward a limit cycle.

Your Turn 11D

Follow logic similar to that in Equation 11.7 to justify the statement just made.

The power of the **linear stability analysis** just given is that we didn't need to know much about the complicated, nonlinear phase portrait. Because the system's only fixed point is unstable, it cannot come to rest at any steady state. Nor can the system run away to infinite values of Δm or θ, on physical grounds. Thus, it must oscillate.

Linearized analysis like the one just given can be applied to any fixed point in a two-dimensional phase space. If both eigenvalues are real and negative, the fixed point is a stable node;[9] if both are real and positive, it's an unstable node. If one is negative but the other is positive, the fixed point is a saddle. If both are complex with negative real part, then the fixed point is a stable spiral.[10] To this menagerie we can now add the case of two purely imaginary eigenvalues (a neutral fixed point or "center," Equation 11.5), and two complex values, each with a positive real part (an "unstable spiral," Equation 11.8). Some other exotic cases are also possible (see Strogatz, 2014).

Similar analyses apply for phase spaces with any number of dimensions. Beyond 2D, however, there is a new, qualitatively distinct option for the motion of a bounded system with an unstable fixed point (see Section 11.4′b).

11.4.1′b Noise-induced oscillation

A system whose dynamical equations have only a stable fixed point can nevertheless oscillate as a result of molecular randomness. For example, the system may be close to a bifurcation to oscillation (see Problem 11.3); then fluctuations can repeatedly kick it over the threshold, creating a series of individual events that resemble truly periodic oscillations (Hilborn et al., 2012).

T_2 | **Track 2**

11.5.2′ Analysis of *Xenopus* mitotic oscillator

To obtain a tractable set of dynamical equations, we'll make some approximations, following Yang & Ferrell (2013). Their model was conceptually similar to the proposal of Novák & Tyson (1993a), but with recent experimental determinations of key parameters.

The approximations made below are rooted in biochemical facts, but the authors also solved the more difficult equations that result when we don't make some of the approximations, confirming that the results were qualitatively unchanged.

We'll make some abbreviations:

Q	cyclin-Cdk1 complex
P	APC-Cdc20 complex
R	Wee1
S	Cdc25
Y	concentration of Q^* (active cyclin-Cdk1)
Z	concentration of Q_0 (inactive cyclin-Cdk1)
X	$= Y + Z$ (total cyclin)

Main negative loop (See Figure 11.8a.)
Yang and Ferrell supposed that Q is continuously created at a rate $\beta_Q = 1$ nM/min, and that all newly synthesized molecules of cyclin immediately form active complexes with Cdk1. Thus, dY/dt has a production term, but dZ/dt does not.

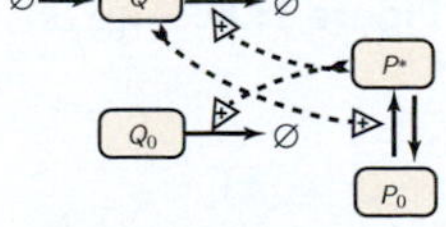

Figure 11.8a (page 287)

[9] See Section 9.7.1′ (page 237).
[10] See Problem 9.8.

The researchers also supposed that P^* responds so quickly to the level of Q^* that we may simply take its concentration to be a function of Y (the concentration of Q^*). Then its relevant activity, which is to degrade both forms of Q, can be expressed as

- A contribution to dY/dt of the form $-g_P(Y)Y$, where g_P is an empirically determined function, and
- A contribution to dZ/dt of the form $-g_P(Y)Z$.

These formulas assume that P^* acts by first-order kinetics on any Q it finds, and that the activation of P is itself a saturating, increasing function of Y. The authors found experimentally that the function g_P could be adequately represented by a basal rate plus a Hill function:

$$g_P(Y) = a_P + b_P \frac{Y^{n_P}}{K_P^{\,n_P} + Y^{n_P}}, \tag{11.10}$$

with approximate parameter values $a_P = 0.01/\mathrm{min}$, $b_P = 0.04/\mathrm{min}$, $K_P = 32\,\mathrm{nM}$, and $n_P = 17$.

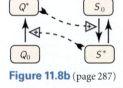

Figure 11.8b (page 287)

Double positive loop (Figure 11.8b)

Next, suppose that S^*, too, responds so quickly to the level of Q^* that we may take its concentration to be a function of Y. Then its relevant activity, which is to convert Q_0 to active form, can be expressed as

- A contribution to dY/dt of the form $+g_S(Y)Z$, where g_S is an empirically determined function, and
- An equal and opposite contribution to dZ/dt.

These formulas assume that S^* acts by first-order kinetics on any Q_0 it finds, and that the activation of S is itself a saturating, increasing function of Y. The authors cited their own and earlier experimental work that found that the empirical function g_S could be adequately represented by a basal rate plus a Hill function:

$$g_S(Y) = a_S + b_S \frac{Y^{n_S}}{K_S^{\,n_S} + Y^{n_S}}, \tag{11.11}$$

with approximate parameter values $a_S = 0.16/\mathrm{min}$, $b_S = 0.80/\mathrm{min}$, $K_S = 35\,\mathrm{nM}$, and $n_S = 11$.

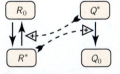

Figure 11.8c (page 287)

Double negative loop (Figure 11.8c)

Finally, suppose that R^* also responds so quickly to the level of Q^* that we may take its concentration to be a function of Y. Then its relevant activity, which is to convert Q^* to inactive form, can be expressed as

- A contribution to dY/dt of the form $-g_R(Y)Y$, where g_R is an empirically determined function, and
- An equal and opposite contribution to dZ/dt.

These formulas assume that R^* acts by first-order kinetics on any Q^* it finds, and that the activation of R is itself a *decreasing* function of Y, because Q^* *deactivates* R^*. The authors cited their own and earlier experimental work that found that the empirical function g_R could be adequately represented by a basal rate plus a Hill function:

$$g_R(Y) = a_R + b_R \frac{K^{n_R}}{K_R{}^{n_R} + Y^{n_R}}, \tag{11.12}$$

with approximate parameter values $a_R = 0.08/\text{min}$, $b_R = 0.40/\text{min}$, $K_R = 30\,\text{nM}$, and $n_R = 3.5$.

Combined system

It is convenient to re-express the variable Z in terms of the total $X = Y + Z$. Then

$$dY/dt = \beta_Q - g_P(Y)Y + g_S(Y)(X - Y) - g_R(Y)(Y) \tag{11.13}$$

$$dX/dt = \beta_Q - g_P(Y)X. \tag{11.14}$$

Figure 11.10a shows the vector field defined by Equations 11.10–11.14, as well as the nullclines and a typical trajectory. Panel (b) shows the time evolution of the variables.

Yang and Ferrell drew particular attention to the remarkably high Hill coefficient in the negative loop. This feature makes one of the nullclines in Figure 11.10a nearly horizontal, and hence similar to the horizontal nullcline in our mechanical analogy (Figure 11.5a). They found that models without high Hill coefficient gave less robust oscillations, and could even fail to oscillate at all.

The authors also performed stochastic simulations, along the lines of Chapter 8, to confirm that their results were qualitatively maintained despite cellular randomness.

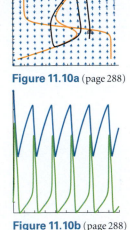

Figure 11.10a (page 288)

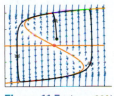

Figure 11.10b (page 288)

Figure 11.5a (page 283)

PROBLEMS

11.1   Relaxation oscillator

a. Create a two-dimensional phase portrait representing the mechanical oscillator without a toggle element (Equations 11.1 and 11.2). Include the vector field and nullclines, and discuss the qualitative behavior. Add a streamline representing a typical trajectory.[11]

b. Repeat for the relaxation oscillator (Equations 11.1 and 11.4) with $\alpha = 1\,\text{s}^{-1}$.

11.2 $\boxed{T_2}$ High-friction regime

Section 11.4.1 claimed that, in the limit of high friction, we may neglect the angular acceleration term in Newton's law, approximating it by the statement that all torques approximately balance. This seems reasonable—the acceleration term is associated with inertia, and if you try to throw a ball in a vat of molasses, it stops moving immediately after leaving your hand, instead of coasting for a while. To investigate further, replace Equation 11.2 (page 281) by the more complete Newton law

$$I\frac{\mathrm{d}^2\theta}{\mathrm{d}t^2} = -\zeta\frac{\mathrm{d}\theta}{\mathrm{d}t} + (\Delta m)gR\cos\theta.$$

In this formula, ζ is a friction constant and I is the oscillator's moment of inertia, whose time dependence we will neglect.

   Following the main text, write approximate versions of this formula and Equation 11.1 for the case where θ is small, and combine them to eliminate Δm, obtaining a generalized form of Equation 11.3. Because this equation is linear with constant coefficients, we can write a trial solution of the form $e^{\beta t}$, obtaining an ordinary algebraic equation for β. Do this, and comment on whether the inertial term matters in the limit where ζ becomes large holding other constants fixed.

11.3 $\boxed{T_2}$ Oscillation bifurcation

a. The dynamical equations for the relaxation oscillator, Equations 11.1 and 11.4, have only one fixed point. Linearize them about this fixed point, find solutions for small deviations, and comment. Is there more than one kind of behavior possible, depending on the values of parameters?

b. Modify the toggle element in the equations, replacing $\alpha(\theta - \theta^3)$ by $\alpha\theta - \gamma\theta^3$, and imagine adjusting only α, holding all other constants fixed. The text considered only the case $\alpha > 0$; instead investigate what happens to the solutions as α is reduced to zero, and beyond it to negative values.

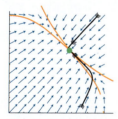

Figure 10.8a (page 255)

The behavior you have found is sometimes called the **Hopf bifurcation**. The lesson is that, as in the toggle switch, a plausible-looking network diagram by itself does not guarantee oscillation; we must also be in the right region of parameter space.

11.4 $\boxed{T_2}$ Linear stability analysis

a. Go back to Equations 9.21 (page 226), which describe the pendulum. Analyze small deviations from the stable and the unstable fixed points by the method of linear stability analysis (Section 11.4.1′, page 291).

b. Go back to Equations 10.7 (page 254), which describe the two-gene toggle. Assume that $\tau_1 = \tau_2$, $\Gamma_1 = \Gamma_2$, and $n_1 = n_2$. The portraits in Figure 10.8 make it seem reasonable

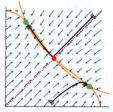

Figure 10.8b (page 255)

---

[11] See Problem 9.5 (page 238).

that there will always be a fixed point with $\bar{c}_1 = \bar{c}_2$. Confirm this. Find this fixed point for the two sets of parameter values shown in the figure, and in each case assess its stability and comment.

11.5 $\boxed{T_2}$ *Xenopus* oscillator

Carry out the analysis outlined in Section 11.5.2′ (page 293), using the parameter values given there, and create figures similar to Figures 11.10a,b.

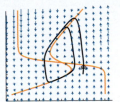

Figure 11.10a (page 288)

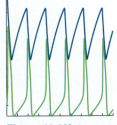

Figure 11.10b (page 288)

Epilog

Eccentric, intervolved, yet regular
Then most, when most irregular they seem;
And in their motion harmony divine.
—John Milton, 1667

So far, this book has skirted a big question: *What is a physical model?* You have seen many examples of an effective approach to scientific problems in the preceding chapters. Did they have anything in common?

The figure below represents one kind of answer: The models we discussed have attempted to find synergy between four different modes of thought and expression, each

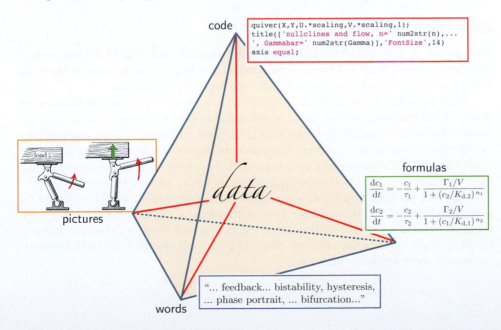

code

```
quiver(X,Y,U.*scaling,V.*scaling,1);
title(['nullclines and flow, n=' num2str(n),...
', Gammabar=' num2str(Gamma)],'FontSize',14)
axis equal;
```

pictures

load

data

formulas

$$\frac{dc_1}{dt} = -\frac{c_1}{\tau_1} + \frac{\Gamma_1/V}{1 + (c_2/K_{d,2})^{n_1}}$$

$$\frac{dc_2}{dt} = -\frac{c_2}{\tau_2} + \frac{\Gamma_2/V}{1 + (c_1/K_{d,1})^{n_2}}$$

"... feedback... bistability, hysteresis, ... phase portrait, ... bifurcation..."

words

illuminating the others. We can start at any point of this tetrahedron, then bounce from point to point as we refine the ideas. We may start with visual imagery, exploring a possible analogy between a new system (in this case bistability of a network) and one we know (in this case the mechanical toggle, Figure 10.5). Words that express particles of experience can sharpen our expectations, and remind us what phenomena to look for. Mathematical formulas can give precise instantiations of those words, and adapt them to known aspects of the problem at hand. Computer coding can help us solve those equations, or even bring to light hidden consequences that we did not expect. Throughout the process, each aspect of the model must be compatible with what we know already (the data); moreover, each may suggest the right experiments to dig out new data.

A model is a reduced description of a real-world system to demonstrate particular features of, or investigate specific questions about, the system. Modeling can help us find what few features of a complex system are really necessary to get certain behaviors; as we have seen, it can also guide the construction of useful artificial systems inspired by natural ones. But what makes a model "physical"? The edges are fuzzy here, but it should be clear that analogies to nonliving systems have been helpful throughout this book. Tactile images, like the toggle, helped us to guess mechanisms hit upon by evolution, presumably because they were evolvable from existing components, and performed adaptive tasks well. And hypotheses rooted in a web of other known facts about the living and nonliving world, even if provisional, often bear fruit, perhaps because those roots confer a high prior probability even before any data have been taken. Other physical ideas, such as the discrete character of light, also proved to be indispensable for imagining new experimental techniques, such as localization microscopy.

It may seem surprising that physical modeling ever works at all, and even miraculous that sometimes we can extrapolate a model successfully to new situations not yet studied experimentally. As scientists, we cannot explain the regularity of Nature; instead, we seek to exploit it, in part by studying cases where it has appeared in the past, and by developing the skills needed to recognize it.

Skills and frameworks

Facts go out of date quickly; individual facts are generally also tied to narrow areas of experience. Certain skills and frameworks, however, can help you make sense of the world as it is now known, as new facts come to light, and across disparate areas.

Working some of this book's many problems has given you practice in the intimate details of the modeling process. Some of the skills involved everyday scientific tasks like dimensional analysis and curve fitting. Others involved randomness, from characterizing distributions to maximum-likelihood inference to stochastic simulation. Along the way, we needed bits of bigger frameworks, such as genetics, dynamical systems, physiology, control theory, biochemistry, statistical inference, physical chemistry, and cell biology.

The goal has been to uncover some aspects of life science, physical science, and the general problem of "how do we know." Beyond knowledge for its own sake, however, many scientists view the second part of their job as seeing how such insights can lead to improvements in health, sustainability, or some other big goal. Here, too, the skills and frameworks mentioned above will be useful.

Like any other tool, physical modeling can mislead us if pushed too far or too uncritically. The attempt to explain many things in terms of a few common agents and mechanisms does not always succeed. But the impulse to search for additional relevant data that could falsify our favorite model, even if we have found some that seem to support it, is itself a skill that can lead to better science.

Vista

Another goal of this book has been to show you that more of science is more interconnected than you may have realized. Some say that mechanistic understanding tarnishes the mysterious beauty of the world. Many scientists counter that we need a considerable amount of mechanistic understanding even to *grasp* what is truly beautiful and mysterious.

As I write these words, birds are calling in the forest around me. I know they want territory, mates, insects to eat. But knowing that I am embedded in a dazzlingly intricate, self-regulating system doesn't detract from the pleasure of hearing them. Knowing a tiny bit about the mechanisms they employ to make their living only sharpens my sense of wonder. Unlike the ancients, we can also begin to appreciate the dense, sentient, and mostly invisible living world in which they live, going right down to the level of the microorganisms in the soil around them, and begin to make sense of that world's complex dance.

A lot of doors now open in every direction. Good luck with your own search.

Philip Nelson
Philadelphia, 2014

Appendix A:
Global List of Symbols

It is not once nor twice but times without number that the same ideas make their appearance in the world.
—Aristotle

A.1 Mathematical Notation

Abbreviated words

var x variance (Section 3.5.2, page 54).
corr(ℓ, s) correlation coefficient (Section 3.5.2′, page 60).
cov(ℓ, s) covariance (Section 3.5.2′, page 60).

Operations

Both $\times$ and $\cdot$ denote ordinary multiplication; no vector cross products are used.
$\mathcal{P}_1 \star \mathcal{P}_2$ convolution of two distributions (Section 4.3.5, page 79).
$\langle f \rangle$ expectation (Section 3.5.1, page 53). $\langle f \rangle_\alpha$, expectation in a family of distributions with parameter α.
$\bar{f}$ sample mean of a random variable (Section 3.5.1, page 53). However, an overbar can have other meanings (see below).

Other modifiers

$\bar{c}$ dimensionless rescaled form of a variable c.
Δ is often used as a prefix: For example, Δx is a small, but finite, change of x. Sometimes this symbol is also used by itself, if the quantity being changed is clear from context.

The subscript 0 appended to a quantity can mean an initial value of a variable, or the center of a small range of specific values for a variable.

The subscript $*$ or $\star$ appended to a quantity can mean any of the following:

An optimal value (for example, the maximally likely value), an extreme value, or some other critical value (for example, an inflection point)

The value at a fixed point of a phase portrait

A value being sent to infinity in some limit (Section 4.3.2, page 75)

The value of some function at a particular time of interest, for example, starting or ending value, or value when some event first occurs

The superscript $*$ or $\star$ can perform these functions:

Distinguish between multiple operators (Section 10.4.1′, page 266)

Indicate the activated form of an enzyme (Section 11.5.2, page 286)

Vectors

Vectors are denoted by $\boldsymbol{v} = (v_x, v_y, v_z)$, or just (v_x, v_y) if confined to a plane.

Relations

The symbol $\overset{?}{=}$ signals a provisional formula, or guess.

The symbol $\approx$ means "approximately equal to."

In the context of dimensional analysis, $\sim$ means "has the same dimensions as."

The symbol $\propto$ means "is proportional to."

Miscellaneous

The symbol $\left. \frac{\mathrm{d}G}{\mathrm{d}x} \right|_{x_0}$ refers to the derivative of G with respect to x, evaluated at the point $x = x_0$.

Inside a probability function, $|$ is pronounced "given" (Section 3.4.1, page 45).

A.2 Graphical Notation

A.2.1 Phase portraits

Each point on a line or plane represents a system state. Arrows indicate how any starting state will evolve. Black curves give examples of the system's possible trajectories. Nullclines are drawn in orange. A separatrix, if any, is drawn in magenta. Stable fixed points appear as ●. Unstable fixed points appear as ◎.

A.2.2 Network diagrams

See Section 9.5.1, page 219.

Each box represents a state variable, usually the inventory of some type of molecule.

Incoming and outgoing solid arrows represent processes (chemical reactions) that increase or decrease a state variable.

If a process transforms one species to another, and both are of interest, then we draw
the solid line joining the two species' boxes. But if a species' precursor is not of
interest to us, for example, because its inventory is maintained constant by some
other mechanism, we can replace it by the symbol ∅, and similarly when the loss of a
particular species creates something not of interest to us.

Each dashed line that ends on a solid arrow represents an interaction in which one type
of molecule modifies the rate of a process (see Figure 8.8b). However, such "influence
lines" are omitted for the common case in which the rate of degradation depends on
the level at which a species is present.

A dashed line may also end on another dashed line, representing the modulation of one
molecule's effect on a process by another molecule (see Figure 10.14b).

Dashed lines terminate with a symbol: A blunt end, $-----\!$, indicates repression, whereas
an open arrowhead, $----\!\!\triangleright$, indicates activation.

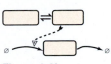

Figure 8.8b (page 192)

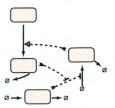

Figure 10.14b (page 261)

A.3 Named Quantities

The lists below act as a glossary for usage adopted in this book. Although symbolic names
for quantities are in principle arbitrary, still it's convenient to use standard names for the
ones that recur the most, and to use them as consistently as possible. But the limited number
of letters in the Greek and Latin alphabets make it inevitable that some letters must be used
for more than one purpose. See Appendix B for explanation of the dimensions, and for the
corresponding units.

Latin alphabet

c number density ("concentration") of a small molecule, for example, a nutrient
(Sections 9.7.2 and 9.4.3) [dimensions $\mathbb{L}^{-3}$].

$C(j)$ autocorrelation function (Section 3.5.2′, page 60) [dimensions depend on those of
the random variable].

d generic name for a distance, for example, the step size for a random walk
[dimensions $\mathbb{L}$].

$\mathcal{D}(x)$ cumulative distribution function (Section 5.4, page 110) [dimensionless].

E generic name for an "event" in probability (Section 3.3.1, page 41) [not a quantity].

f generic name for a function, or specifically the gene regulation function
(Equation 9.10, page 218).

G function defining a transformation of a continuous random variable (Section 5.2.5,
page 104).

$\hbar$ reduced Planck's constant (Section B.6, page 313) [dimensions $\mathbb{M}\mathbb{L}^2\mathbb{T}^{-1}$].

i generic name for an integer quantity, for example, a subscripted index that counts
items in a list [dimensionless].

I moment of inertia (Section 11.4.1, page 279) [dimensions $\mathbb{M}\mathbb{L}^2$].

j generic name for an integer quantity, for example, a subscripted index that counts
items in a list. Specifically, the discrete waiting time in a Geometric distribution
(Section 3.4.1.2, page 47) [dimensionless].

k a rate constant, for example, $k_\emptyset$, the degradation rate constant; k_I, clearance rate
of infected T cells (Section 1.2.3, page 12); k_V, clearance rate of virus particles
(Section 1.2.3, page 12) [dimensions depend on the order of the reaction].

k_e electric force constant (Section B.6, page 313) [dimensions $\mathbb{ML}^3\mathbb{T}^{-2}$].

k_g bacterial growth rate constant [dimensions $\mathbb{T}^{-1}$]; $k_{g,max}$, its maximum value.

K generic name for an equilibrium constant; K_d, dissociation equilibrium constant (Section 9.4.3, page 214) [dimensions depend on the reaction].

K half-maximum parameter in a saturating rate function (Section 9.7.2, page 227) [dimensions $\mathbb{L}^{-3}$].

ℓ generic name for a discrete random variable. Modified forms, such as ℓ_*, may instead represent constants [dimensionless].

m mass; m_e, mass of electron (Section B.6, page 313) [dimensions $\mathbb{M}$].

m generic name for an integer quantity, or specifically, number of resistant bacteria in the Luria-Delbrück experiment, or number of mRNA molecules in a transcriptional burst [dimensionless].

M generic name for an integer quantity, or specifically, total number of coin flips summed to obtain a Binomial distribution [dimensionless].

n cooperativity parameter ("Hill parameter"), a real number ≥ 1 (Section 9.4.3, page 214) [dimensionless].

N generic name for an integer quantity, or specifically, the number of times a particular outcome has been measured in a random system [dimensionless]; N_I number of infected T cells in blood; N_V, number of free virus particles (virions) in blood (Section 1.2.3, page 12).

$\wp_x(x)$ probability density function for a continuous random variable x, sometimes abbreviated $\wp(x)$ (Section 5.2.1, page 98) [dimensions match those of $1/x$]. $\wp(x \mid y)$, conditional pdf.

$\mathcal{P}(\mathsf{E})$ probability of event E [dimensionless]. $\mathcal{P}_\ell(\ell)$, probability mass function for the discrete random variable ℓ, sometimes abbreviated $\mathcal{P}(\ell)$ (Section 3.3.1, page 41). $\mathcal{P}(\mathsf{E} \mid \mathsf{E}')$, $\mathcal{P}(\ell \mid s)$, conditional probability (Section 3.4.1, page 45).

$\mathcal{P}_{name}(\ell; p_1, \dots)$ or $\wp_{name}(x; p_1, \dots)$ mathematical function of ℓ or x, with parameter(s) $p_1, \dots$, that specifies a particular idealized distribution, for example, $\mathcal{P}_{unif}$ (Section 3.3.2, page 43), $\wp_{unif}$ (Equation 5.6, page 99), $\mathcal{P}_{bern}$ (Equation 3.5, page 43), $\mathcal{P}_{binom}$ (Equation 4.1, page 71), $\mathcal{P}_{pois}$ (Equation 4.6, page 76), $\wp_{gauss}$ (Equation 5.8, page 100), $\mathcal{P}_{geom}$ (Equation 3.13, page 47), $\wp_{exp}$ (Equation 7.5, page 159), $\wp_{cauchy}$ (Equation 5.9, page 101).

P APC-Cdc20 complex (Section 11.5.2, page 286).

Q fluid flow rate (Section 9.7.2, page 227) [dimensions $\mathbb{L}^3\mathbb{T}^{-1}$].

Q cyclin-Cdk1 complex (Section 11.5.2, page 286).

R Wee1 (Section 11.5.2, page 286).

s generic outcome label for a discrete distribution. In some cases, these may be outcomes without any natural interpretation as a number, for example, a Bernoulli trial ("coin flip").

S Cdc25 (Section 11.5.2, page 286).

t_w waiting time between events in a random process (Section 7.3.2.1, page 158) [dimensions $\mathbb{T}$]. $t_{w,stop}$, duration of a transcriptional burst (waiting time to turn off); $t_{w,start}$, waiting time to turn on (Section 8.4.2, page 189).

V electric potential [dimensions $\mathbb{ML}^2\mathbb{T}^{-2}\mathbb{Q}^{-1}$].

W dynamical vector field on a phase portrait (Section 9.2.2, page 204).

X total cyclin inventory (Section 11.5.2′, page 293) [dimensionless].

Y inventory of Q^* (active cyclin-Cdk1) (Section 11.5.2′, page 293) [dimensionless].

Z inventory of Q_0 (inactive cyclin-Cdk1) (Section 11.5.2′, page 293) [dimensionless].

Greek alphabet

α generic name for a subscripted index that counts items in a list.

α parameter describing a power-law distribution (Section 5.4, page 110) [dimensionless].

α_g mutation probability per doubling (Section 4.4.4, page 84) [dimensionless].

β probability per unit time, appearing, for example, in a Poisson process or its corresponding Exponential distribution (Section 7.3.2, page 157) [dimensions $\mathbb{T}^{-1}$]. In the continuous, deterministic approximation, β can also represent the rate of the corresponding zeroth-order reaction. (Higher-order rate constants are generally denoted by k.) β_s, $\beta_\emptyset$, synthesis and degradation rates in a birth-death process (Section 8.3.1, page 182). β_{start}, β_{stop}, probabilities per unit time for a gene to transition to the "on" or "off" transcription state.

γ parameter appearing in the chemostat equation (page 230) [dimensionless].

γ mean rate of virus particle production per infected T cell (Section 1.2.3, page 12).

Γ maximum protein production rate of a gene (Equation 9.10, page 218) [dimensions $\mathbb{T}^{-1}$]; Γ_{leak}, "leakage" protein production rate of a gene (Equation 10.13, page 269).

Δ amount by which some quantity changes. Usually used as a prefix: Δx denotes a small change in x.

ζ friction constant (Section 9.7.1, page 226) [dimensions $\mathbb{MLT}^{-1}$].

η width parameter of a Cauchy distribution (Equation 5.9, page 101) [same dimensions as its random variable].

μ parameter describing a Poisson distribution (Equation 4.6, page 76) [dimensionless].

μ_x parameter setting the expectation of a Gaussian or Cauchy distribution in x (Equation 5.8, page 100) [same dimensions as its random variable x].

ν frequency (cycles per unit time) [dimensions $\mathbb{T}^{-1}$].

ν number of molecules of a critical nutrient required to create one new bacterium (Equation 9.23, page 229) [dimensionless].

ξ parameter describing a Bernoulli trial ("probability to flip <u>heads</u>") (Section 3.2.1, page 36) [dimensionless]. ξ_{thin}, thinning factor applied to a Poisson process (Section 7.3.2.2, page 160).

ρ number density, for example, of bacteria (Section 9.7.2, page 227) [dimensions $\mathbb{L}^{-3}$].

σ variance parameter of a Gaussian distribution (Equation 5.8, page 100) [same dimensions as its random variable].

τ_{tot} e-folding time scale for concentration of repressor in a cell (Section 9.4.5, page 218) [dimensions $\mathbb{T}$]. τ_e, e-folding time for cell growth (Equation 9.12, page 218).

Appendix B: Units and Dimensional Analysis

The root cause for the loss of the spacecraft was the failure to use metric units in the coding of a ground software file . . . . The trajectory modelers assumed the data was provided in metric units per the requirements.
—Mars Climate Orbiter Mishap Investigation Board

Physical models discuss physical quantities. Some physical quantities are integers, like the number of cells in a culture. But most are continuous, and most continuous quantities carry *units*. This book will generally use the Système Internationale, or **SI units**, but it's essential to be able to convert, accurately, when reading other works or even when speaking to other scientists. (Failure to convert units led to the loss of the \$125 million Mars Climate Orbiter spacecraft.) Units and their conversions are part of a larger framework called **dimensional analysis**.

Students sometimes don't take dimensional analysis too seriously because it seems trivial, but it's a very powerful method for catching algebraic errors. *Much more importantly,* it gives a way to organize and classify numbers and situations, and even to guess new physical laws, as we'll see below. When faced with an unfamiliar situation, dimensional analysis is usually step one. We can use dimensional analysis to (*i*) instantly spot a formula that must contain an error, (*ii*) recall formulas that we have partially forgotten, and even (*iii*) construct promising new hypotheses for further checking.

Our point of view will be that every useful thing about units can be systematized by using a simple maxim:

*Most physical quantities should be regarded as the **product** of a pure number times one or more "units." A unit acts like a symbol representing an unknown quantity.*

(A few physical quantities, for example, those that are intrinsically integers, have no units and are called **dimensionless**.) We carry the unit symbols along throughout our calculations. They behave just like any other multiplicative factor; for example, a unit can cancel if it appears in the numerator and denominator of an expression.[1] Although they behave like unknowns, we do know relations among certain units; for example, we know that 1 inch $\approx$ 2.54 cm. Dividing both sides of this formula by the numeric part 2.54, we find 0.39 inch $\approx$ 1 cm, and so on.

B.1 Base Units

The SI begins by choosing arbitrary "base" units for length, time, and mass: Lengths are measured in meters (abbreviated m), masses in kilograms (kg), time in seconds (s), and electric charge in coulombs. We also create related units by attaching prefixes giga ($=10^9$, or billion), mega ($=10^6$, or million), kilo ($=10^3$, or thousand), milli ($=10^{-3}$, or thousandth), micro ($=10^{-6}$, or millionth), nano ($=10^{-9}$, or billionth), or pico ($=10^{-12}$). In writing, we abbreviate these prefixes to G, M, k, m, μ, n, and p, respectively. Thus, 1 μg is a microgram (or 10^{-9} kg), 1 ms is a millisecond, and so on.

In addition, there are some traditional though non-SI prefixes, such as centi ($= 10^{-2}$, abbreviated c), and deci ($= 10^{-1}$, abbreviated d).

B.2 Dimensions versus Units

Other quantities, such as force, derive their standard units from the base units. But it is useful to think about force in a way that is less strictly tied to a particular unit system. Thus, we define abstract **dimensions**, which tell us *what kind of thing* a quantity represents.[2] For example,

- We define the symbol $\mathbb{L}$ to denote the *dimension* of length. The SI assigns it a base *unit* called "meters," but other units exist with the same dimension (for example, miles or centimeters). Having chosen a unit of length, we then also get a derived unit for volume, namely, cubic meters, or m^3, which has dimensions $\mathbb{L}^3$.
- We define the symbol $\mathbb{M}$ to denote the dimension of mass. Its SI base unit is the kilogram.
- We define the symbol $\mathbb{T}$ to denote the dimension of time. Its SI base unit is the second.
- We define the symbol $\mathbb{Q}$ to denote the dimension of electric charge. Its SI base unit is the coulomb.
- Velocity has dimensions of $\mathbb{L}\mathbb{T}^{-1}$. The SI assigns it a standard unit called "meter per second," written m/s or m s^{-1}.
- Force has dimensions $\mathbb{M}\mathbb{L}\mathbb{T}^{-2}$. The SI assigns it a standard unit kg m/s^2, also called "newton" and abbreviated N.
- Energy has dimensions $\mathbb{M}\mathbb{L}^2\mathbb{T}^{-2}$. The SI assigns it a standard unit kg m^2/s^2, also called "joule" and abbreviated J.
- Power (energy per unit time) has dimensions $\mathbb{M}\mathbb{L}^2\mathbb{T}^{-3}$. The SI assigns it a standard unit kg m^2/s^3, also called "watt" and abbreviated W.

---

[1] One exception involves temperatures expressed using the Celsius and Fahrenheit scales of temperature, which differ from the absolute (Kelvin) scale by an offset as well as a multiplier.

[2] This distinction, and many other points in this section, were made in a seminal paper by J. C. Maxwell and F. Jenkins.

The answers to a quantitative exam problem will also have some appropriate dimensions, which you can use to check your work. Suppose that you are asked to compute a force. You work hard and write down a formula made out of various given quantities. To check your work, write down the dimensions of each of the quantities in your answer, cancel whatever cancels, and make sure the result is $\mathbb{M}\mathbb{L}\mathbb{T}^{-2}$. If it's not, you probably forgot to copy something from one step to the next. It's easy, and it's amazing how quickly you can spot and fix errors in this way.

When you multiply two quantities, the dimensions just pile up: Force ($\mathbb{M}\mathbb{L}\mathbb{T}^{-2}$) times length ($\mathbb{L}$) has dimensions of energy ($\mathbb{M}\mathbb{L}^2\mathbb{T}^{-2}$). On the other hand, you can never add or subtract terms with different dimensions in a valid equation, any more than you can add dollars to kilograms. Equivalently, an equation of the form $A = B$ cannot be valid if A and B have different dimensions.[3] For example, suppose that someone gives you a formula for the mass m of a sample as $m = aL$, where a is the cross-sectional area of a test tube and L the height in the tube. One side is $\mathbb{M}$, the other is $\mathbb{L}^3$; the disagreement is a clue that another factor is missing from the equation (in this case the mass density of the substance). It is possible that a formula like this may be valid with a certain special choice of units. For example, the author of such a formula may mean "(mass in grams) = (area in cm^2) $\times$ (length in cm)," which might better be written

$$\frac{m}{1\,\text{g}} = \frac{a}{1\,\text{cm}^2}\frac{L}{1\,\text{cm}}.$$

In this form, the formula equates two dimensionless quantities, so it's valid in any set of units; indeed it says $m = \rho_w aL$, where $\rho_w = 1\,\text{g}\,\text{cm}^{-3}$ is the mass density (of water).

You *can* add dollars to rupees, with the appropriate conversion factor, and similarly meters to miles. Meters and miles are different units that both have the same dimensions. We can automate unit conversions, and eliminate errors, if we restate 1 mile $\approx$ 1609 m in the form

$$1 \approx \frac{1609\,\text{m}}{\text{mile}}.$$

Because we can freely insert a factor of 1 into any formula, we may introduce as many factors of the above expression as we need to cancel all the mile units in that expression. This simple prescription ("multiply or divide by 1 as needed to cancel unwanted units") eliminates confusion about whether to place the pure number 1609 in the numerator or denominator. For example,

$$230\,\text{m} + 2.6\,\text{mile} \approx 230\,\text{m} + 2.6\,\cancel{\text{mile}} \times \frac{1609\,\text{m}}{\cancel{\text{mile}}} \approx 4400\,\text{m}.$$

Functions applied to dimensional quantities

If $x = 1$ m, then we understand expressions like $2\pi x$ (with dimensions $\mathbb{L}$), and even x^3 (with dimensions $\mathbb{L}^3$). But what about $\exp(x)$, $\cos(x)$, or $\ln x$? These expressions are meaningless; more precisely, they don't transform in any simple multiplicative way when we change units, unlike $2\pi x$ or x^3. (One way to see why such expressions are meaningless is to use the Taylor

---

[3]There is an exception: If a formula sets a dimensional quantity equal to *zero*, then we may omit the units on the zero without any ambiguity. Thus, a statement like, "The potential difference is $\Delta V = 0$," is legitimate, and many authors will omit the unnecessary units in this way.

series expansion of $\exp(x)$, and notice that it involves adding terms with incompatible units.)

Additional SI units

This book occasionally mentions electrical units (volt V, ampere A, and derived units like μV, pA, and so on), but does not make essential use of their definitions. They involve the dimension $\mathbb{Q}$.

Traditional but non-SI units

length: One Ångstrom unit (Å) equals 0.1 nm.

volume: One liter (L) equals 10^{-3} m^3. Thus, 1 mL = 10^{-6}m^3.

number density: A 1 M solution has a number density of 1 mole L^{-1} = 1000 mole m^{-3}, where "mole" represents the dimensionless number $\approx 6.02 \cdot 10^{23}$.

energy: One calorie (cal) equals 4.184 J. An electron volt (eV) equals $e \times (1\,\text{V}) = 1.60 \cdot 10^{-19}$ J = 96 kJ/mole. An erg (erg) equals 10^{-7} J. Thus, 1 kcal mole^{-1} = 0.043 eV = $6.9 \cdot 10^{-21}$ J = $6.9 \cdot 10^{-14}$ erg = 4.2 kJ mole^{-1}.

B.3 Dimensionless Quantities

Sometimes a quantity is stated as a multiple of some other quantity with the same units. For example, "concentration relative to time zero," means $c(t)/c(0)$. Such relative quantities are dimensionless; they carry no units. Other quantities are intrinsically dimensionless, for example, angles (see below).

B.4 About Graphs

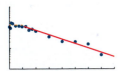

Figure 0.3 (page 4)

When graphing a continuous quantity, it's usually essential to state the units, to give meaning to the labels on the axes. For example, if the axis label says `length [m]` then we understand that a point aligned with the tick mark labeled `1.5` represents a measured length that, when divided by 1 m, yields the pure number 1.5.

The same interpretation applies to logarithmic axes. If the axis label says `length [m]`, and the tick marks are unequal, as they are on the vertical axis in Figure 0.3, then we understand that a point aligned with the first tick after the one labeled `1000` represents a measured length that when divided by 1 m, yields the pure number 2000. Alternatively, we can make an ordinary graph of the logarithm of a quantity x, indicating this in the axis label, which says "$\log_{10} x$" or "ln x" instead of "x." The disadvantage of the latter system is that, if x carries units, then strictly speaking we must instead write something like "$\log_{10}(x/(1\,\text{m}))$," because the logarithm of a quantity with dimensions has no meaning.

B.4.1 Arbitrary units

Sometimes a quantity is stated in some unknown or unstated unit. It may not be necessary to be more specific, but you should alert your reader by saying something like `virus concentration, arbitrary units`. Many authors abbreviate this as "a.u."

When using arbitrary units on one axis, it's usually a good practice to make sure the other axis crosses it at the value 0 (which should be labeled), rather than at some other value.[4] (Otherwise, your reader won't be able to judge whether you have exaggerated an insignificant effect by blowing up the scale of the graph!)

B.5 About Angles

Angles are dimensionless. After all, we get the angle between two intersecting rays by drawing a circular arc of any radius r between them and dividing the circumference of that arc (with dimensions $\mathbb{L}$) by r (with dimensions $\mathbb{L}$). Another clue is that if θ carried dimensions, then trigonometric functions like sine and cosine wouldn't be defined (see Section B.2).

What about degrees versus radians? We can think of deg as a convenient or traditional unit with *no* dimensions: It's just an abbreviation for the pure number $\pi/180$. The radian represents the pure number 1; we can omit it. Stating it explicitly as rad is just a helpful reminder that we're *not* using degrees. Similarly, when phrases like "cycles per second" or "revolutions per minute" are regarded as angular frequencies, we can think of the words "cycles" and "revolutions" as dimensionless units (pure numbers), both equal to 2π.

B.6 Payoff

Dimensional analysis has other uses. Let's see how it actually helps us to discover new science.

An obsolete physical model of an atom regarded it as a miniature solar system, with electrons orbiting a heavy nucleus. Like our solar system, an atom was known to consist of a massive, positive kernel (the nucleus) surrounded by lighter electrons. The force law between charges was known to have the same $1/r^2$ form as the Sun's gravitational pull on Earth. The analogy failed, however, when scientists noted that Newtonian physics doesn't determine the size of planetary orbits, and only partially determines their shapes. (Unfriendly aliens could change the size and eccentricity of Earth's orbit just by giving it a push.) In contrast, somehow all hydrogen atoms have exactly the *same size and shape* in their ground state. Indeed *all* kinds of atoms have similar sizes, about a tenth of a nanometer. What could determine that size?

The problem can be restated succinctly by using dimensional analysis. The force between a proton and an electron is k_e/r^2, where $k_e = 2.3 \cdot 10^{-28}$ J m is a constant of Nature involving the charge. All we need to know is that its dimensions are $\mathbb{M}\mathbb{L}^3/\mathbb{T}^2$. The only other relevant constant in the problem is the electron mass $m_e = 9.1 \cdot 10^{-31}$ kg. Play around with the constants k_e/r^2 and m_e for a while, and show that there is no way to put them together to get a number with dimensions of length. The problem is that there's no way to get rid of the $\mathbb{T}$'s. No theory in the world can get a well-defined atomic size without some *new constant of Nature*.[5]

Early in the 20th century, Niels Bohr knew that Max Planck had recently discovered a new constant of Nature in a different context, involving *light*. We will call Planck's constant $\hbar = 1.05 \cdot 10^{-34}$ kg m^2/s. Bohr suspected that this same constant played a role in atomic physics.

---

[4] Except when using log axes, which cannot show the value 0. But on a logarithmic axis, changing units simply shifts the graph, without changing its shape, so the reader can always tell whether a variation is fractionally significant or not.

[5] The speed of light is a constant of Nature that involves time dimensions, but it is not relevant to this situation.

Let's see how far we can go without any real theory. Can we construct a length scale using k_e, m_e, and $\hbar$? We are looking for a formula for the size of atoms, so it must have dimensions $\mathbb{L}$. We put the relevant constants together in the most general way by considering the product $(k_e)^a(m_e)^b(\hbar)^c$, and try to choose the exponents a, b, c to get the dimensions right:

$$\left(\mathbb{M}\mathbb{L}^3/\mathbb{T}^2\right)^a \mathbb{M}^b \left(\mathbb{M}\mathbb{L}^2/\mathbb{T}\right)^c = \mathbb{L}.$$

We must choose $c = -2a$ to get rid of $\mathbb{T}$. We must choose $b = a$ to get rid of $\mathbb{M}$. Then $a = -1$; there's no freedom whatever.

Does it work? The proposed length scale is

$$(k_e)^{-1}(m_e)^{-1}(\hbar)^2 \approx 0.5 \times 10^{-10}\text{m}, \tag{B.1}$$

which is right in the tenth-of-a-nanometer ballpark. *Atoms really are that size!*

What did we really accomplish here? This isn't the end; it's the beginning: We don't have a *theory* of atoms. But we have found that *any* theory that predicts an atomic size, using only the one new constant $\hbar$, must give a value similar to Equation B.1 (maybe times some factors of 2, π, or something similar). The fact that this estimate does coincide with the actual size scale of atoms strengthens the hypothesis that there is an atomic theory based only on these constants, motivating us to go find such a theory.

Appendix C: Numerical Values

Human knowledge will be erased from the world's archives
before we possess the last word that a gnat has to say to us.
— Jean-Henri Fabre

(See also http://bionumbers.hms.harvard.edu/)

C.1 Fundamental Constants

reduced Planck's constant, $\hbar = 1.05 \cdot 10^{-34}$ J s.
electric force constant, $k_e = e^2/(4\pi\varepsilon_0) = 2.3 \cdot 10^{-28}$ J m.
electron mass, $m_e = 9.1 \cdot 10^{-31}$ kg.
Avogadro's number, $N_{\text{mole}} = 6.02 \cdot 10^{23}$.

Acknowledgments

Whoever is in search of knowledge, let him fish for it where it dwells.
—Michel de Montaigne

This book is an emergent phenomenon—it arose from the strongly coupled dynamics of many minds. It's a great pleasure to remember the friends and strangers who directly taught me things, replied to my queries, created some of the graphics, and willingly subjected themselves to the draftiest of drafts. Many others have entered the book though their writing and through lectures I attended.

Some of the big outlines of this book were inspired by two articles (Bialek & Botstein, 2004; Wingreen & Botstein, 2006), and I've benefited from discussions with those authors. Bruce Alberts gave several useful suggestions, but a single sentence from him literally wrenched the project off its foundations and reoriented it. Nily Dan tirelessly discussed the scope and purposes of this book, and the skills that students will need in the future. The topics I have chosen have also been guided by those techniques I found I needed in my own research, many of which I've learned from my collaborators, including John Beausang, Yale Goldman, Timon Idema, Andrea Liu, Rob Phillips, Jason Prentice, and particularly Vijay Balasubramanian.

In an intense, year-long exchange, Sarina Bromberg helped channel a chaotic sequence of ideas into a coherent story. She also supplied technical expertise, tactfully pointed out errors small and large, and explained nuances of expression that I had obtusely missed.

Much of the art in this book sprang from the visual imagination of Sarina Bromberg, David Goodsell, and Felice Macera, as well as individual scientists named below. Steven Nelson offered expert counsel on photographic reproduction. William Berner, who is my own Physical Model for vivid, tactile exposition, created several of the classroom demonstrations.

Ideas and inspiration alone don't create a book; they must be supplemented by more concrete support (see page 321). The US National Science Foundation has steadfastly supported my ideas about education for many years; here I'd especially like to thank Krastan Blagoev, Neocles Leontes, Kamal Shukla, and Saran Twombly, both for support, and for their patience as this project's timeline exceeded all my expectations. At Penn, Dawn Bonnell and A. T. Johnson have also recognized this project as an activity of the Nano-Bio Interface Center, which in addition has been an endless source of great science—some of it reflected in these pages. I am also grateful to the University of Pennsylvania for extraordinary help. Larry Gladney unhesitatingly made the complex arrangements needed for two scholarly leaves, Dennis Deturck supplied some funding, and all my colleagues undertook big and small duties that I might have done. Tom Lubensky asked me to create two new courses, each of which contributed to material in this book.

I am also lucky to have found a publisher, W. H. Freeman and Company, that remains so committed to creating really new textbooks to the highest standards. Over several years of gestation, Alicia Brady, Christine Buese, Taryn Burns, Jessica Fiorillo, Richard Fox, Jeanine Furino, Lisa Kinne, Tracey Kuehn, Courtney Lyons, Matt McAdams, Philip McCaffrey, Kate Parker, Amy Thorne, Vicki Tomaselli, Susan Wein, and particularly Elizabeth Widdicombe maintained unflagging enthusiasm. In the final stages, Kerry O'Shaughnessy, Blythe Robbins, and Teresa Wilson held all the threads of the project, and didn't let any of them go. Their professionalism and good grace made this intricate process bearable, and even at times joyful.

Yaakov Cohen, Raghuveer Parthasarathy, and particularly Ann Hermundstad committed to the long haul of reading absolutely everything; their invisible contributions improved nearly every page. John Briguglio, Edward Cox, Jennifer Curtis, Mark Goulian, Timon Idema, Kamesh Krishnamurthy, Natasha Mitchell, Rob Phillips, Kristina Simmons, Daniel Sussman, and Menachem Wanunu also read multiple chapters and made incisive comments. In addition, my teaching assistants over the years have made countless suggestions, including solving, and even writing, many of the exercises: They are Ed Banigan, Isaac Carruthers, David Chow, Tom Dodson, Jan Homann, Asja Radja, and most of all, ex officio, André Brown and Jason Prentice.

For seven consecutive years, the students in my class have somehow managed to teach themselves this material using inscrutable drafts of this book. Each week, each of them was asked to pelt me with questions, which often exposed my ignorance. Students at other institutions also used preliminary versions, including Earlham College, Emory University, Harvard University, MIT, University of Chicago, University of Florida, University of Massachusetts, and University of Michigan. They, and their instructors (Jeff Gore, Maria Kilfoil, Michael Lerner, Erel Levine, Ilya Nemenman, Stephanie Palmer, Aravinathan Samuel, and Kevin Wood), flagged many rough patches. Special thanks to Steve Hagen for using a very early version, and for his extensive advice.

Many reviewers generously read and commented on the book's initial plan, including Larry Abbott, Murat Acar, David Altman, Russ Altman, John Bechhoefer, Meredith Betterton, David Botstein, André Brown, Anders Carlsson, Paul Champion, Horace Crogman, Peter Dayan, Markus Deserno, Rhonda Dzakpasu, Gaute Einevoll, Nigel Goldenfeld, Ido Golding, Ryan Gutenkunst, Robert Hilborn, K. C. Huang, Greg Huber, Maria Kilfoil, Jan Kmetko, Alex Levine, Anotida Madzvamuse, Jens-Christian Meiners, Ethan Minot, Simon Mochrie, Liviu Movileanu, Daniel Needleman, Ilya Nemenman, Julio de Paula, Rob Phillips, Thomas Powers, Thorsten Ritz, Steve Quake, Aravinathan Samuel, Ronen Segev, Anirvan Sengupta, Sima Setayeshgar, John Stamm, Yujie Sun, Dan Tranchina, Joe Tranquillo, Joshua Weitz, Ned Wingreen, Eugene Wong, Jianghua Xing, Haw Yang, Daniel Zuckerman, as well as other, anonymous, referees. Several of these people also gave suggestions on drafts of the book. In the end game, Andrew Belmonte, Anne Caraley, Venkatesh Gopal, James Gumbart, William Hancock, John Karkheck, Michael Klymkowsky, Wolfgang Losert, Mark Matlin, Kerstin Nordstrom, Joseph Pomerening, Mark Reeves, Erin Rericha, Ken Ritchie, Hanna Salmon, Andrew Spakowitz, Megan Valentine, Mary Wahl, Kurt Wiesenfeld, and other, anonymous, referees reviewed chapters. I couldn't include all the great suggestions for topics that I got, but all highlighted the diversity of the field, and the passion of those who engage with it.

Many colleagues read sections, answered questions, supplied graphics, discussed their own and others' work, and more, including: Daniel Andor, Bill Ashmanskas, Vijay Balasubramanian, Mark Bates, John Beausang, Matthew Bennett, Bill Bialek, Ben Bolker, Dennis Bray, Paul Choi, James Collins, Carolyn A. Cronin, Tom Dodson, Michael Elowitz, James Ferrell, Scott Freeman, Noah Gans, Timothy Gardner, Andrew Gelman, Ido Golding, Yale Goldman, Siddhartha Goyal, Urs Greber, Jeff Hasty, David Ho, Farren Isaacs, Randall Kamien, Hiroaki Kitano, Mark Kittisopikul, Michael Laub, David Lubensky, Louis Lyons, Will Mather, Will McClure, Thierry Mora, Alex Ninfa, Liam Paninski, Johan Paulsson, Alan Perelson, Josh Plotkin, Richard Posner, Arjun Raj, Devinder Sivia, Lok-Hang So, Steve Strogatz, Gürol Süel, Yujie Sun, Tatyana Svitkina, Alison Sweeney, Gasper Tkacik, Tony Yu-Chen Tsai, John Tyson, Chris Wiggins, Ned Wingreen, Qiong Yang, and Ahmet Yildiz. Mark Goulian was always ready, seemingly at any hour of day or night, with an expert clarification, no matter how technical the question. Scott Weinstein and Peter Sterling gave me strength.

Many of the key ideas in this book first took shape in the inspiring atmosphere of the Aspen Center for Physics, and in the urban oases of Philadelphia's Fairmount Park system and the Free Library of Philadelphia. The grueling revisions also benefited from the warm hospitality of the Nicolás Cabrera Institute of the Universidad Autónoma de Madrid and the American Philosophical Society.

Lastly, I think that everyone who ever encountered Nicholas Cozzarelli or Jonathan Widom learned something about kindness, rigor, and passion. They, and everyone else on these pages, have my heartfelt thanks.

Credits

Protein Data Bank entries

Several images in this book are based on data obtained from the RCSB Protein Data Bank (`http://www.rcsb.org/`; Berman et al., 2000), which is managed by two members of the RCSB (Rutgers University and UCSD) and funded by NSF, NIGMS, DOE, NLM, NCI, NINDS, and NIDDK.

The entries below include both the PDB ID code and, if published, a Digital Object Identifier (DOI) or PubMed citation for the original source:

Fig. 1.1: RT enzyme: `1hys` (DOI: `10.1093/emboj/20.6.1449`); protease: `1hsg` (PubMed: `7929352`); *gag* polyprotein: `116n` (DOI: `10.1038/nsb806`).

Fig. 1.4: Wildtype: `1hxw` (PubMed: `7708670`); mutant: `1rl8`.

Fig. 7.1: `1m8q` and `2dfs` (PubMed: `12160705` and `16625208`).

Fig. 8.4: RNA polymerase: `1i6h` (DOI: `10.1126/science.1059495`); ribosome: `2wdk` and `2wdl` (DOI: `10.1038/nsmb.1577`); tRNA + EF–Tu: `1ttt` (PubMed: `7491491`); EFG: `1dar` (PubMed: `8736554`); EF–Tu + EF–Ts `1efu` (DOI: `10.1038/379511a0`); aminoacyl tRNA synthetases `1ffy`, `1eiy`, `1ser`, `1qf6`, `1gax`, `1asy` (PubMed: `10446055`, `9016717`, `8128220`, `10319817`, `11114335`, `2047877`).

Fig. 9.4: `1hw2` (DOI: `10.1074/jbc.M100195200`).

Software

This book was built with the help of several pieces of freeware and shareware, including TeXShop, TeX Live, LaTeXiT, and DataThief.

Grant support

This book is partially based on work supported by the United States National Science Foundation under Grants EF–0928048 and DMR–0832802. The Aspen Center for Physics, which is supported by NSF grant PHYS-1066293, also helped immeasurably with the conception, writing, and production of this book. Any opinions, findings, conclusions, silliness, or recommendations expressed in this book are those of the author and do not necessarily reflect the views of the National Science Foundation.

The University of Pennsylvania Research Foundation provided additional support for this project.

Trademarks

MATLAB is a registered trademark of The MathWorks, Inc. *Mathematica* is a registered trademark of Wolfram Research, Inc.

Bibliography

For oute of olde feldys, as men sey,
Comyth al this newe corn from yere to yere;
And out of old bokis, in good fey,
Comyth al this newe science that men lere.
—Geoffrey Chaucer, *Parlement of Fowles*

Many of the articles listed below are published in high-impact scientific journals. It is important to know that frequently such an article is only the tip of an iceberg: Many of the technical details (generally including specification of any physical model used) are relegated to a separate document called Supplementary Information, or something similar. The online version of the article will generally contain a link to that supplement.

Acar, M, Mettetal, J T, & van Oudenaarden, A. 2008. Stochastic switching as a survival strategy in fluctuating environments. *Nat. Genet.*, **40**(4), 471–475.

Ahlborn, B. 2004. *Zoological physics.* New York: Springer.

Alberts, B, Johnson, A, Lewis, J, Raff, M, Roberts, K, & Walter, P. 2008. *Molecular biology of the cell.* 5th ed. New York: Garland Science.

Alberts, B, Bray, D, Hopkin, K, Johnson, A, Lewis, J, Raff, M, Roberts, K, & Walter, P. 2014. *Essential cell biology.* 4th ed. New York: Garland Science.

Allen, L J S. 2011. *An introduction to stochastic processes with applications to biology.* 2d ed. Upper Saddle River NJ: Pearson.

Alon, U. 2006. *An introduction to systems biology: Design principles of biological circuits.* Boca Raton FL: Chapman and Hall/CRC.

Amador Kane, S. 2009. *Introduction to physics in modern medicine.* 2d ed. Boca Raton FL: CRC Press.

American Association for the Advancement of Science. 2011. *Vision and change in undergraduate biology education.* http://www.visionandchange.org.

American Association of Medical Colleges. 2014. *The official guide to the MCAT exam.* 4th ed. Washington DC: AAMC.

American Association of Medical Colleges / Howard Hughes Medical Institute. 2009. *Scientific foundations for future physicians.* Washington DC. https://members.aamc.org/eweb/DynamicPage.aspx?webcode=PubByTitle&Letter=S.

Andresen, M, Stiel, A C, Trowitzsch, S, Weber, G, Eggeling, C, Wahl, M C, Hell, S W, & Jakobs, S. 2007. Structural basis for reversible photoswitching in Dronpa. *Proc. Natl. Acad. Sci. USA*, **104**(32), 13005–13009.

Atkins, P W, & de Paula, J. 2011. *Physical chemistry for the life sciences.* 2d ed. Oxford UK: Oxford Univ. Press.

Barrangou, R, Fremaux, C, Deveau, H, Richards, M, Boyaval, P, Moineau, S, Romero, D A, & Horvath, P. 2007. CRISPR provides acquired resistance against viruses in prokaryotes. *Science*, **315**(5819), 1709–1712.

Bates, M, Huang, B, Dempsey, G T, & Zhuang, X. 2007. Multicolor super-resolution imaging with photo-switchable fluorescent probes. *Science*, **317**(5845), 1749–1753.

Bates, M, Huang, B, & Zhuang, X. 2008. Super-resolution microscopy by nanoscale localization of photo-switchable fluorescent probes. *Curr. Opin. Chem. Biol.*, **12**(5), 505–14.

Bates, M, Jones, S. A, & Zhuang, X. 2013. Stochastic optical reconstruction microscopy (STORM): A method for superresolution fluorescence imaging. *Cold Spring Harbor Protocols*, **2013**(6), 498–520.

Bechhoefer, J. 2005. Feedback for physicists: A tutorial essay on control. *Rev. Mod. Phys.*, **77**, 783–836.

Becskei, A, & Serrano, L. 2000. Engineering stability in gene networks by autoregulation. *Nature*, **405**(6786), 590–593.

Beggs, J M, & Plenz, D. 2003. Neuronal avalanches in neocortical circuits. *J. Neurosci.*, **23**(35), 11167–11677.

Benedek, G B, & Villars, F M H. 2000. *Physics with illustrative examples from medicine and biology.* 2d ed. Vol. 2. New York: AIP Press.

Berendsen, H J C. 2011. *A student's guide to data and error analysis.* Cambridge UK: Cambridge Univ. Press.

Berg, H C. 2004. *E. coli in motion.* New York: Springer.

Berg, J M, Tymoczko, J L, & Stryer, L. 2012. *Biochemistry.* 7th ed. New York: WH Freeman and Co.

Berman, H M, Westbrook, J, Feng, Z, Gilliland, G, Bhat, T N, Weissig, H, Shindyalov, I N, & Bourne, P E. 2000. The Protein Data Bank. *Nucl. Acids Res.*, **28**, 235–242.

Betzig, E. 1995. Proposed method for molecular optical imaging. *Opt Lett*, **20**(3), 237–9.

Betzig, E, Patterson, G H, Sougrat, R, Lindwasser, O W, Olenych, S, Bonifacino, J S, Davidson, M W, Lippincott-Schwartz, J, & Hess, H F. 2006. Imaging intracellular fluorescent proteins at nanometer resolution. *Science*, **313**(5793), 1642–1645.

Bialek, W. 2012. *Biophysics: Searching for principles.* Princeton NJ: Princeton Univ. Press.

Bialek, W, & Botstein, D. 2004. Introductory science and mathematics education for 21st-century biologists. *Science*, **303**(5659), 788–790.

Bloomfield, V. 2009. *Computer simulation and data analysis in molecular biology and biophysics: An introduction using R.* New York: Springer.

Boal, D. 2012. *Mechanics of the cell.* 2d ed. Cambridge UK: Cambridge Univ. Press.

Bobroff, N. 1986. Position measurement with a resolution and noise-limited instrument. *Rev. Sci. Instrum.*, **57**, 1152–1157.

Bolker, B M. 2008. *Ecological models and data in R.* Princeton NJ: Princeton Univ. Press.

Boyd, I A, & Martin, A R. 1956. The end-plate potential in mammalian muscle. *J. Physiol. (Lond.)*, **132**(1), 74–91.

Bray, D. 2009. *Wetware: A computer in every living cell.* New Haven: Yale Univ. Press.

Burrill, D R, & Silver, P A. 2011. Synthetic circuit identifies subpopulations with sustained memory of DNA damage. *Genes and Dev.*, **25**(5), 434–439.

Calo, S, Shertz-Wall, C, Lee, S C, Bastidas, R J, Nicolás, F E, Granek, J A, Mieczkowski, P, Torres-Martínez, S, Ruiz-Vázquez, R M, Cardenas, M E, & Heitman, J. 2014. Antifungal drug resistance evoked via RNAi-dependent epimutations. *Nature*, **513**(7519), 555–558.

Cheezum, M K, Walker, W F, & Guilford, W H. 2001. Quantitative comparison of algorithms for tracking single fluorescent particles. *Biophys. J.*, **81**(4), 2378–2388.

Cherry, J L, & Adler, F R. 2000. How to make a biological switch. *J. Theor. Biol.*, **203**(2), 117–133.

Choi, P J, Cai, L, Frieda, K, & Xie, X S. 2008. A stochastic single-molecule event triggers phenotype switching of a bacterial cell. *Science*, **322**(5900), 442–446.

Clauset, A, Shalizi, C R, & Newman, M E J. 2009. Power-law distributions in empirical data. *SIAM Rev.*, **51**, 661–703.

Cosentino, C, & Bates, D. 2012. *Feedback control in systems biology.* Boca Raton FL: CRC Press.

Coufal, N G, Garcia-Perez, J L, Peng, G E, Yeo, G W, Mu, Y, Lovci, M T, Morell, M, O'shea, K S, Moran, J V, & Gage, F H. 2009. L1 retrotransposition in human neural progenitor cells. *Nature*, **460**(7259), 1127–1131.

Cowan, G. 1998. *Statistical data analysis.* Oxford UK: Oxford Univ. Press.

Cronin, C A, Gluba, W, & Scrable, H. 2001. The *lac* operator-repressor system is functional in the mouse. *Genes and Dev.*, **15**(12), 1506–1517.

Dayan, P, & Abbott, L F. 2000. *Theoretical neuroscience.* Cambridge MA: MIT Press.

Denny, M, & Gaines, S. 2000. *Chance in biology.* Princeton NJ: Princeton Univ. Press.

DeVries, P L, & Hasbun, J E. 2011. *A first course in computational physics.* 2d ed. Sudbury MA: Jones and Bartlett.

Dickson, R M, Cubitt, A B, Tsien, R Y, & Moerner, W E. 1997. On/off blinking and switching behaviour of single molecules of green fluorescent protein. *Nature*, **388**(6640), 355–8.

Dill, K A, & Bromberg, S. 2010. *Molecular driving forces: Statistical thermodynamics in biology, chemistry, physics, and nanoscience.* 2d ed. New York: Garland Science.

Dillon, P F. 2012. *Biophysics: A physiological approach.* Cambridge UK: Cambridge Univ. Press.

Echols, H. 2001. *Operators and promoters: The story of molecular biology and its creators.* Berkeley CA: Univ. California Press.

Efron, B, & Gong, G. 1983. A leisurely look at the bootstrap, the jackknife, and cross-validation. *Amer. Statistician*, **37**, 36–48.

Eldar, A, & Elowitz, M B. 2010. Functional roles for noise in genetic circuits. *Nature*, **467**(7312), 167–173.

Ellner, S P, & Guckenheimer, J. 2006. *Dynamic models in biology.* Princeton NJ: Princeton Univ. Press.

Elowitz, M B, & Leibler, S. 2000. A synthetic oscillatory network of transcriptional regulators. *Nature*, **403**(6767), 335–338.

English, B P, Min, W, van Oijen, A M, Lee, K T, Luo, G, Sun, H, Cherayil, B J, Kou, S C, & Xie, X S. 2006. Ever-fluctuating single enzyme molecules: Michaelis-Menten equation revisited. *Nat. Chem. Biol.*, **2**(2), 87–94.

Epstein, W, Naono, S, & Gros, F. 1966. Synthesis of enzymes of the lactose operon during diauxic growth of Escherichia coli. *Biochem. Biophys. Res. Commun.*, **24**(4), 588–592.

Ferrell, Jr., J E. 2008. Feedback regulation of opposing enzymes generates robust, all-or-none bistable responses. *Curr. Biol.*, **18**(6), R244–5.

Ferrell, Jr., J E, Tsai, T Y-C, & Yang, Q. 2011. Modeling the cell cycle: Why do certain circuits oscillate? *Cell*, **144**(6), 874–885.

Franklin, K, Muir, P, Scott, T, Wilcocks, L, & Yates, P. 2010. *Introduction to biological physics for the health and life sciences.* Chichester UK: John Wiley and Sons.

Freeman, S, & Herron, J C. 2007. *Evolutionary analysis.* 4th ed. Upper Saddle River NJ: Pearson Prentice Hall.

Gardner, T S, Cantor, C R, & Collins, J J. 2000. Construction of a genetic toggle switch in *Escherichia coli*. *Nature*, **403**(6767), 339–342.

Gelles, J, Schnapp, B J, & Sheetz, M P. 1988. Tracking kinesin-driven movements with nanometre-scale precision. *Nature*, **331**(6155), 450–453.

Gelman, A, Carlin, J B, Stern, H S, Dunson, D B, Vehtari, A, & Rubin, D B. 2014. *Bayesian data analysis.* 3d ed. Boca Raton FL: Chapman and Hall/CRC.

Gerstner, W, Kistler, W M, Naud, R, & Paninski, L. 2014. *Neuronal dynamics: From single neurons to networks and models of cognition.* Cambridge UK: Cambridge Univ. Press.

Gigerenzer, G. 2002. *Calculated risks: How to know when numbers deceive you.* New York: Simon and Schuster.

Gillespie, D T. 2007. Stochastic simulation of chemical kinetics. *Annu. Rev. Phys. Chem.*, **58**, 35–55.

Gireesh, E D, & Plenz, D. 2008. Neuronal avalanches organize as nested theta- and beta/gamma-oscillations during development of cortical layer 2/3. *Proc. Natl. Acad. Sci. USA*, **105**(21), 7576–7581.

Golding, I, Paulsson, J, Zawilski, S M, & Cox, E C. 2005. Real-time kinetics of gene activity in individual bacteria. *Cell*, **123**(6), 1025–1036.

Haddock, S H D, & Dunn, C W. 2011. *Practical computing for biologists.* Sunderland MA: Sinauer Associates.

Hand, D J. 2008. *Statistics: A very short introduction.* Oxford UK: Oxford Univ. Press.

Hasty, J, Pradines, J, Dolnik, M, & Collins, J J. 2000. Noise-based switches and amplifiers for gene expression. *Proc. Natl. Acad. Sci. USA*, **97**(5), 2075–2080.

Hell, S W. 2007. Far-field optical nanoscopy. *Science*, **316**(5828), 1153–1158.

Hell, S W. 2009. Microscopy and its focal switch. *Nat. Methods*, **6**(1), 24–32.

Herman, I P. 2007. *Physics of the human body: A physical view of physiology.* New York: Springer.

Hess, S T, Girirajan, T P K, & Mason, M D. 2006. Ultra-high resolution imaging by fluorescence photoactivation localization microscopy. *Biophys. J.*, **91**(11), 4258–4272.

Hilborn, R C, Brookshire, B, Mattingly, J, Purushotham, A, & Sharma, A. 2012. The transition between stochastic and deterministic behavior in an excitable gene circuit. *PLoS ONE*, **7**(4), e34536.

Hinterdorfer, P, & van Oijen, A (Eds.). 2009. *Handbook of single-molecule biophysics.* New York: Springer.

Ho, D D, Neumann, A U, Perelson, A S, Chen, W, Leonard, J M, & Markowitz, M. 1995. Rapid turnover of plasma virions and CD4 lymphocytes in HIV-1 infection. *Nature*, **373**(6510), 123–126.

Hoagland, M, & Dodson, B. 1995. *The way life works.* New York: Random House.

Hobbie, R K, & Roth, B J. 2007. *Intermediate physics for medicine and biology*. 4th ed. New York: Springer.

Hoffmann, P M. 2012. *Life's ratchet: How molecular machines extract order from chaos*. New York: Basic Books.

Hoogenboom, J P, den Otter, W K, & Offerhaus, H L. 2006. Accurate and unbiased estimation of power-law exponents from single-emitter blinking data. *J. Chem. Phys.*, **125**, 204713.

Huang, B, Bates, M, & Zhuang, X. 2009. Super-resolution fluorescence microscopy. *Annu. Rev. Biochem.*, **78**, 993–1016.

Ingalls, B P. 2013. *Mathematical modeling in systems biology: An introduction*. Cambridge MA: MIT Press.

Ioannidis, J P A. 2005. Why most published research findings are false. *PLoS Med.*, **2**(8), e124.

Isaacs, F J, Hasty, J, Cantor, C R, & Collins, J J. 2003. Prediction and measurement of an autoregulatory genetic module. *Proc. Natl. Acad. Sci. USA*, **100**(13), 7714–7719.

Iyer-Biswas, S, Hayot, F, & Jayaprakash, C. 2009. Stochasticity of gene products from transcriptional pulsing. *Phys. Rev. E*, **79**, 031911.

Jacobs, K. 2010. *Stochastic processes for physicists*. Cambridge UK: Cambridge Univ. Press.

Jaynes, E T, & Bretthorst, G L. 2003. *Probability theory: The logic of science*. Cambridge UK: Cambridge Univ. Press.

Jones, O, Maillardet, R, & Robinson, A. 2009. *Introduction to scientific programming and simulation using R*. Boca Raton FL: Chapman and Hall/CRC.

Karp, G. 2013. *Cell and molecular biology: Concepts and experiments*. 7th ed. Hoboken NJ: John Wiley and Sons.

Katz, B, & Miledi, R. 1972. The statistical nature of the acetylcholine potential and its molecular components. *J. Physiol. (Lond.)*, **224**(3), 665–699.

Keener, J, & Sneyd, J. 2009. *Mathematical physiology I: Cellular physiology*. 2d ed. New York: Springer.

Klipp, E, Liebermeister, W, Wierling, C, Kowald, A, Lehrach, H, & Herwig, R. 2009. *Systems biology: A textbook*. New York: Wiley-Blackwell.

Koonin, E V, & Wolf, Y I. 2009. Is evolution Darwinian or/and Lamarckian? *Biol. Direct*, **4**, 42.

Lacoste, T D, Michalet, X, Pinaud, F, Chemla, D S, Alivisatos, A P, & Weiss, S. 2000. Ultrahigh-resolution multicolor colocalization of single fluorescent probes. *Proc. Natl. Acad. Sci. USA*, **97**(17), 9461–9466.

Laughlin, S, & Sterling, P. 2015. *Principles of neural design*. Cambridge MA: MIT Press.

Laurence, T A, & Chromy, B A. 2010. Efficient maximum likelihood estimator fitting of histograms. *Nat. Methods*, **7**(5), 338–339.

Lea, D, & Coulson, C. 1949. The distribution of the numbers of mutants in bacterial populations. *J. Genetics*, **49**, 264–285.

Leake, M C. 2013. *Single-molecule cellular biophysics*. Cambridge UK: Cambridge Univ. Press.

Lederberg, J, & Lederberg, E M. 1952. Replica plating and indirect selection of bacterial mutants. *J. Bacteriol.*, **63**(3), 399–406.

Le Novère, N, et al. 2009. The systems biology graphical notation. *Nat. Biotechnol.*, **27**(8), 735–741.

Lewis, M. 2005. The *lac* repressor. *C. R. Biol.*, **328**(6), 521–548.

Lidke, K, Rieger, B, Jovin, T, & Heintzmann, R. 2005. Superresolution by localization of quantum dots using blinking statistics. *Opt. Express*, **13**(18), 7052–7062.

Linden, W von der, Dose, V, & Toussaint, U von. 2014. *Bayesian probability theory: Applications in the physical sciences.* Cambridge UK: Cambridge Univ. Press.

Little, J W, & Arkin, A P. 2012. Stochastic simulation of the phage *lambda* gene regulatory circuitry. *In:* Wall, M E (Ed.), *Quantitative biology: From molecular to cellular systems.* Boca Raton FL: Taylor and Francis.

Lodish, H, Beck, A, Kaiser, C A, Krieger, M, Bretscher, A, Ploegh, H, Amon, A, & Scott, M P. 2012. *Molecular cell biology.* 7th ed. New York: W H Freeman and Co.

Luria, S E. 1984. *A slot machine, a broken test tube: An autobiography.* New York: Harper and Row.

Luria, S E, & Delbrück, M. 1943. Mutations of bacteria from virus sensitivity to virus resistance. *Genetics*, **28**, 491–511.

Mantegna, R N, & Stanley, H E. 2000. *Introduction to econophysics: Correlations and complexity in finance.* Cambridge UK: Cambridge Univ. Press.

María-Dolores, R, & Martínez-Carrión, J M. 2011. The relationship between height and economic development in Spain, 1850–1958. *Econ. Hum. Biol.*, **9**(1), 30–44.

Marks, F, Klingmüller, U, & Müller-Decker, K. 2009. *Cellular signal processing: An introduction to the molecular mechanisms of signal transduction.* New York: Garland Science.

McCall, R P. 2010. *Physics of the human body.* Baltimore MD: Johns Hopkins Univ. Press.

Mertz, J. 2010. *Introduction to optical microscopy.* Greenwood Village, CO: Roberts and Co.

Mills, F C, Johnson, M L, & Ackers, G K. 1976. Oxygenation-linked subunit interactions in human hemoglobin. *Biochemistry*, **15**, 5350–5362.

Mlodinow, L. 2008. *The drunkard's walk: How randomness rules our lives.* New York: Pantheon Books.

Monod, J. 1942. *Recherches sur la croissance des cultures bactérienne.* Paris: Hermann et Cie.

Monod, J. 1949. The growth of bacterial cultures. *Annu. Rev. Microbiol.*, **3**(1), 371–394.

Mora, T, Walczak, A M, Bialek, W, & Callan, C G. 2010. Maximum entropy models for antibody diversity. *Proc. Natl. Acad. Sci. USA*, **107**(12), 5405–5410.

Mortensen, K I, Churchman, L S, Spudich, J A, & Flyvbjerg, H. 2010. Optimized localization analysis for single-molecule tracking and super-resolution microscopy. *Nat. Methods*, **7**(5), 377–381.

Müller-Hill, B. 1996. *The* lac *operon: A short history of a genetic paradigm.* Berlin: W. de Gruyter and Co.

Murray, J D. 2002. *Mathematical biology.* 3d ed. New York: Springer.

Myers, C J. 2010. *Engineering genetic circuits.* Boca Raton FL: CRC Press.

Nadeau, J. 2012. *Introduction to experimental biophysics.* Boca Raton FL: CRC Press.

National Research Council. 2003. *Bio2010: Transforming undergraduate education for future research biologists.* Washington DC: National Academies Press.

Nelson, P. 2014. *Biological physics: Energy, information, life—With new art by David Goodsell.* New York: W. H. Freeman and Co.

Newman, M. 2013. *Computational physics.* Rev. and expanded ed. Amazon CreateSpace.

Nordlund, T. 2011. *Quantitative understanding of biosystems: An introduction to biophysics.* Boca Raton FL: CRC Press.

Novák, B, & Tyson, J J. 1993a. Modeling the cell division cycle: M-phase trigger, oscillations, and size control. *J. Theor. Biol.*, **165**, 101–134.

Novák, B, & Tyson, J J. 1993b. Numerical analysis of a comprehensive model of M-phase control in *Xenopus* oocyte extracts and intact embryos. *J. Cell Sci.*, **106 (Pt 4)**, 1153–1168.

Novák, B, & Tyson, J J. 2008. Design principles of biochemical oscillators. *Nat. Rev. Mol. Cell Biol.*, **9**(12), 981–991.

Novick, A, & Weiner, M. 1957. Enzyme induction as an all-or-none phenomenon. *Proc. Natl. Acad. Sci. USA*, **43**(7), 553–566.

Nowak, M A. 2006. *Evolutionary dynamics: Exploring the equations of life.* Cambridge MA: Harvard Univ. Press.

Nowak, M A, & May, R M. 2000. *Virus dynamics.* Oxford UK: Oxford Univ. Press.

Ober, R J, Ram, S, & Ward, E S. 2004. Localization accuracy in single-molecule microscopy. *Biophys. J.*, **86**(2), 1185–1200.

Otto, S P, & Day, T. 2007. *Biologist's guide to mathematical modeling in ecology and evolution.* Princeton NJ: Princeton Univ. Press.

Ozbudak, E M, Thattai, M, Lim, H N, Shraiman, B I, & van Oudenaarden, A. 2004. Multistability in the lactose utilization network of *Escherichia coli. Nature*, **427**(6976), 737–740.

Pace, H C, Lu, P, & Lewis, M. 1990. *lac* repressor: Crystallization of intact tetramer and its complexes with inducer and operator DNA. *Proc. Natl. Acad. Sci. USA*, **87**(5), 1870–1873.

Paulsson, J. 2005. Models of stochastic gene expression. *Physics of Life Reviews*, **2**(2), 157–175.

Perelson, A S. 2002. Modelling viral and immune system dynamics. *Nat. Rev. Immunol.*, **2**(1), 28–36.

Perelson, A S, & Nelson, P W. 1999. Mathematical analysis of HIV-1 dynamics in vivo. *SIAM Rev.*, **41**, 3–44.

Perrin, J. 1909. Mouvement brownien et réalité moléculaire. *Ann. Chim. Phys.*, **8**(18), 5–114.

Pevzner, P, & Shamir, R. 2009. Computing has changed biology—Biology education must catch up. *Science*, **325**(5940), 541–542.

Phillips, R, Kondev, J, Theriot, J, & Garcia, H. 2012. *Physical biology of the cell.* 2d ed. New York: Garland Science.

Pomerening, J R, Kim, S Y, & Ferrell, Jr., J E. 2005. Systems-level dissection of the cell-cycle oscillator: Bypassing positive feedback produces damped oscillations. *Cell*, **122**(4), 565–578.

Pouzat, C, Mazor, O, & Laurent, G. 2002. Using noise signature to optimize spike-sorting and to assess neuronal classification quality. *J. Neurosci. Meth.*, **122**(1), 43–57.

Press, W H, Teukolsky, S A, Vetterling, W T, & Flannery, B P. 2007. *Numerical recipes: The art of scientific computing.* 3d ed. Cambridge UK: Cambridge Univ. Press.

Ptashne, M. 2004. *A genetic switch: Phage* lambda *revisited.* 3d ed. Cold Spring Harbor NY: Cold Spring Harbor Laboratory Press.

Raj, A, & van Oudenaarden, A. 2008. Nature, nurture, or chance: Stochastic gene expression and its consequences. *Cell*, **135**(2), 216–226.

Raj, A, & van Oudenaarden, A. 2009. Single-molecule approaches to stochastic gene expression. *Annu. Rev. Biophys.*, **38**, 255–270.

Raj, A, Peskin, C S, Tranchina, D, Vargas, D Y, & Tyagi, S. 2006. Stochastic mRNA synthesis in mammalian cells. *PLoS Biol.*, **4**(10), e309.

Rechavi, O, Minevich, G, & Hobert, O. 2011. Transgenerational inheritance of an acquired small RNA-based antiviral response in *C. elegans. Cell*, **147**(6), 1248–1256.

Rechavi, O, Houri-Ze'evi, L, Anava, S, Goh, W S S, Kerk, S Y, Hannon, G J, & Hobert, O. 2014. Starvation-induced transgenerational inheritance of small RNAs in *C. elegans. Cell*, **158**(2), 277–287.

Ro, D-K, Paradise, E M, Ouellet, M, Fisher, K J, Newman, K L, Ndungu, J M, Ho, K A, Eachus, R A, Ham, T S, Kirby, J, Chang, M C Y, Withers, S T, Shiba, Y, Sarpong, R, & Keasling, J D. 2006. Production of the antimalarial drug precursor artemisinic acid in engineered yeast. *Nature*, **440**(7086), 940–943.

Roe, B P. 1992. *Probability and statistics in experimental physics.* New York: Springer.

Rosche, W A, & Foster, P L. 2000. Determining mutation rates in bacterial populations. *Methods*, **20**(1), 4–17.

Rosenfeld, N, Elowitz, M B, & Alon, U. 2002. Negative autoregulation speeds the response times of transcription networks. *J. Mol. Biol.*, **323**(5), 785–793.

Rosenfeld, N, Young, Jonathan W, Alon, U, Swain, P S, & Elowitz, M B. 2005. Gene regulation at the single-cell level. *Science*, **307**(5717), 1962–1965.

Ross, S M. 2010. *A first course in probability.* 8th ed. Upper Saddle River NJ: Pearson Prentice Hall.

Rossi-Fanelli, A, & Antonini, E. 1958. Studies on the oxygen and carbon monoxide equilibria of human myoglobin. *Arch. Biochem. Biophys.*, **77**, 478–492.

Rust, M J, Bates, M, & Zhuang, X. 2006. Sub-diffraction-limit imaging by stochastic optical reconstruction microscopy (STORM). *Nat. Methods*, **3**(10), 793–795.

Santillán, M, Mackey, M C, & Zeron, E S. 2007. Origin of bistability in the *lac* operon. *Biophys. J.*, **92**(11), 3830–3842.

Savageau, M A. 2011. Design of the *lac* gene circuit revisited. *Math. Biosci.*, **231**(1), 19–38.

Schiessel, H. 2013. *Biophysics for beginners: A journey through the cell nucleus.* Boca Raton FL: CRC Press.

Segrè, G. 2011. *Ordinary geniuses: Max Delbrück, George Gamow and the origins of genomics and Big Bang cosmology.* New York: Viking.

Selvin, P R, Lougheed, T, Tonks Hoffman, M, Park, H, Balci, H, Blehm, B H, & Toprak, E. 2008. *In vitro* and *in vivo* fiona and other acronyms for watching molecular motors walk. *Pages 37–72 of:* Selvin, P R, & Ha, T (Eds.), *Single-molecule techniques: A laboratory manual.* Cold Spring Harbor NY: Cold Spring Harbor Laboratory Press.

Sha, W, Moore, J, Chen, K, Lassaletta, A D, Yi, C-S, Tyson, J J, & Sible, J C. 2003. Hysteresis drives cell-cycle transitions in *Xenopus laevis* egg extracts. *Proc. Natl. Acad. Sci. USA*, **100**(3), 975–980.

Shahrezaei, V, & Swain, P S. 2008. Analytical distributions for stochastic gene expression. *Proc. Natl. Acad. Sci. USA*, **105**(45), 17256–17261.

Shankar, R. 1995. *Basic training in mathematics: A fitness program for science students.* New York: Plenum.

Sharonov, Alexey, & Hochstrasser, Robin M. 2006. Wide-field subdiffraction imaging by accumulated binding of diffusing probes. *Proc Natl Acad Sci USA*, **103**(50), 18911–6.

Shonkwiler, R W, & Herod, J. 2009. *Mathematical biology: An introduction with Maple and MATLAB.* 2d ed. New York: Springer.

Silver, N. 2012. *The signal and the noise.* London: Penguin.

Sivia, D S, & Skilling, J. 2006. *Data analysis: A Bayesian tutorial.* 2d ed. Oxford UK: Oxford Univ. Press.

Small, A R, & Parthasarathy, R. 2014. Superresolution localization methods. *Annu. Rev. Phys. Chem.*, **65**, 107–125.

Sneppen, K, & Zocchi, G. 2005. *Physics in molecular biology.* Cambridge UK: Cambridge Univ. Press.

So, L-H, Ghosh, A, Zong, C, Sepúlveda, L A, Segev, R, & Golding, I. 2011. General properties of transcriptional time series in *Escherichia coli*. *Nat. Genet.*, **43**(6), 554–560.

Stinchcombe, A R, Peskin, C S, & Tranchina, D. 2012. Population density approach for discrete mRNA distributions in generalized switching models for stochastic gene expression. *Phys. Rev. E*, **85**, 061919.

Stricker, J, Cookson, S, Bennett, M R, Mather, W H, Tsimring, L S, & Hasty, J. 2008. A fast, robust and tunable synthetic gene oscillator. *Nature*, **456**(7221), 516–519.

Strogatz, S. 2003. *Sync: The emerging science of spontaneous order.* New York: Hyperion.

Strogatz, S H. 2012. *The joy of x: A guided tour of math, from one to infinity.* Boston MA: Houghton Mifflin Harcourt.

Strogatz, S H. 2014. *Nonlinear dynamics and chaos with applications in physics, biology, chemistry, and engineering.* 2d ed. San Francisco: Westview Press.

Suter, D M, Molina, N, Gatfield, D, Schneider, K, Schibler, U, & Naef, F. 2011. Mammalian genes are transcribed with widely different bursting kinetics. *Science*, **332**(6028), 472–474.

Taniguchi, Y, Choi, P J, Li, G-W, Chen, H, Babu, M, H, Jeremy, Emili, A, & Xie, X S. 2010. Quantifying *E. coli* proteome and transcriptome with single-molecule sensitivity in single cells. *Science*, **329**(5991), 533–538.

Thomas, C M, & Nielsen, K M. 2005. Mechanisms of, and barriers to, horizontal gene transfer between bacteria. *Nat. Rev. Microbiol.*, **3**(9), 711–721.

Thompson, R E, Larson, D R, & Webb, W W. 2002. Precise nanometer localization analysis for individual fluorescent probes. *Biophys. J.*, **82**(5), 2775–2783.

Toprak, E, Kural, C, & Selvin, P R. 2010. Super-accuracy and super-resolution: Getting around the diffraction limit. *Meth. Enzymol.*, **475**, 1–26.

Tyson, J J, & Novák, B. 2010. Functional motifs in biochemical reaction networks. *Annu. Rev. Phys. Chem.*, **61**, 219–240.

Tyson, J J, & Novák, B. 2013. Irreversible transitions, bistability and checkpoint controls in the eukaryotic cell cycle: A systems-level understanding. *In:* Walhout, A J M, Vidal, M, & Dekker, J (Eds.), *Handbook of systems biology: Concepts and insights.* Amsterdam: Elsevier/Academic Press.

Tyson, J J, Chen, K C, & Novák, B. 2003. Sniffers, buzzers, toggles and blinkers: Dynamics of regulatory and signaling pathways in the cell. *Curr. Opin. Cell Biol.*, **15**(2), 221–231.

Vecchio, D Del, & Murray, R M. 2014. *Biomolecular feedback systems.* Princeton NJ: Princeton Univ. Press.

Voit, E O. 2013. *A first course in systems biology.* New York: Garland Science.

Walton, H. 1968. *The how and why of mechanical movements.* New York: Popular Science Publishing Co./E. P. Dutton and Co.

Weber, W, & Fussenegger, M (Eds.). 2012. *Synthetic gene networks: Methods and protocols.* New York: Humana Press.

Wei, X, Ghosh, S K, Taylor, M E, Johnson, V A, Emini, E A, Deutsch, P, Lifson, J D, Bonhoeffer, S, Nowak, M A, Hahn, B H, Saag, M S, & Shaw, G M. 1995. Viral dynamics in human immunodeficiency virus type 1 infection. *Nature*, **373**(6510), 117–122.

Weinstein, J A, Jiang, N, White III, R A, Fisher, D S, & Quake, S R. 2009. High-throughput sequencing of the zebrafish antibody repertoire. *Science*, **324**(5928), 807–810.

Weiss, R A. 1993. How does HIV cause AIDS? *Science*, **260**(5112), 1273–1279.

Wheelan, C J. 2013. *Naked statistics: Stripping the dread from the data.* New York: W. W. Norton and Co.

White, E P, Enquist, B J, & Green, J L. 2008. On estimating the exponent of power-law frequency distributions. *Ecology*, **89**(4), 905–912.

Wilkinson, D J. 2006. *Stochastic modelling for systems biology.* Boca Raton FL: Chapman and Hall/CRC.

Winfree, A T. 2001. *The geometry of biological time.* 2d ed. New York: Springer.

Wingreen, N, & Botstein, D. 2006. Back to the future: Education for systems-level biologists. *Nat. Rev. Mol. Cell Biol.*, **7**, 829–832.

Woodworth, G G. 2004. *Biostatistics: A Bayesian introduction.* Hoboken NJ: Wiley-Interscience.

Woolfson, M M. 2012. *Everyday probability and statistics: Health, elections, gambling and war.* 2d ed. London: Imperial College Press.

Yanagida, T, & Ishii, Y (Eds.). 2009. *Single molecule dynamics in life science.* Weinheim: Wiley-VCH.

Yang, Q, & Ferrell, Jr., J E. 2013. The Cdk1-APC/C cell cycle oscillator circuit functions as a time-delayed, ultrasensitive switch. *Nat. Cell Biol.*, **15**(5), 519–525.

Yildiz, A, Forkey, J N, McKinney, S A, Ha, T, Goldman, Y E, & Selvin, P R. 2003. Myosin V walks hand-over-hand: Single fluorophore imaging with 1.5-nm localization. *Science*, **300**(5628), 2061–2065.

Zeng, L, Skinner, S O, Zong, C, Sippy, J, Feiss, M, & Golding, I. 2010. Decision making at a subcellular level determines the outcome of bacteriophage infection. *Cell*, **141**(4), 682–691.

Zenklusen, D, Larson, D R, & Singer, R H. 2008. Single-RNA counting reveals alternative modes of gene expression in yeast. *Nat. Struct. Mol. Biol.*, **15**(12), 1263–1271.

Zimmer, C. 2011. *A planet of viruses.* Chicago IL: Univ. Chicago Press.

Index

Bold references are the defining instance of a key term. Symbol names and mathematical notations are defined in Appendix A.